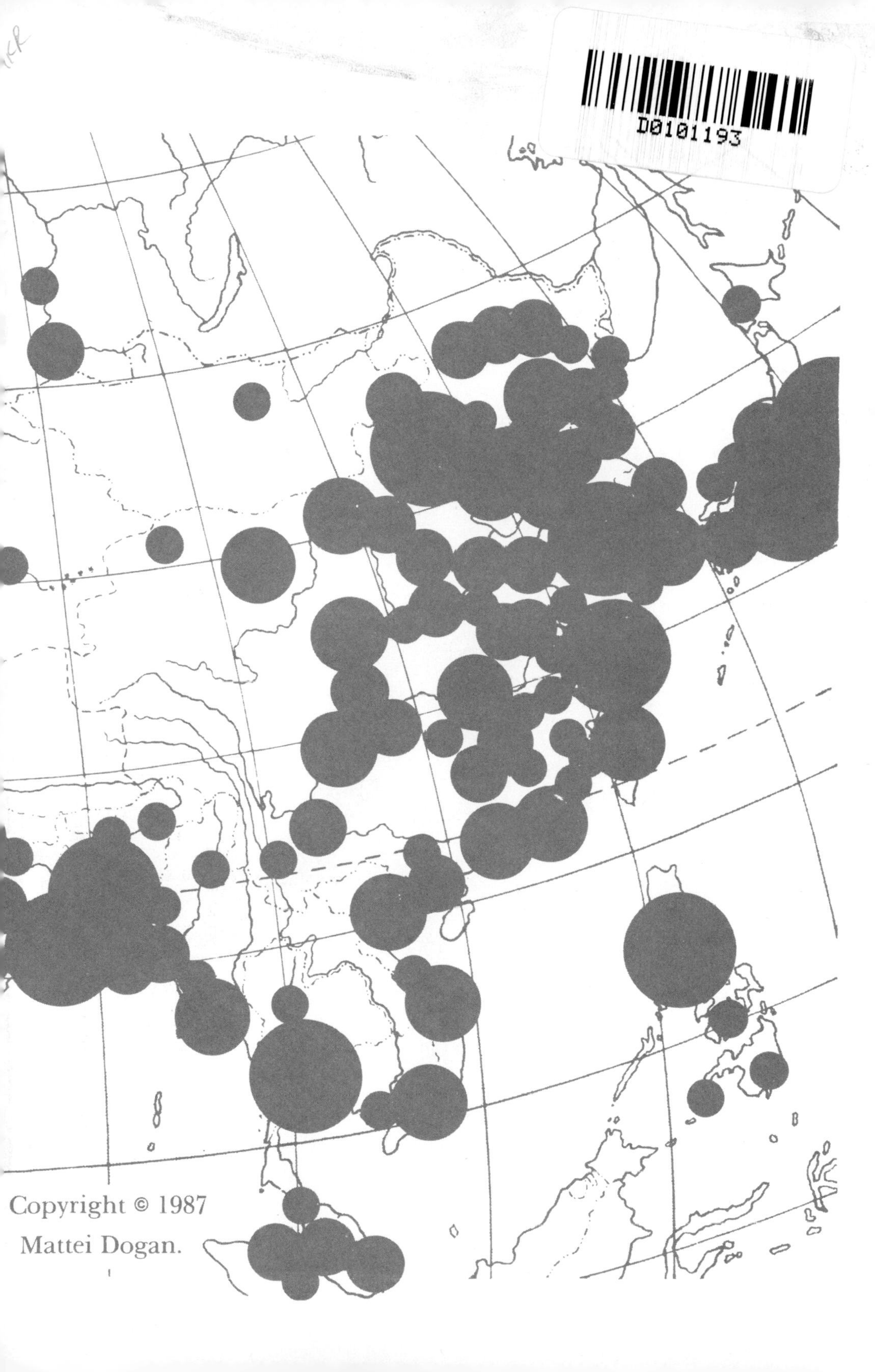

A World of Giant Cities

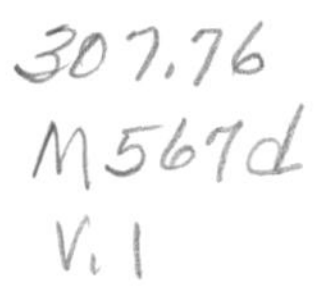

The Metropolis Era

Volume 1

A World of Giant Cities

EDITORS

Mattei Dogan

John D. Kasarda

SAGE PUBLICATIONS
The Publishers of Professional Social Science
Newbury Park Beverly Hills London New Delhi

This book is dedicated to the beautiful metropolis of Barcelona

For information address:

SAGE Publications, Inc.
2111 West Hillcrest Drive
Newbury Park, California 91320

SAGE Publications Inc.
275 South Beverly Drive,
Beverly Hills
California 90212

SAGE Publications Ltd.
28 Banner Street
London EC1Y 8QE
England

SAGE PUBLICATIONS India Pvt. Ltd.
M-32 Market
Greater Kailash I
New Delhi 110 048 India

Printed in the United States of America

Library of Congress Cataloging-in-Publication Data

Main entry under title:

The metropolis era.

Contents: v. 1. A world of giant cities—v. 2. Mega-cities.
Includes bibliographies and indexes.
1. Metropolitan areas. 2. Cities and towns—Growth. I. Dogan, Mattei. II. Kasarda, John D. III. Title: Mega-cities.
HT330.M45 1987 307.7′64 87-23247
ISBN 0-8039-2602-2 (v. 1)
ISBN 0-8039-2603-0 (v. 2)

Contents

Foreword

Luis Ramallo
International Social Science Council

At the invitation of the Corporacio Metropolitana de Barcelona (Greater Barcelona Council), the International Social Science Council in close cooperation with the Research Committee on Social Ecology of the International Sociological Association organized a conference on giant cities of the world in Barcelona from 25 February to 1 March 1985. The main objective of the Barcelona Conference was examining how huge urban conglomerates and metropolitan areas can confront the malfunctions which stem from their very expansion as population centers and apply forward-looking solutions to these problems. The conference was designed and conceived as a breakthrough experience in analyzing and discussing urban problems in a cross-cultural, worldwide approach.

The conference focus was not only on urban problems (using the medical analogy of pathology), but also on the opportunities giant cities provide for social mobility of the underprivileged as well as functions and potentials of each city in national and international economic orders. At the conference efforts of local, state, and national governments were assessed to reduce the most pressing problems facing giant cities.

During the course of the conference, problems and potentials of the world's largest cities, the courses they have taken in recent decades, and their foreseeable development were discussed from the double viewpoints of science and knowledge and the political and administrative affirmation of the urban reality.

Mayors of the metropolitan area of Barcelona attended the conference and participated in the closing session which agreed on the conclusions and recommendations on future prospects and policies for giant cities. The main points of the "Declaration of Barcelona" follow:

- First, referring to the Declaration of Rome adopted at the International Conference on Population and the Urban Future in 1980, participants acknowledged that today planned and orderly conditions for urbanization do not exist and stressed the combined effect of the economic crises and criticisms of planning concepts and techniques which have caused a serious loss of confidence in society's ability to construct efficient cities.
- Second, particularly in less-developed countries, large cities continue and will continue to grow. The problems characterizing urbanization in these countries to date, such as inadequate urban infrastructures and housing deficits, poverty and alienation of a large part of the population, and environmental degradation will worsen every day. These problems are also found in cities of more developed countries, although to a lesser extent.
- Third, large cities still hold important attraction factors, particularly in less-developed countries where they continue taking in population that cannot survive in the rural areas. Presently in large cities there exists a set of important factors for technological and economic reconversion and implementation of new patterns of social development.
- Fourth, the large city must be at the head of economic reconversion processes and, therefore, the leading seat for development of new technologies while assuming the social challenge of this economic reconversion.
- Fifth, it is fundamental to reinforce the autonomy and representativeness of local government. Creating participatory mechanisms is an indispensable condition for efficacy of city management and for democratizing governments.

The mayors of the metropolitan area of Barcelona formally endorsed this declaration. They stressed the benefits from such international scholarly exchange and expressed the hope that such dialogues could be pursued.

Participants expressed their deep gratitude to the Metropolitan Corporation of Barcelona for its most generous invitation and particularly hospitable welcome.

Preface

Comparative urban research confronts a dual phenomenon: the rapid growth of densely compacted giant cities in the Third World and, because of new technologies, enormous urban deconcentration in more developed countries. For this reason the Research Committee on Social Ecology of the International Sociological Association has given high priority to large cities at the last two World Congress of Sociology meetings (1982 and 1986).

At our suggestion the International Social Science Council agreed to sponsor a conference on "Giant Cities of the World and the Future." The I.S.S.C. organized the conference with generous support from the metropolitan municipality of Barcelona, in February 1985.

The International Social Science Council commissioned well-known scholars—many of them of international fame—to carry out a number of research studies. While the methodology applied in these studies differed according to the individual objectives, analytical tools, available information, and the researchers' own theoretical frame of reference, all contributors were invited to provide replies to a series of issues:

What roles does each giant city play in national, regional, and international economies?

What are the main "pathologies" of these cities and what are the demographic, social, political, economic, and technological causes and consequences of these pathologies?

To what extent are these phenomena peculiar to the city studied or similar to those revealed in other cities?

To what extent can these phenomena be controlled? What are the political and administrative mechanisms that have been successfully implemented to address particular problems?

What is the future outlook regarding both the persistence of these malfunctions (and even their possible aggravation and the appearance of new ones) and creative solutions to these problems?

What conclusions can be drawn about the future of the giant cities and the quality of life there?

Most of the chapters contained in these two volumes are revisions of papers presented at the Barcelona Conference. Some chapters were solicited specifically for these volumes.

We believe that these two volumes fill an important gap in the literature. The urban world is immensely diverse. To cover such diversity, our approach was to ask scholars across ranges of disciplines to apply their insights and knowledge to particular cities and world regions for which they held expertise. No single scholar possesses the information, understanding, and experience to cover thoroughly the variety of cities and regions assessed herein.

Among the 28 contributors to these two volumes are sociologists, geographers, demographers, political scientists, economists, historians, and urbanists. They belong to 16 different countries. About 15 other scholars participated at the Barcelona Conference.

Most of the 13 chapters of the first volume have a comparative approach. Combined, they cover virtually all major urban regions of the world: Western Europe, Eastern Europe, North America, South America, Africa, South-East Asia, China, and India. This broadly cross-cultural approach is also manifest in the giant cities selected for examination in Volume II: four from industrially advanced countries (New York, London, Los Angeles, and Tokyo) and six from developing countries (Shanghai, Cairo, Mexico City, Delhi, São Paulo and Lagos). Each of the chapters on these ten giant cities has been written by a scholar who lives and teaches in the city that he or she describes—except two, who sojourned for long times in the respective city. Each of these contributors is considered as an expert of "his" or "her" city. All but rapidly growing Lagos are among the world's largest concentrations of human beings. We would have liked to have included chapters on other metropolises, such as Paris, Moscow, Milan, Buenos Aires, Beijing, Seoul, Lima, Calcutta, Teheran, Bangkok. Doing so would have required an additional volume.

These volumes are especially pertinent to three audiences: first, to scholars and students of urban life who may use them for classroom texts as well as for their research; second, to municipal officials of large cities who are grappling with decisions to improve the quality of life in their cities; third, to a concerned public who wish to be better informed about challenges facing the world's largest cities.

These two volumes would not have been possible without the help we received from various persons and institutions. We are particularly grateful to the International Social Science Council and its general secretary, Luis Ramallo, and to the deputy general secretary, Evelyn Blamont, who generously helped us. We are heavily indebted to the Mayor of Barcelona, Pasqual Maragall; to Luis Carreno Piera, Jordi Borja, the Corporacio Metropolitana de Barcelona, the Deputacio de Barcelona, and to the Generalitat de Catalunya, for their kind, friendly, and efficient support.

The editors received invaluable assistance from the University of North Carolina at Chapel Hill, the University of California at Los Angeles, and the National Center for Scientific Research in Paris. We are grateful to Anna K. Tyndall of the UNC-CH Department of Sociology for invaluable assistance.

The Carolina Population Center (UNC-CH) was most gracious in facilitating the technical preparation of the manuscript. In particular, we wish to express our gratitude to Lynn Moody Igoe for her skillful copyediting.

We express our thanks also to our publishers, Sara and George McCune, who did not lose their patience during the sometimes tortuous and prolonged preparation of the final manuscript. We are grateful to Lisa Freeman-Miller, our earlier Sage editor and especially to Ann West, our final editor, who worked graciously in determining the final shape of these volumes.

—*Mattei Dogan*
John D. Kasarda

Introduction:

How Giant Cities Will Multiply and Grow

Mattei Dogan and John D. Kasarda

The rapid growth of population in most developing countries is a well-known fact and continuing projection. Concurrent with such growth is the multiplication and growth of giant cities. The world is becoming more and more a world of giant cities, and these cities are increasingly located in less-developed countries.

As illustrated in tables 1.1 and 1.2, 23 of the 35 world's largest urban agglomerations in 1950 were located in more developed countries. By 1985, the distribution had completely reversed with 23 of the 35 largest urban agglomerations found in less developed countries. United Nations projections indicate that at the turn of the 21st century, 17 of the world's 20 largest metropolises will be in developing nations.

Not only has the locus of giant city growth geographically shifted in recent decades, so has its scale and pace. Mexico City, which contained 3.1 million residents in 1950, is expected to exceed 26 million residents during the coming decade. Likewise, São Paulo, a city of 2.8 million in 1950, is projected to reach 24 million by the year 2000. In terms of pace, New York (the world's largest metropolitan area in 1950) took nearly 150 years to expand by eight million residents; Mexico City and São Paulo are currently growing at a rate that will add eight million to their population bases in fewer than 15 years.

Though not on quite the same scale, the dramatic growth represented by Mexico City and São Paulo is being echoed in developing nations around the world. Unfortunately, much of this urban growth is occurring in world regions whose economic systems are least equipped to sustain giant cities. For example, Dennis A. Rondinelli's chapter in this volume notes that in 1950 only two cities on the African continent had more than one million residents. By 1990, there will be 37 such cities, containing nearly 40 percent of Africa's urban population. Some African cities are experiencing annual growth rates approaching 6 percent, which means they are doubling in size every 12 years.

Worldwide Metropolitan Trends and Projections

The number of metropolises with over a million inhabitants has tripled during the past 35 years. Table 1.3 reveals that in 1950 only 78 metropolises exceeded this size. According to United Nations estimates, there were 258 metropolises around the world with over one million inhabitants in 1985. Among these, 42 had more than four million, 56 between two and four million, and 158 between one and two million. Three hundred metropolises had between half a million and one million inhabitants.

U.N. projections indicate that there will be 511 metropolises exceeding a million inhabitants by 2010. Thereafter, more than 40 such metropolises will be added every five years, so that in the year 2025, there will be 639 metropolises exceeding one million residents. Before the children born in 1985 become adults, half of the world's population will be urban, and half of this half will be located in metropolises with over a million inhabitants.

The largest of metropolises, those containing over four million inhabitants, will more than double in number in the quarter century from 1985 to 2010, reaching 135 in 2025. The total population residing in these metropolises will aggregate to 1.26 billion in 2025, compared to approximately 342 million in 1985 and just 88 million in 1950. Similar trends are projected for metropolises with between two and four million inhabitants and those with between one and two million inhabitants.

Trends and projections are strikingly different for the developed and developing regions of the world, however. In Europe and the United States, major metropolitan areas are growing slowly with many of their core (central city) areas actually losing population, as chapters in this volume by Manuel Castells, Peter Hall, and John D. Kasarda describe. The number of metropolitan areas in developed countries with populations over a million is expected to increase modestly from 110 in 1985 to 128 in 2000, and 153 in 2025.

Conversely, in less-developed countries, only 31 metropolises in 1950 had over a million residents. In the ensuing 35 years, the number grew over fourfold to 146 metropolises. They will almost double again in the next 15 years, numbering 279 in the year 2000, becoming 374 a decade later in 2010, reaching 486 by 2025. Thus, in the 75 years between 1950 and 2025, the number of metropolises in developing countries exceeding a million inhabitants will have multiplied sixteenfold, from 31 to 486.

Giant metropolises of over four million inhabitants are also quickly forming within developing nations. In 1950, there were only five metropolises

TABLE 1.1
World's 35 Largest Metropolises Ranked by Population Size (in millions), 1950-2000

Rank	Metropolis	*1950* *Size*	Metropolis	*1985* *Size*	Metropolis	*2000* *Size*
1	New York/Northeastern NJ, USA	12.4	Mexico City, Mexico	18.1	Mexico City, Mexico	26.3
2	London, United Kingdom	10.4	Tokyo/Yokohama, Japan	17.2	São Paulo, Brazil	24.0
3	Shanghai, China	10.3	São Paulo, Brazil	15.9	Tokyo/Yokohama, Japan	17.1
4	Rhein-Ruhr, Federal Republic of Germany	6.9	New York/Northeastern NJ, USA	15.3	Calcutta, India	16.6
5	Tokyo/Yokohama, Japan	6.7	Shanghai, China	11.8	Greater Bombay, India	16.0
6	Beijing (Peking), China	6.7	Calcutta, India	11.0	New York/Northeastern NJ, USA	15.5
7	Paris, France	5.5	Greater Buenos Aires, Argentina	10.9	Seoul, Republic of Korea	13.5
8	Tianjin, China	5.4	Rio de Janeiro, Brazil	10.4	Shanghai, China	13.5
9	Greater Buenos Aires, Argentina	5.3	Seoul, Republic of Korea	10.2	Rio de Janeiro, Brazil	13.3
10	Chicago/Northwestern IN, USA	5.0	Greater Bombay, India	10.1	Delhi, India	13.3
11	Moscow, USSR	4.8	Los Angeles/Long Beach, CA, USA	10.0	Greater Buenos Aires, Argentina	13.2
12	Calcutta, India	4.4	London, United Kingdom	9.8	Cairo/Giza/Imbaba, Egypt	13.2
13	Los Angeles/Long Beach, CA, USA	4.1	Beijing (Peking), China	9.2	Jakarta, Indonesia	12.8
14	Osaka/Kobe, Japan	3.8	Rhein-Ruhr, Federal Republic of Germany	9.2	Baghdad, Iraq	12.8
15	Milan, Italy	3.6	Paris, France	8.9	Teheran, Iran	12.7
16	Rio de Janeiro, Brazil	3.5	Moscow, USSR	8.7	Karachi, Pakistan	12.2
17	Mexico City, Mexico	3.1	Cairo/Giza/Imbaba, Egypt	8.5	Istanbul, Turkey	11.9
18	Philadelphia/Western NJ, USA	3.0	Osaka/Kobe, Japan	8.0	Los Angeles/Long Beach, CA, USA	11.2

19	Greater Bombay, India	2.9	Jakarta, Indonesia	7.9	Dacca, Bangladesh	11.2
20	Detroit, MI, USA	2.8	Tianjin, China	7.8	Manila, Philippines	11.1
21	São Paulo, Brazil	2.8	Delhi, India	7.4	Beijing (Peking), China	10.8
22	Naples, Italy	2.8	Baghdad, Iraq	7.2	Moscow, USSR	10.1
23	Leningrad, USSR	2.6	Teheran, Iran	7.2	Bangkok/Thonburi, Thailand	9.5
24	Manchester, United Kingdom	2.5	Manila, Philippines	7.0	Tianjin, China	9.2
25	Birmingham, United Kingdom	2.5	Milan, Italy	7.0	Paris, France	9.2
26	Cairo/Giza/Imbaba, Egypt	2.5	Chicago/Northwestern IN, USA	6.8	Lima/Callo, Peru	9.1
27	Boston, MA, USA	2.3	Istanbul, Turkey	6.8	London, United Kingdom	9.1
28	Shenyang (Mukden), China	2.2	Karachi, Pakistan	6.8	Kinshasa, Zaire	8.9
29	West Berlin, Federal Republic of Germany	2.2	Lima/Callo, Peru	5.7	Rhein-Ruhr, Federal Republic of Germany	8.6
30	San Francisco/Oakland, CA, USA	2.0	Bangkok/Thonburi, Thailand	5.5	Lagos, Nigeria	8.3
31	Leeds-Bradford, United Kingdom	1.9	Madras, India	5.2	Madras, India	8.2
32	Glasgow, United Kingdom	1.9	Hong Kong, Hong Kong	5.1	Bangalore, India	8.0
33	Jakarta, Indonesia	1.8	Madrid, Spain	5.1	Osaka/Kobe, Japan	7.7
34	Hamburg, Federal Republic of Germany	1.8	Leningrad, USSR	5.1	Milan, Italy	7.5
35	Vienna, Austria	1.8	Dacca, Bangladesh	4.9	Chicago/Northwestern IN, USA	7.2

SOURCE: United Nations (1985: Table A-12).

TABLE 1.2
Population of Metropolises Appearing in Table 1.1 by Major Area (in millions), 1950-2000

Metropolis/Country	*1950*	*1955*	*1960*	*1965*	*1970*	*1975*	*1980*	*1985*	*1990*	*1995*	*2000*
More Developed Regions											
Birmingham, United Kingdom	2.5	2.6	2.7	2.8	2.8	2.8	2.8	2.8	2.8	2.8	2.8
Boston, MA, USA	2.3	2.3	2.4	2.6	2.7	2.7	2.7	2.7	2.7	2.8	2.9
Chicago/Northwestern IN, USA	5.0	5.5	6.0	6.4	6.8	6.8	6.8	6.8	6.9	7.0	7.2
Detroit, MI, USA	2.8	3.2	3.6	3.8	4.0	3.9	3.8	3.8	3.8	3.8	3.9
Glasgow, United Kingdom	1.9	1.9	1.9	1.9	1.9	1.9	1.8	1.8	1.8	1.7	1.7
Hamburg, Federal Republic of Germany	1.8	1.9	2.1	2.2	2.2	2.2	2.2	2.2	2.1	2.1	2.1
Katowice, Poland	1.8	2.1	2.4	2.6	2.8	2.9	3.1	3.2	3.5	3.7	3.9
Leeds-Bradford, United Kingdom	1.9	1.9	1.9	2.0	2.0	2.0	2.0	2.0	2.0	2.0	1.9
Leningrad, USSR	2.6	3.0	3.5	3.8	4.0	4.3	4.7	5.1	5.4	5.7	6.0
London, United Kingdom	10.4	10.5	10.7	10.8	10.6	10.3	10.0	9.8	9.5	9.3	9.1
Los Angeles/Long Beach, CA, USA	4.1	5.2	6.6	7.6	8.4	9.0	9.5	10.0	10.5	10.9	11.2
Madrid, Spain	1.7	2.0	2.3	2.8	3.3	4.0	4.6	5.1	5.5	5.8	5.9
Manchester, United Kingdom	2.5	2.5	2.5	2.6	2.5	2.5	2.4	2.4	2.4	2.4	2.3
Milan, Italy	3.6	4.0	4.5	5.0	5.6	6.2	6.6	7.0	7.3	7.5	7.5
Moscow, USSR	4.8	5.5	6.3	6.8	7.1	7.6	8.2	8.7	9.2	9.7	10.1
Naples, Italy	2.8	3.0	3.2	3.4	3.6	3.9	4.0	4.1	4.2	4.3	4.4
New York/Northeastern NJ, USA	12.4	13.3	14.2	15.4	16.3	16.0	15.6	15.3	15.2	15.3	15.5
Osaka/Kobe, Japan	3.8	4.7	5.7	6.7	7.6	8.0	8.0	8.0	7.8	7.7	7.7
Paris, France	5.5	6.3	7.2	8.0	8.3	8.6	8.8	8.9	9.0	9.1	9.2
Philadelphia/Western NJ, USA	3.0	3.3	3.7	3.9	4.0	4.1	4.1	4.2	4.2	4.3	4.5
Rhein-Ruhr, Federal Republic of of Germany	6.9	7.7	8.7	9.2	9.3	9.3	9.3	9.2	9.1	8.9	8.6
San Francisco/Oakland, CA, USA	2.0	2.2	2.5	2.7	3.0	3.1	3.2	3.3	3.4	3.5	3.6
Tokyo/Yokohama, Japan	6.7	8.6	10.7	12.6	14.9	16.4	17.0	17.2	17.2	17.1	17.1
Vienna, Austria	1.8	1.8	1.8	1.8	1.8	1.8	1.7	1.7	1.6	1.6	1.6
West Berlin, Federal Republic of Germany	2.2	2.2	2.2	2.2	2.1	2.0	2.0	1.9	1.9	1.8	1.8

Less Developed Regions											
Africa											
Cairo/Giza/Imbaba, Egypt	2.5	3.0	3.7	4.6	5.4	6.2	7.3	8.5	10.0	11.5	13.2
Kinshasa, Zaire	0.2	0.3	0.6	0.9	1.4	2.2	3.2	4.3	5.8	7.3	8.9
Lagos, Nigeria	0.4	0.5	0.7	1.0	1.4	2.1	2.8	3.6	4.8	6.4	8.3
Asia											
Baghdad, Iraq	0.6	0.7	1.0	1.6	2.5	3.8	5.7	7.2	8.9	10.8	12.8
Bangalore, India	0.8	0.9	1.2	1.4	1.7	2.2	3.0	4.0	5.2	6.5	8.0
Bangkok/Thonburi, Thailand	1.4	1.8	2.2	2.7	3.3	3.9	4.6	5.5	6.5	7.8	9.5
Beijing (Peking), China	6.7	7.1	7.3	7.6	8.3	8.0	9.1	9.2	9.5	9.9	10.8
Calcutta, India	4.4	4.9	5.5	6.2	7.1	8.2	9.5	11.0	12.6	14.5	16.6
Dacca, Bangladesh	0.4	0.5	0.7	1.0	1.5	2.4	3.4	4.9	6.5	8.6	11.2
Delhi, India	1.4	1.8	2.3	2.9	3.6	4.6	5.9	7.4	9.2	11.1	13.2
Greater Bombay, India	2.9	3.4	4.0	4.9	5.9	7.2	8.5	10.1	11.9	13.8	16.0
Hong Kong, Hong Kong	1.7	2.2	2.7	3.3	3.5	4.0	4.6	5.1	5.6	6.0	6.4
Istanbul, Turkey	1.0	1.2	1.5	2.0	2.8	3.9	5.3	6.8	8.4	10.2	11.9
Jakarta, Indonesia	1.8	2.2	2.8	3.5	4.5	5.5	6.7	7.9	9.3	11.0	12.8
Karachi, Pakistan	1.0	1.4	1.8	2.4	3.1	4.0	5.2	6.8	8.2	10.0	12.1
Madras, India	1.4	1.5	1.7	2.3	3.1	3.8	4.4	5.2	6.1	7.1	8.2
Manila, Philippines	1.6	1.9	2.3	2.9	3.6	5.0	6.0	7.0	8.3	9.7	11.1
Seoul, Republic of Korea	1.1	1.6	2.4	3.5	5.4	7.0	8.5	10.2	11.5	12.7	13.5
Shanghai, China	10.3	10.6	10.7	10.8	11.4	11.6	11.8	11.8	12.0	12.5	13.5
Shenyang (Mukden), China	2.2	2.3	2.5	2.6	2.8	2.9	3.0	2.9	2.9	3.1	3.3
Teheran, Iran	0.9	1.3	1.8	2.4	3.3	4.3	5.6	7.2	9.0	10.9	12.7
Tianjin, China	5.4	5.8	6.0	6.3	6.9	7.4	7.7	7.8	8.0	8.5	9.2
Latin America											
Greater Buenos Aires, Argentina	5.3	6.1	7.0	7.7	8.5	9.3	10.1	10.9	11.7	12.5	13.2
Lima/Callo, Peru	1.1	1.4	1.7	2.3	2.9	3.7	4.6	5.7	6.8	8.0	9.1
Mexico City, Mexico	3.1	4.0	5.2	7.0	9.2	12.1	15.0	18.1	21.3	24.2	26.3
Rio de Janeiro, Brazil	3.5	4.2	5.1	6.1	7.2	8.2	9.2	10.4	11.4	12.3	13.3
São Paulo, Brazil	2.8	3.7	4.9	6.4	8.3	10.3	12.8	15.9	18.8	21.5	24.0

SOURCE: United Nations (1985: Table A-13).
NOTE: The more developed regions include Northern America, Japan, all regions of Europe, Australia-New Zealand, and the Union of Soviet Socialist Republics. Less developed regions include all regions of Africa, Latin America, and South Asia; China, other East Asia, Melanesia, and Micronesia-Polynesia.

TABLE 1.3
Total Population and Number of Cities by Urban Size Category

	Total Population in Size Category (millions)			*Number of Cities in Size Category*			
Year	*4+*	*2-4*	*1-2*	*4+*	*2-4*	*1-2*	*Total*
1950	87.8	47.0	65.5	13	17	48	78
1960	136.3	67.0	95.9	19	26	69	114
1970	186.9	109.2	136.1	23	39	98	160
1980	281.4	140.9	182.3	35	51	136	222
1985	341.6	154.4	219.1	42	56	158	256
1990	405.6	198.0	246.7	48	72	178	298
2000	587.3	290.3	326.4	66	106	236	408
2010	826.7	380.0	399.0	90	139	282	511
2025	1255.0	493.2	457.1	135	182	322	639

SOURCE: United Nations (1985).

of this size in the Third World. By 1985 there were 28. There will be 50 in the year 2000 and 72 in 2010. By 2025, 114 of the world's 135 urban agglomerations exceeding 4 million in size will be in developing countries and their 1.1 billion inhabitants (up from 222 million in 1985) will account for 88 percent of the world's total population residing in these giant metropolises.

The dramatic growth of large urban agglomerations in less-developed regions of the world will parallel anticipated global population trends. For the world as a whole, population is projected to expand from 4.9 billion in 1985 to 8.2 billion in the year 2025. Of this projected 3.3 billion world population increase, a 3.1 billion increase is expected to occur in less-developed countries; 2.7 billion of this 3.1 billion increase will occur in these countries' urban areas (as nationally defined). Thus, in the 40-year period between 1985 and 2025, urban areas of less-developed countries will *add* more inhabitants than the total size of the world's population through the year 1960.

Nearly 30 percent of the urban population in developing nations will be concentrating in metropolises exceeding four million inhabitants in 2025 (up from 19 percent in 1985). For the more developed regions of the world, the percentage of urban population in metropolises of four million plus inhabitants will actually decline from 14 percent in 1985 to 12.8 percent in 2025.

The metropolitan explosion in the Third World will be greatest in Africa and Latin America. By the year 2025, both continents will have nearly half of their urban populations concentrated in metropolises exceeding a million inhabitants and approximately one-third living in metropolises of over four million. Africa will have 36 metropolises of over four million each (compared to none in 1960 and just 2 in 1985), and Latin America will have 21 metropolises exceeding this size (compared to 4 in 1960 and 8 in 1985). In Africa, the total population residing in metropolises of 4 million or more inhabitants will expand from 12.9 million in 1985 to 325 million 40 years later. In Latin America, residents in these giant metropolitan areas will increase from 74 million in 1985 to 197 million in 2025. At this time, the average size of the 36 giant metropolises on the African continent will be just over 9 million, while the 21 in Latin America will average nearly 9.4 million inhabitants.

Forces Behind Giant City Growth

The growth of giant metropolises in Africa, Latin America, and other developing regions is resulting in a multitude of problems of seemingly unmanageable proportions. These include, among others:

(1) high rates of unemployment and underemployment as urban labor markets are unable to absorb the expanding numbers of urban job seekers;
(2) insufficient housing and shelter;
(3) health and nutrition problems;
(4) inadequate sanitation and water supplies;
(5) overloaded and congested transportation systems;
(6) air, water, and noise pollution;
(7) municipal budget crises;
(8) rising crime and other social malaise; and
(9) a general deterioration of the perceived quality of urban life.

Despite these problems, the flood of migrants to the cities continues apace. Why? The answer lies in natural population increase in rural areas, limited rural economic development, and the decision-making calculus of urban migrants. Demographic forces are now well known: declining mortality rates in rural areas of most developing nations have not been matched with corresponding fertility declines. The resulting increase of population cannot be sustained by stagnating rural economies which leads to growing demographic-employment opportunity imbalances in the countryside. Migration becomes the only mechanism to relieve this imbalance. Rural migrants pour into the cities, exacerbating already overcrowded conditions in urban subareas. The age selectivity of rural migrants (largely teenagers and young adults) further contributes to city growth through new family formation and natural increase. What this all means, of course, is that the primary cause of what some have termed "overurbanization" (more urban residents than the economies of cities can sustain) is increasingly severe "overruralization" (more rural residents than the economies of rural areas can sustain). Let us elaborate this important point.

For rural areas to absorb greater numbers of younger people, they must either industrialize, extend agricultural land, or intensify cultivation of presently used acreage. The first option (industrialization) will, at best, proceed slowly. An extension of agricultural land is possible in some Third World countries yet for many it seems that a limit of agricultural density has already been reached. Available statistics, although they do not distinguish clearly between arable land, land under permanent crops, and permanent meadows and pastures, give useful indications of these limits (USAID, 1974:1).

In Asia, where irrigation, labor-intensive farming, and double cropping are common, agricultural density is already very high. Excluding Japan and including China, about two-thirds of the labor force is in agriculture and the average peasant cultivates only one acre. Even if technological inputs were raised, the number of people in agriculture could hardly increase and might even decrease in a country like Bangladesh where there are, on the average, three acres for every ten peasants. In the Philippines or Indonesia, more land could be converted to agriculture, and, consequently, there is still room to absorb additional rural labor, perhaps reducing, but certainly not stopping, substantial emigration to cities.

In Latin America, the number of acres of agricultural land per capita varies from 0.5 in Haiti, 0.6 in Jamaica, 2.2 in Ecuador to 17 in Argentina, 11 in Paraguay, 13 in Uruguay. Thus, in the temperate countries agricultural density is lower than in the tropical countries. In Brazil, which covers about 40 percent of the surface of Latin America, 16 percent of the land is used for agriculture with 3.3 acres per capita. Nevertheless, vegetation is luxuriant in most of the country. Why, then, are millions of Brazilian peasants pushed to cities? Why is agricultural activity so poor in the green Amazon basin? We can look at research on Africa where the pace of urban growth is greatest for answers.

In Africa, the tropical climate can be humid or dry. The number of agricultural acres per capita in dry Africa varies from 78 in Mauritania to 32 in Chad, and the proportion of agricultural land from 5 percent in Libya to 65 percent in Ethiopia (USAID, 1974:5). These figures are approximations, particularly in the desert margins of the West African Sahel where a trend toward increasing aridity has been observed, "a natural desiccation of the region that man can do nothing to stop" (Glantz and Orlovsky, 1986:220). "Desertification is a self-accelerating process, feeding on itself, and as it advances, rehabilitation costs rise exponentially" (p. 214).

In humid Africa, a large part of the territory is not used for agriculture: 70 percent in Zaire, 60 percent in Niger, half in Zambia. Why, then, are the agricultural areas not expanding in these areas?

In his book, *The Tropics and Economic Development*, Andrew M. Kamarck, although rejecting Ellsworth Huntington's old theory on climate as a factor in economic development, explains persuasively why agricultural productivity is so low in the three types of tropical countries—humid, dry, and alternatively wet and dry—and "why tropical countries have lagged during the last two hundred years in the process of modern economic growth" (Kamarck, 1976:10), and will continue to do so for some time. The hazards of annual rainfall and heat, the enemies of people and livestock (bilharzia, trypanosomiasis, malaria, river blindness, parasitic worms, hookworms, yellow fever, dengue, leishmaniasis, etc.) constitute a serious handicap to agricultural productivity.

> Trypanosomiasis is one of the most important obstacles to the economic development of Africa. . . . For centuries, by killing transport animals, it abetted the isolation of tropical Africa

> from the rest of the world and the isolation of the African peoples from one another. . . . Over ten million square kilometers of Africa—an area greater than that of the United States—are thereby denied to cattle (Kamarck, pp. 38-39).

Animal trypanosomiasis prevents the raising of domestic animals and limits the availability of meat, particularly in the most humid regions. "There are still extensive areas of relatively underdeveloped land, often unused because of the presence of trypanosomiasis" (Lee and Maurice, 1983:1). Mechanical plowing and transport could replace animals and human labor. But this would likely result in even less need for agricultural labor—assuming that these countries were financially able and willing to invest in agricultural mechanization.

A reduction of these handicaps would require enormous investments over several decades and there is the likelihood agricultural land extension and intensification would be subject to diminishing marginal returns (United Nations, 1974:269). For the moment it costs less to import cereals than to produce them. Even a country endowed with agricultural resources, like Morocco, imports more than half its wheat needs (Payne, 1986:154), partly in exchange for citrus fruits. Thus, with a few exceptions, the agricultural sector of African and most other Third World regions will not be able to absorb a significant part of their rural demographic surpluses. Emigration to cities is unavoidable.

There is also mounting evidence that bias in Third World public policies toward investing government resources in giant cities (at the expense of smaller cities and rural areas) contributes to the flood of urban migrants. Not only are economic opportunities more abundant in the giant cities (in both their formal and informal sectors) but these cities also further afford migrants access to schools, health clinics, and other public facilities and services unavailable in most of the countryside and many smaller cities. In addition, large urban agglomerations provide myriad stimuli, consumer goods, and cultural attractions. Under such circumstances, the utility function of individual migrants would logically be quite different from that of the city to which they are migrating. Whereas additional numbers of migrants may bring more costs than benefits to the city, migration improves the conditions of the rural migrants who view the economic and social benefits of moving to a large city as substantially outweighing the costs.

The point above was frequently overlooked in the early literature on urban growth in developing nations where giant cities were often depicted as abnormal and unhealthy and life-styles of the residents were cast in appalling terms. Even today, some popular press and academic writers appear surprised when they observe the high degrees of optimism and satisfaction elicited by recent migrants residing in the most densely compacted, impoverished Third World cities. Residents of these cities, in turn, often are puzzled by the Western, middle-class perspectives writers apply to the economic, social, and environmental circumstances confronting the inhabitants. As one participant from Mexico City stated at the Barcelona Conference on Giant Cities: "We didn't feel we were in such dire straits until you told us how awful our conditions were." Perceived deprivation, no doubt, is relative.

Our comments should not be construed as blissful naivete. For sure, tens of millions of residents of giant cities in the developing world live in squalor, poverty, and human depravity. The cities and national economies (on which the overall health of the cities rests) cannot cope with their rates of population increase—either in terms of providing sufficient job opportunities or basic public services. Unlike Europe, which adapted to its periodic demographic-employment opportunity imbalances during the 19th and early 20th century by exporting surplus population to the Americas and other New World colonies, Third World nations do not have such demographic outlets. They only have inlets—their giant cities—which simultaneously serve as demographic reservoirs and islands of hope for those seeking economic opportunity and a better life.

As Ignacy Sachs insightfully observes in this volume, there is a common feature demographically driving large Third World cities:

> they act like a gigantic Las Vegas in the sense that the bulk of their population consists of gamblers. Except that games are different... the rewards may look insignificant when compared with the excessively high price of the tickets in terms of daily life difficulties.... The important thing is that there are rewards for some. The large cities are places of hope, while the drudgery of rural life looks hopeless.

Three Illustrative Cases

The impact of demographic growth in the Third World on the expansion of giant cities can be further illustrated by examining three selected countries: India, Nigeria, and Mexico—one for each of the developing continents.

India

India may be as heterogeneous as Europe from cultural, religious, or linguistic points of view; nevertheless, it is a political entity, the second most populous country in the world and one which is experiencing rapid growth. Between 1975 and 1985 the Indian population grew by 142 million people (from 619 to 761 million). This growth is more than the equivalent of the entire Brazilian population in 1985. In the coming 15 years, India will add 200 million (as many people as the current populations of Japan, Canada, Australia, Scandinavia, and Chile combined) to exceed 960 million by the year 2000. Five years later, it will have expanded by 50 million more, exceeding a billion in 2005. During the following decade and a half, India will gain another 150 million, surpassing the combined populations of North America, western Europe, Australia, and Japan today. By the year 2025, India will likely count nearly 1.2 billion people in a country a third the area of the United States.

More than two-thirds of the Indian labor force is currently in agriculture, and half the territory is cultivated. An Indian peasant has, on the average, only

0.7 acres at his/her disposal. It is not by fragmenting even further the small land parcel ownership that more people will find jobs in agriculture. An extension of agricultural areas seems difficult in this ancient land; there is no "western frontier" in India. As a consequence, cities will have to absorb much of the population growth in the rural areas. In 1985, one-quarter of the Indian population lived in cities; in the year 2000 the urban population will represent one-third; by 2025 more than one-half. In absolute figures, the urban population will grow from 194 million in 1985 to 329 million in 2000, and to 637 million in 2025—an additional 443 million urbanites over the next 40 years. (See table 1.4)

Many of these 443 million will live in relatively smaller cities of less than a million, but the major part will be attracted to giant metropolises. If past trends continue, one might imagine, in the year 2025, one dozen or more urban agglomerations of over ten million each, with a few possibly exceeding 20 million. Also emerging will be 30 to 40 metropolises of between four and ten million residents, surrounded by dozens of cities of over one million each. Bombay, Calcutta, Delhi, Madras, Hyderabad, Ahmedabad, and others may well become the epicenters of enormous megalopolises. Hans Nagpaul, in his chapter on India's giant cities, testifies that, "No scholar or writer from any field wants to predict future needs and basic requirements of food, shelter, clothing, education, and health for a population of about 50 million likely to inhabit these [four] cities [Bombay, Calcutta, Delhi, and Madras] in the year 2000. The task is staggering with bewildering implications."

Nigeria

Nigeria, Africa's largest country, totaled 95 million people in 1985, its population having increased by 28 million in ten years. The population is expected to reach 162 million in the year 2000, a growth of 67 million people in just 15 years or nearly the equivalent of the entire Mexican population in 1980. United Nations projections indicate that by 2010 Nigeria, a country twice the area of the state of California, will contain 228 million people—more people than lived in the entire United States in 1980. If the birth rate does not substantially decline in the near future, the Nigerian population will reach 302 million in 2020. Placed in comparative perspective, the five largest European countries combined (Britain, France, West Germany, Italy, and Spain) have fewer total people today: only about 270 million. By 2025, Nigeria will reach 338 million people, roughly equivalent to the projected population of the entire North American continent in the same year.

About two-thirds of the Nigerian population works in agriculture, encompassing about half the territory. As a result, agricultural density is very high—only two acres per capita. A substantial increase of the rural population seems possible only if accompanied by an extension of the arable land, but this is feasible only to a limited extent, since the richest areas, in terms of soil and climate, are already in use. Even if Nigeria's rural population increased by 50 percent of the projected amount in the next 20 years, the urban population,

TABLE 1.4
Population Projections (in millions)—Asia, 1985-2025

Selected Countries	*Total*			*Urban*[a]		
	1985	*2000*	*2025*	*1985*	*2000*	*2025*
China	1063	1256	1460	224	334	665
India	761	962	1189	194	329	637
Indonesia	165	204	255	42	75	143
Bangladesh	101	146	219	12	27	79
Pakistan	102	143	213	30	54	121
Vietnam	59	78	105	12	21	49
Philippines	55	75	102	22	37	68
Thailand	52	66	86	8	15	37
Turkey	50	68	99	24	40	72
Iran	55	66	96	25	43	75
Burma	39	55	82	12	23	49

SOURCE: United Nations (1985: Tables A-3, A-7).
a. Urban population as nationally defined.

which amounted to 23 percent of the total in 1985, would expand to a third of the country's population by the year 2000 and more than half in 2025 as displaced rural residents flee to the cities.

In absolute figures, the urban population will probably grow from 22 million in 1985 to 54 million in 2000 and to 179 million in 2025. In 40 years Nigeria's urban population will have increased by nearly 160 million. How many of the new urbanites will be located in cities of over five million, and how many in cities of between one and five million? It is highly probable that Lagos, now the largest seaport in Africa, will be the largest metropolis in Africa, competing with Cairo. Given the ethnic heterogeneity of this country, one might expect to see several metropolises attracting millions of rural emigrants. Although Nigeria is growing particularly fast, table 1.5 illustrates that it is representative of many African countries.

Mexico

The population of Mexico will nearly double between 1985 and 2025, unless an unanticipated rapid birthrate decline occurs (see table 1.6). The 79 million Mexicans of 1985 will become 154 million in 2025; the 75 additional million equals the combined population of East and West Germany today.

In this arid country, lacking abundant rivers, the increase of the work force engaged in agriculture is severely limited by water supply. In many areas, pumping water is already too costly.

Half the Mexican territory is used for agriculture. Considering that a large part of this 761,604-square mile country is mountainous or arid, it would be difficult to expand the arable land. Even if a significant part of the increased population remains in rural areas and continues to be occupied in agriculture, a large portion will migrate to the cities. The urban population is expected to increase from 55 million in 1985 to 131 million in 2025, the equivalent of 13

TABLE 1.5
Population Projections (in millions)—Africa, 1985-2025

Selected Countries	*Total*			*Urban*[a]		
	1985	*2000*	*2025*	*1985*	*2000*	*2025*
Nigeria	95	162	338	22	54	179
Egypt	47	65	97	22	36	68
Ethiopia	36	58	112	6	16	54
Zaire	33	52	104	15	29	75
Morocco	24	36	60	10	20	42
Tanzania	22	39	84	3	10	37
Sudan	22	33	55	6	14	34
Algeria	22	35	57	15	27	49
Kenya	21	39	83	3	10	38

SOURCE: See Table 1.4.
NOTE: See Table 1.4.

cities of ten million each. Is there a limit to Mexico City's growth, this bowl constrained by a circle of pollutant-trapping mountains?

Guadalajara and Veracruz might become enormous metropolises. How many new cities of over a million will spring up in the Mexican "desert"? How many Mexicans will head for California, either legally or clandestinely? Much of Los Angeles's downtown area is already predominantly Mexican, as Ivan Light describes in his chapter in volume 2.

The projections we've discussed for these three countries are what the United Nations Population Division refers to as standard projections, based on medium variants of fertility and mortality. Different estimates have been developed by other organizations such as the World Bank, but they do not challenge the basic trends.

Possible Corrections of Projections

The population of a country will continue to grow for some time after fertility declines because the complete effects of such a decline are often staggered by at least two generations. That population can grow even while the fertility rate declines is an established fact for demographers, but a stumbling block for many nonspecialists. For instance, if the average woman has four children instead of five or six, population will still increase spectacularly even with this fertility rate decline. In the extreme case where the birth rate drops to fewer than two children born per couple, the population will continue to grow if the population is predominantly young, as is the case of China today.

The scope of the problem was pointed out by the United Nations over a decade ago: "A typical poor country achieving replacement reproduction by the year 2000 will stabilize shortly after the middle of the next century at about 2.5 times its present population." (United Nations, 1974, p. 269)

Thus, even with greater acceptance of family planning and corresponding

TABLE 1.6
Population Projections (in millions)—Tropical Latin America, 1985-2025

Selected Countries	Total			Urban[a]		
	1985	*2000*	*2025*	*1985*	*2000*	*2025*
Brazil	136	179	246	99	148	219
Mexico	79	109	154	55	84	131
Colombia	29	38	52	19	29	43
Peru	20	28	41	13	21	34
Venezuela	18	27	43	16	24	40

SOURCE: See Table 1.4.
NOTE: See Table 1.4.

reductions in fertility in developing regions, their absolute population growth will be immense through at least the year 2025. At this time, the world will surely be a world of giant cities.

Unfortunately, as a number of chapters to follow will show, the word *developing* is far more appropriate from a demographic than from an economic or social standpoint. Huge population increases could lead to an explosive situation, particularly in the largest cities. Where there is famine in rural areas people tend to die of starvation slowly and silently; in giant cities they tend to revolt! It would be interesting to record, from the *New York Times* or *Le Monde*, the number of riots in various giant cities during the last decade or so. In the past few years alone, the rise in the price of bread was sufficient to engender riots in several cities in Morocco, Tunisia, and Egypt.

Considering the increasing concentration of so many people in giant cities of the Third World, one might be seduced by the concept of overshoot, suggested by Lester Milbrath:

> The fundamental difficulty of our modern large cities is that they are in overshoot. The overshoot concept comes from biology. Think about a species occupying a niche in a given habitat. This habitat has a carrying capacity for that species which is the maximum population level that the habitat can sustain over a long period of time. Normally, population levels of given species are stabilized by such factors as available food supply, activity of predators, incidence of disease, and so forth. It occasionally happens that a species gains access to an abundance of the resources it requires to maintain life; in that event, the species is likely to experience an irruption which is a rapid exponential increase in population. In an irruption condition, reproduction is so swift that it is common for the species to go into overshoot having exceeded the carrying capacity of the habitat. When the species goes into overshoot, it inevitably dies back to a level that habitat can sustain. In an overshoot condition, the population members frantically grasp for any resource that will keep them alive; this generally results in the destruction of some portion of the habitat. It follows that only a smaller population of the species can now be supported in the reduced carrying capacity of that habitat (Milbrath, p. 7).

The World Bank estimates that the gap between food supply and food demand will widen in the developing countries; in the year 2000, 29 such countries may be unable to feed themselves. Already by 1975, 14 sub-Saharan countries lacked enough agricultural production to support their populations entirely. As a group, these 14 countries account for a third of the land area of sub-Saharan Africa and about half of its 1981 population (World Bank Staff,

1984:125). Will Mexico City, Seoul, Teheran, Bangkok, Bombay, Lagos, and other giant cities go into overshoot?

Unanticipated factors may intervene. Malthus was a false prophet because he neglected the impact of technological advances in agriculture and in contraceptive practice. Apropos the latter, some recent discoveries in medical technology and contraception may well have major consequences. Concerned government officials may even take advantage of cultural patterns and emerging techniques to accelerate the decline of the birthrate. For example, amniocentesis and echoradiography, which permit knowledge at an early stage of the sex of the fetus, offer the possibility of responding to cultural aspirations. In some cultures there is a strong preference for sons, particularly in China, India, and the Muslim countries, where traditionally the heir is the son, who is also expected to support parents in their old age.

As a result of this cultural trait, demographers have discovered abnormalities in the demographic pyramids of these countries and pointed to female infanticide. In India, for instance, there were only 935 women for each thousand men in 1981. This figure must be interpreted in the light of there being in almost all countries more women than men because of higher male mortality rates. The disparity in India strongly suggests the practice of female feticide and infanticide, or the deliberate lack of adequate care of female infants. In Bombay nearly 40 thousand abortions were conducted in 1985 to eliminate the fetus identified by prenatal diagnosis as being female.

In China, the demographic pyramid also deviates from its expected shape for similar reasons, a condition more or less tolerated by the authorities. Prenatal diagnosis of sex for purposes of female feticide may expand even more than in India because, according to official policy, the Chinese family is required to have only a single child. If the one child is a girl, transmission of the family name and property are jeopardized. Feticide or infanticide, a painful solution, is culturally accepted.

If prenatal diagnosis of the sex of the fetus becomes a common practice in the Chinese, Indian, and possibly in the Islamic cultures, a rapid decline of the fertility rate may well result one generation later when there will be fewer women who might give birth to even fewer girls. As a result, the projections presented above would have to be modified for the 21st century.

A related technological step has been taken in the recognition in vitro of the y chromosome, which exists only in the male embryo. The technique consists of taking a swab from the embryo. Such a procedure is already applied in France and other countries for cattle breeding. In India, where consumption of beef is forbidden by religion, adoption of this method would favor the birth of cows for milk and avoid the birth of bullocks except as needed for plowing. Apropos humans, and supposing that the strong cultural bias toward sons persists, such a technique might be used to create a disequilibrium between the sexes and ultimately reduce the rate of population growth.

China, India, and the Muslim Arab countries together represent more than half the world's population. If their traditional cultures accepted in a relatively short period the most recent scientific advances for population control, there will be fewer gigantic cities in this part of the world in the 21st century, and these will be smaller than currently projected.

One final possible demographic adjustment (aside from war) should be mentioned. During the 14th century the great plague killed over a third of the population of Europe, disproportionately striking inhabitants of cities. Cholera also decimated many cities. Similar epidemics are not implausible for the future, particularly in giant cities of the Third World where inadequate sanitation and poor health care practices can rapidly diffuse contagious diseases among the concentrated masses. The outcome could be a dramatic reduction in the size of many giant cities of the Third World, accompanied by unparalleled human suffering.

A View of Giant Cities

The thirteen chapters which follow in this volume take diverse views of our world of giant cities. In chapter 1, Mattei Dogan examines the intricacies of giant cities as maritime gateways, considering the world's 195 largest ocean and riverine ports. Chapters 2 and 3 focus on the United States. John D. Kasarda explores the transformation of U.S. cities from centers of goods processing to centers of information processing and derives policy implications for urban revitalization and the underclass. Manuel Castells relates urban dynamics in the United States to high technology and describes the role industrial restructuring plays in uneven spatial development. In two chapters devoted to the urban situation in Europe, Peter Hall discusses urban growth and decline in Western Europe and Jürgen Friedrichs compares and contrasts the spatial ecology of four large cities in Eastern Europe—Budapest, Moscow, Prague, and Warsaw. Four chapters have an Asian focus. In chapter 6, Yeu-man Yeung discusses problems, policies, and prospects for the largest cities of eastern Asia. Sidney Goldstein examines the most recent census data on urbanization in Mainland China in chapter 7 and Xiangming Chen assesses the role of China's giant cities in the nation's urban hierarchy in chapter 8. Hans Nagpaul's chapter investigates problems of some of India's immense cities, focusing on their slums and squatter settlements and assessing policy options. Dennis A. Rondinelli writes on patterns of urbanization in Africa and the need to direct growth away from primate cities to secondary cities, while Guillermo Geisse and Francisco Sabantini look at alternative development plans in the large cities of Latin America, emphasizing problems of shelter and housing.

The closing chapters of this volume are international in their perspective. Ignacy Sachs concentrates on the vulnerability of giant cities and proposes a theory based on a life lottery concept. Henry Teune's concluding chapter deals with the growth and pathologies of giant cities, a theme central to the Barcelona Conference, within the context of the new global economy.

A companion volume, *The Metropolis Era: Mega-Cities*, contains detailed case studies of ten giant cities: Mexico City, São Paulo, Shanghai, Delhi,

Cairo, Lagos, Tokyo, New York, Los Angeles, and London. Together, these two volumes provide broad and thorough coverage of the dynamics of giant city developments as well as their problems and future prospects.

Bibliography

Glantz, Michael H., and Nicolai S. Orlovsky. 1986. "Desertification: Anatomy of a Complex Environmental Process." In *Natural Resources and People: Conceptual Issues in Interdisciplinary Research,* edited by Kenneth A. Dahlberg and John W. Bennett. Boulder, CO: Westview Press.

Kamarck, Andrew M. 1976. *The Tropics and Economic Development: A Provocative Inquiry into the Poverty of Nations.* Baltimore: Johns Hopkins University Press for the World Bank.

Lee, C. W., and J. M. Maurice, eds. 1983. *The African Trypanosomiases: Methods and Concepts of Control and Eradication in Relation to Development.* World Bank Technical Paper no. 4. Washington, DC: World Bank.

Milbrath, W. Lester. 1985. "Pathologies of Giant Cities." Paper presented at the Barcelona Conference on Giant Cities, 25 February-2 March, Barcelona.

Payne, Rhys. 1986. "Food Deficits and Political Legitimacy: The Case of Morocco." In *Africa's Agrarian Crisis: The Roots of Famine,* edited by Stephen K. Commins, Michael F. Lofchie, and Rhys Payne. Boulder, CO: Lynne Rienner.

United Nations. 1974. *Report on the World Social Situation, 1974.* (E/CN.5/512/Rev.1; ST/ESA/24). New York.

United Nations. Department of International Economic and Social Affairs. 1985. *Estimates and Projections of Urban, Rural and City Populations, 1950-2025: The 1982 Assessment.* (ST/ESA/SER.R/58). New York.

USAID. See U.S. Agency for International Development.

U.S. Agency for International Development. 1974. *Selected Economic Data for the Less Developed Countries.* (RC-W-136). Washington, DC.

World Bank Staff. 1984. *Population Change and Economic Development.* New York: Oxford University Press.

1

Giant Cities as Maritime Gateways

Mattei Dogan

How many of the world's giant cities are located on waterways? The main difficulty in analysis does not come from the ubiquity of the phenomenon across all continents, but from its historical dimension. Giant cities do not fall from the sky as giants. All great cities of today were born as villages or towns a long time ago. They grew over decades, centuries, or millenia. Thousands of fishermen's villages have *not* become giant cities.

Why did one site grow rather than another? For many cities of today, the reason remains a mystery; for many others the reasons are known. The Romans built a bridge at the Thames' lowest bridgeable point, so it was there, and not 20 miles farther east or west that London arose. Bishop Absalon built a castle on a little island in 1167 for protection against a Slavic tribe. He was unaware that his fortress was destined to become Copenhagen. Often the official history of a city bears little resemblance to the actual circumstances of its founding. Thus, the site of some ancient cities is attributed to a mythological founder guided by angels in the selection process. The histories of cities too old to have reliable records are steeped with imagination. For example, Romulus, founder of Rome, was said to have been suckled by a she-wolf. Those cities which appear to have recorded their origins accurately, such as colonial cities, reveal what was important to their founders in the way of location.

By location, we should understand not only the site where the old core of the city was established, but also the larger ecological context, considering an area of a diameter of two or three hundred kilometers surrounding the original site. Such a large environment is best perceived in images taken by satellites like the Landsat spacecrafts, flying at a height of 917 kilometers (570 miles). A single Landsat scene covers a ground area of more than 33,000 square kilometers and more than 30 million numbers must be processed before the image can be created (Sheffield, 1983:5). Looking to these spectacular images, we understand more rapidly than by using classical maps the rationality of the location of Cairo at the head of the delta which includes nearly half of the agricultural area of Egypt. No description could be as explicit as the Landsat scene of Rangoon, Shanghai, Stockholm, Jakarta, Tokyo, Montevideo, Khartoum, Istanbul, Leningrad, or Montreal (Sheffield, 1981, 1983; Short et al., 1976).

These large ecological contexts help us explain why a particular fishing village became a giant port and why thousands of other villages did not. If the time dimension is taken into consideration (i.e., technological development), most of the cities of the world appear to be built in the "right" place and not, as Charles Sheffield believes, in the "wrong" place (Sheffield, 1981:44), except in Africa.

For most giant cities, a variety of factors played different roles at different moments. What was important in growth of a city at a certain moment might cease to be the reason for continued growth later. Even if it were possible to analyze the factors of growth for each city—and there is an enormous literature on this topic—it is more difficult to make comparative generalizations.

In addition to the historical dimension, there are difficulties in interpreting seemingly straightforward, physical facts. A physical fact is not a sociological fact. It is not obvious, for instance, what defines *navigable waterway* or *access* to navigable water. The Seine and the Mississippi were more "navigable" in earlier periods than they are today. There are numerous examples of rivers that have lost some of their navigable quality. Navigation at the time of the Roman Empire implied a different technology than that in ancient Greece. At the time of Socrates, the "civilized" world, in the Western tradition, was limited to the Aegean Sea. Later, the Mediterranean Sea became the *mare nostrum*. Nearly two thousand years later, the Atlantic became a lake that could be crossed in less time than was required earlier to cross the Mediterranean from Syracuse to Carthage. Thus, *navigable water* does not have the same meaning at all times.

What is a port? From a geographical point of view, a port is a harbor. But a deep harbor is not enough for growth of a giant city. Hundreds of spectacular harbors are devoid of significant populations because they lack rich hinterlands. Many giant cities are ports without being on sea coasts: They have satellite ports. One could not conceive of these cities as not being seaports. Their economies would decline or collapse if they lost access to the sea. Already in ancient time, Athens had the port of Piraeus. Today the two cities form one giant agglomeration. And, although Rome is located 20 miles inland, the Roman Empire was a maritime power, since the nearby port of

Ostia was able to receive the sailing ships of that time. Tokyo has the port of Yokohama; Seoul has Inchon; São Paulo engendered Santos; Lima integrated Callao; and Caracus, Guaira. I consider as a seaport any giant city having functional access to a harbor of sufficient size to serve its needs. In some cases, the giant city and its port do not constitute a single metropolis; between the two, there is a rural area. For instance, Santiago and Valparaiso are not a compact agglomeration.

Giant cities are defined, for purposes of this analysis, as cities or metropolitan areas with more than one million inhabitants. In 1985, there were, according to the most recent estimations by the United Nations, 285 such cities. More recent statistical information for some countries requires us to raise the number of giant cities to nearly 300 by 1986 (see table 1.1).

Estuaries, Deltas, Bays, Isthmuses, and Archipelagos

From ancient times, history has navigated—so to speak—on great and prestigious rivers. Everyone knows that Egypt and Sudan—particularly Cairo and Khartoum—are gifts of the Nile. The Euphrates and Tigris rivers were the playground of ancient civilization. Western European history is inconceivable without taking into consideration the transportation function played by the Rhine. The same may be said of the Mississippi in American history. The Danube ties together four capitals: Vienna, Budapest, Bratislava, and Belgrade; at the end of the 19th century, it was the grand avenue of central Europe. The Dnieper and Volga played a role in unifying Russia. The Ganges, Amazon, and the Yangtze are the arteries of continents. The symbol of many great cities, Paris and London among them, is a ship on a river.

Many giant ports are located on estuaries (where an ocean's tide meets a river current). Geologically, an estuary is usually associated with a submergent coastline in tectonically stable areas. For example, many more estuaries are on the east coast of the United States than on the west. If they are deep enough, estuaries make fine harbors, as evidenced by the high proportion of large cities located on them. Within an estuary, cities tend to be located where water deep enough for navigation meets firm land. Most of the factors conditioning where in the estuary the city is located are local in origin, making it difficult to reach generalizations. Land near the mouth of estuaries is usually low, swampy, and saturated with saltwater.

Most of the largest ports in Europe are on estuaries: London, Glasgow, Liverpool, Hamburg, Kiel, Rotterdam, Antwerp, Amsterdam, and others from Lisbon to Leningrad. Many great Asian ports are deltaic: Karachi, Rangoon, Calcutta, Bangkok, Canton, Jakarta, Shanghai. Other ports are located on archipelagos: Vancouver, Rio, Singapore, Sidney, Stockholm, Izmir, Copenhagen, and Lagos among many others.

Cities with riverine harbors tend to be smaller than coastal cities. Although

TABLE 1.1
Cities of Over One Million Inhabitants (1985)

	Navigable Waterfront	*Others*	*Total*
Old Countries			
Western Europe	32	15	47
India	3	21	24
China	23	3	26
Japan	10	1	11
USSR	11	10	21
Eastern Europe	3	3	6
Middle East and South Mediterranean	12	6	18
Subtotal	94	59	153
New Countries with Early Urbanization			
United States	26	10	36
Canada, Australia, New Zealand	7	—	7
Mexico	1	4	5
Subtotal	34	14	48
Former Colonial Countries			
Africa	25	5	30
Southeast Asia	15	2	17
South America	15	9	24
Central America	9	—	9
Other (Korea, Sri Lanka, etc.)	3	1	4
Subtotal	67	17	84
Total	195	90	285

often closer to their hinterland sources, their finished products must travel farther to reach the sea lanes. Because it is often easier to transport finished products than the raw materials required to manufacture them, an inland location minimizes transport costs. An industrial city which produces bulky goods tends to locate closer to its markets and farther away from its source of raw materials. Obviously these remarks apply only to cities primarily engaged in manufacturing. Only cities engaged in providing goods or services for regions besides their own reach large size. Transportation is essential for them, and water provided the earliest means of bulk transport.

Cities producing only for their own surrounding area are limited in their growth to the needs of the area they serve.

It is useful to distinguish four types of countries. First are countries without access to the sea, as in Europe: Switzerland, Czechoslovakia, Hungary, and Austria; in Latin America: Bolivia and Paraguay; in Africa: Sudan, Nigeria, Mali, Chad, Burkina Faso (formerly Upper Volta), the Central African Republic, Zimbabwe, Zambia; in the Middle East: Jordan and Afghanistan.

Some of these countries benefit from important navigable rivers. Only 16 among the 150 independent nations (excluding ministates) belong to this category of countries deprived of a seacoast.

Second are the large continental countries, with enormous populations and coastlines relatively limited in comparison with their surface: China,

India, the Soviet Union, and the United States. In these four countries, many giant cities began as riverine ports. Mexico is also a continental country, but without navigable rivers.

The third category is comprised of the coastal countries that, in contrast to the continental states, are open to the sea: the two great islands, Britain and Japan; the less populated islands: Madagascar, Sri Lanka, Cuba; the countries elongated on the coast: Portugal, Chile, Norway, Denmark, Vietnam, and Senegal; and the archipelagic countries, like Indonesia and the Philippines.

The fourth category includes all other countries where there is a kind of equilibrium between coastline and land surface.

Let us take an imaginary round-the-world trip, like Jules Verne, and sail from continent to continent, collecting empirical evidence.

The first evidence, so patent that it tends to be neglected, is that Europe is the only continent without an empty interior. All others have enormous deserts at their centers: the Sahara, the Gobi, the Gibson Desert, the Rocky Mountains, the Cordilleras, the frozen interior of Canada, the Arabian desert, and two green, tropical deserts: the Amazon and the Congo. On a world scale, this configuration contributes among many other factors of a different kind to explain the coastal orientation of so many giant cities.

Western Europe

Among the 47 European cities with more than one million inhabitants in 1985, 18 are seaports and 14 riverine ports. If many contemporary European giant cities are not ports today, this does not imply that some of them have not had a navigable waterfront in the past, either by river or canal. Berlin, Frankfurt, and Hanover are examples of cities with historical ports which are no longer active.

The proportion of ports is higher when one considers only the 20 largest cities: London, Athens, Barcelona, Glasgow, Hamburg, Lisbon, Marseille-Fos, Naples, Stockholm, and Pôrto. To this list we could add several riverine ports on the Rhine and Danube, and Paris, which by tonnage is the third largest French port. The Rhine, together with its effluents and associated network of canals, is the artery of the Rhine-Ruhr megalopolis (nine million inhabitants). Nevertheless, in comparison with other continents, Western Europe appears to be centripetal, in the sense that the largest portion of its population lives in the central part of the continent, not along the coasts.

To understand the social geography of contemporary Western Europe it is necessary to take into consideration the past. As Paul White (1984:3) observes, "the origins of most European cities are unknown and cannot be dated. Few cities have been created as urban places on virgin territory: most have evolved from rural settlements through long processes of economic and population growth and the accretion of new functions." The ancient Greeks founded tiny cities which are today the giant cities of Naples, Bari, Marseilles, and Valencia, to name a few. Other pre-Roman settlements include Barcelona,

Milan, Trieste, Bologna, and Palermo. "By the end of the Roman Empire a high proportion of Western Europe's largest present-day cities were already in existence" (White, 1984:3). Of the major cities of Italy, only Venice is a post-Roman creation, and in Spain, only Madrid and Bilbao.

Not all the towns of previous centuries or millenia later became giant cities, but most of today's giant cities were already in existence before the industrial revolution. A town is like a tree—it grows where the seed was planted; only in exceptional cases can a previous population center be shifted elsewhere.

The European map of great cities can be read not only in ancient history, but also, and to an even greater extent, in the economic history of the last century and a half. Coalfields and river, road, and railway transport played a role in the growth of many cities. Even today, and in contrast to the United States, railroads and motor roads—in part because of the smaller size of the European countries—are serious competitors to waterways (except for oil transportation).

Some cities are giant today because for a long period they were capitals of smaller states, before the unification of territory into nation-states, like Naples or Munich.

Many historians have demonstrated the importance of ports and maritime trade in Europe's economic development. But Europe has been a continent of colonizers, not a continent to be colonized. Europeans have founded colonial ports in many parts of the world. Their own ports have benefited from trade with other continents, but modernization and innovation did not penetrate Europe through its ports. Europe is the only continent which never had colonial ports.

Today, trade among European countries represents the largest part of the foreign trade of each European country. This fact was so even before creation of the European Common Market. Consequently, Europe does not need seaports to the same degree as other continents. Most of the traffic is inland.

The United States

Among the 36 cities in the United States with over one million inhabitants in 1980, only 10 have no waterfront, and only one of them has a population surpassing two million (Dallas). All metropolises over four million are seaports or on the Great Lakes: New York, Los Angeles, Chicago, Philadelphia, Detroit, and San Francisco.

Many of the 26 ports are riverine ports, founded before the advent of rail transportation. At that time the Mississippi-Ohio water route system comprised "more than 35 rivers aggregating over 12,000 miles of navigable water, spread like a tree across the most fertile part of the nation" (Mackenzie, 1933:130). Another important waterway system was the Hudson River-Erie Canal-Great Lakes.

American cities grew at points of intersection between water and land transportation. The railways and the motorways have not determined the initial location of the giant cities, even if they have contributed to their continuing growth. The most famous case is Chicago, strategically located near the southern tip of Lake Michigan and the terminal point of railways reaching into the interior. Chicago is one of the best examples of how a break in transportation favors growth of a city. Cities at junction points became gateway cities. As McKenzie (1933) notes, the motor vehicle and motor highway have not created new large cities. They have only contributed to further growth of existing large cities, born during the era of waterways and growing during the development of railway transportation.

It is not necessary to review here the impressive literature on most of the American giant ports, particularly because two chapters of this book focus on the United States—those by John D. Kasarda and Manuel Castells. Volume 2 contains case studies of New York City and Los Angeles.

Mexico

Mexico is a mountainous country, crossed along the Atlantic and Pacific by two chains of mountains which meet in a volcanic axis. Between the Sierra Madre Occidental and Sierra Madre Oriental lies the Mexican Plateau, the country's most densely populated region: 90 percent of the Mexican population lives on the plateau, at more than 500 meters altitude, where water is scarce. Areas of temperate climate are determined by altitude rather than by latitude. The presence of mountain ranges is one reason why Mexico lacks large rivers; no stream system has yet eroded a sizable watershed. Climatically the upland areas are called the *tierra templada* (temperate land). It is not surprising, then, that we find Mexico's three largest cities located not on the seaboard or lowlands, but in the *tierra templada*. The climate explains why, in spite of 10 thousand km of coastline, Mexico has few important ports.

Located on the southern end of the Mexican plateau, Mexico City, with its suburbs, is one of only two megalopolises in the world without direct access to the sea or an important river passing through it (the other is Teheran). Historically, the Aztecs founded the first settlement on an island in a shallow lake. The lake's location, at an altitude of 7,350 feet, gave it a moderate climate. Well before European colonization, the upland Mexican plateau areas had been the country's most productive region. The Aztecs exercised hegemony over the other aboriginal groups. Their selection of the central plain freed them from many of the unwholesome attributes of the tropical lowlands and must be considered influential in their creation of a great civilization. In contrast, rival groups inhabiting the humid lowlands did not achieve the level of society that the Aztecs did (except the Maya of Yucatan). Consequently the only Indian concentrations which could be termed cities before the arrival of the Spanish were high above sea level in the *tierra templada*. When the Spanish conquistadores invaded Mexico, they quickly recognized the benefits of residing on the plateau rather than in the coastal lowlands. On the coast the soil was leached, an unsatisfactory condition for growing crops. The humidity was extremely uncomfortable and fostered the presence of disease-carrying insects and parasites. For the same reasons, subsequent invading armies, such as the French and the American, also sought to avoid the coastal lowlands as much as possible. During the warm season, yellow fever would ravage any army foolish enough to remain in the lowlands. Mexico is unique because most nations of the tropics do not possess easily accessible highlands, and are therefore forced to build their cities in the coastal lands.

The port city of Vera Cruz, on the Gulf of Mexico, prospered as Mexico City grew, but its generally unhealthful climate in the yellow fever-ridden lowlands prevented it from reaching a great size. Moreover, it is located relatively far from the plateau, connected by a road which traverses two mountain passes. Today Vera Cruz ranks as one of Mexico's leading ports, but still remains relatively small.

In contrast, Guadalajara and Monterrey were not established cities prior to the Spanish Conquest. They are located in the *tierra templada*, although they are both on the seaward side of the crest of the mountains. Only after construction of a railroad terminus from Laredo did Monterrey show signs of growth. The completion of the Pan-American highway in 1930 further benefited the city, until its growth was checked by lack of an adequate water supply. Today Monterrey is the chief beneficiary of the Falco Dam on the lower Rio Grande, and for the present sufficient water is available. The city has industrialized.

Guadalajara, on the other hand, grew immediately following its founding, chiefly as a center for the Indian slave trade. The lands surrounding Guadalajara are among Mexico's most fertile and productive, enabling the city to grow as an agricultural center. During the postwar period, the Mexican government encouraged capital investment outside Mexico City, and as a result Guadalajara has enjoyed rapid industrialization. The city is near enough to Mexico City to receive capital, but far enough away to avoid being swallowed into the megalopolis.

In her chapter on Mexico City in volume 2, Martha Schteingart describes the problems this mega-city deprived of a waterfront must face.

South America

Of the 24 cities of South America with a population of more than one million inhabitants in 1985, 14 are seaports (considering São Paulo-Santos and Santiago-Valparaiso as metropolitan areas at the sea). Only Bogota, Medellin, and Cali in Colombia; Rosario and Cordoba in Argentina; Brasilia, Belo Horizonte, and Curitiba in Brazil; Quito in Ecuador, and La Paz, capital of a country without seacoast, are not ports. Of these inland cities, only three have three million or more inhabitants: Bogota (3.8), Medellin (3.0), and Belo Horizonte (3.8). (See Figures 1.1 and 1.2.)

Five of the ten capitals of this continent are seaports: Buenos Aires,

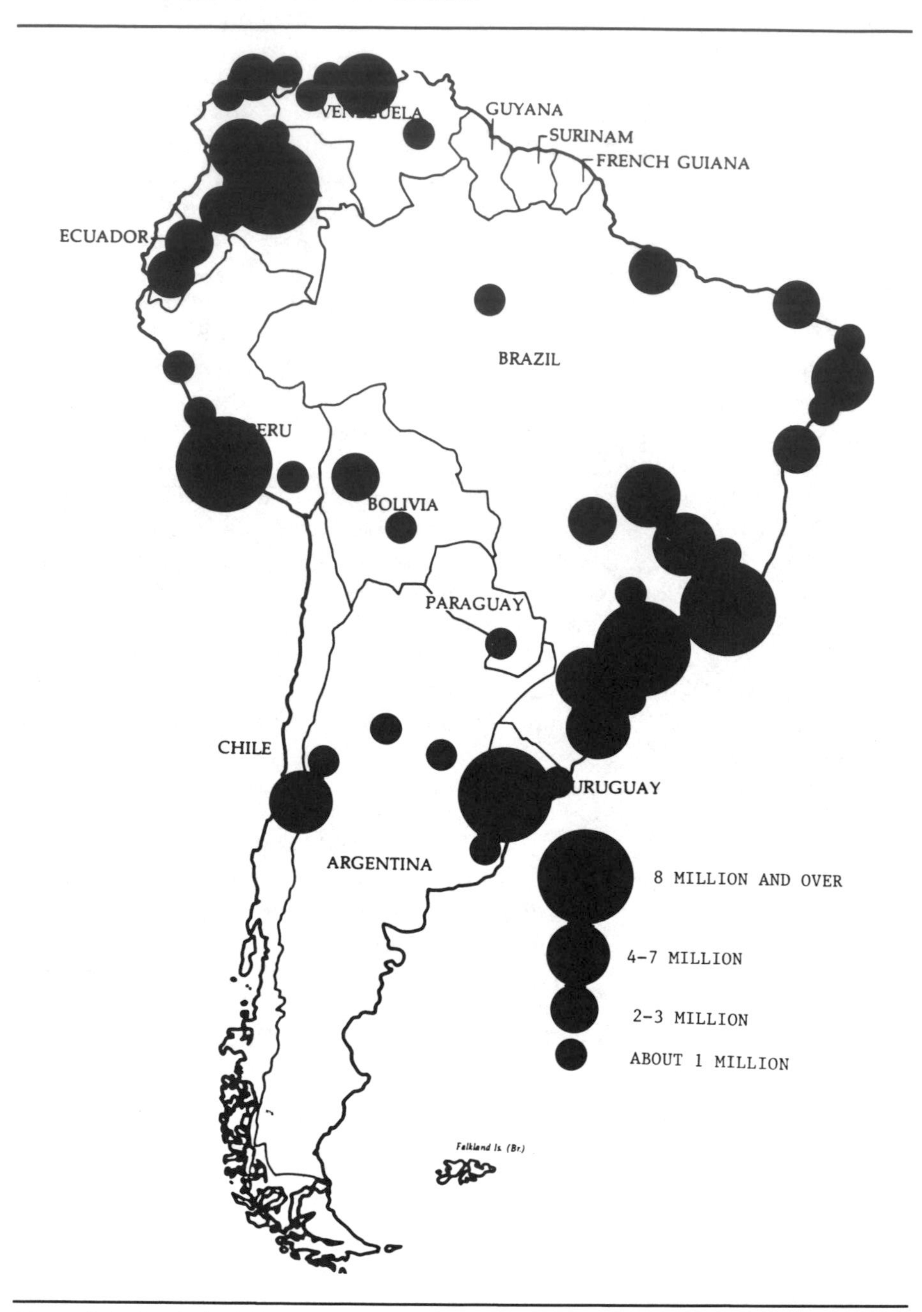

Figure 1.1: South America

SOURCE: Copyright held by Mattei Dogan, 1987.

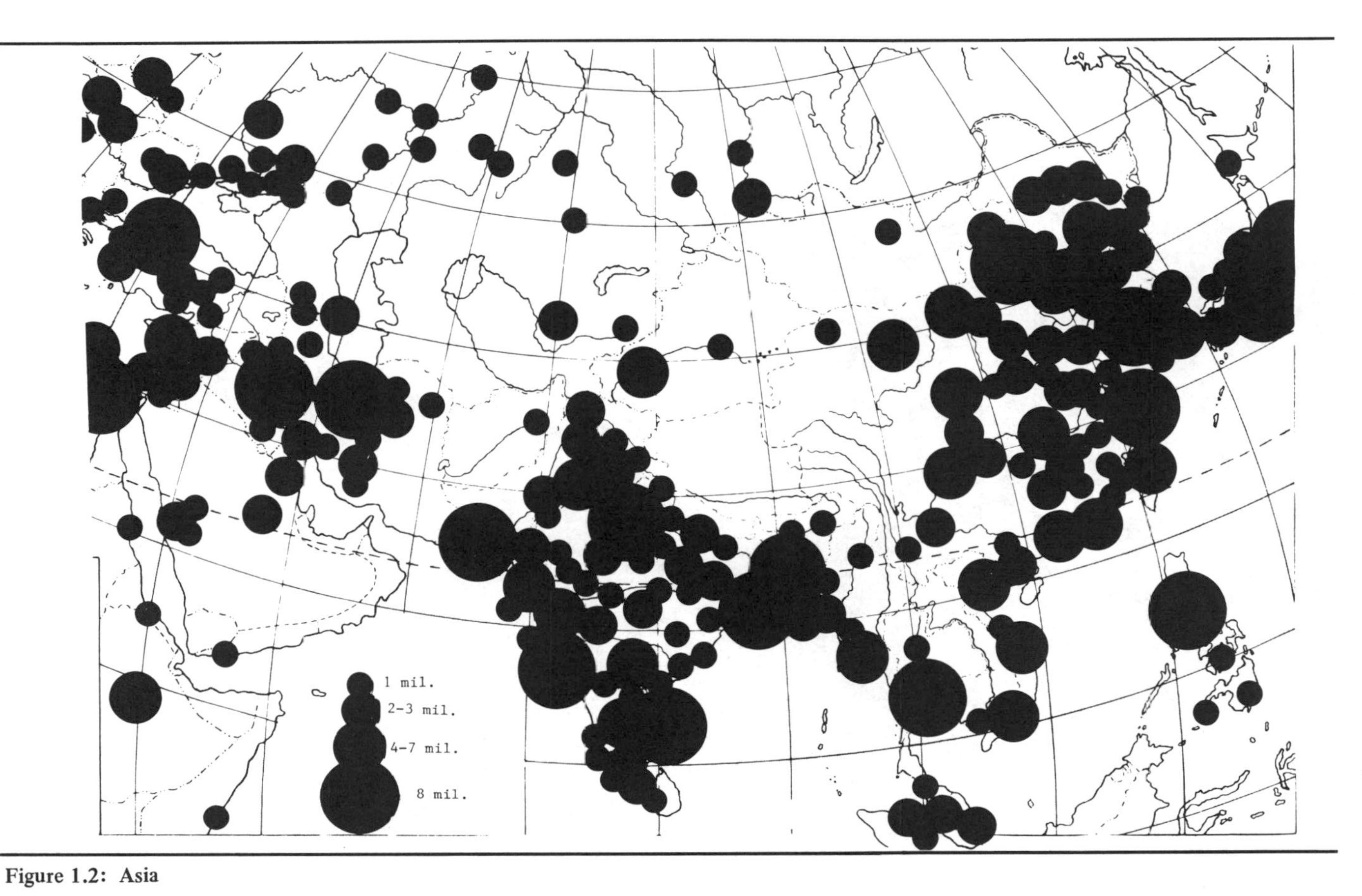

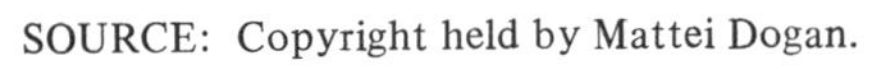

Figure 1.2: Asia

SOURCE: Copyright held by Mattei Dogan.

Santiago, Lima, Caracas, and Montevideo. All except the last one are enormous metropolises. Only one of the capitals that are not ports is a metropolis (Bogota); the others have just over one million inhabitants (Quito, La Paz) or less (Asuncion, river port capital of a country without a coast). It is significant than in Ecuador, the seaport of Guayaquil is larger and growing faster than Quito, the capital.

One case in this picture merits particular consideration: São Paulo, one of the greatest metropolises in the world, located 30 miles from the ocean on the coastal ridges of the Brazilian highlands. By choosing a location above the coastal lowlands (at about 2,600 feet) the Portuguese settlers sought to avoid the humid coastal climate. São Paulo grew without any plan into a congested colonial city. As the population of Brazil grew, São Paulo kept pace until the 1930s, at which time its population began to increase far more rapidly than Brazil's. Such a booming metropolis needed a port, and it effectively engulfed the port of Santos, now a suburb of São Paulo, just as Tientsin is becoming a suburb of Peking, as, in the past, Yokohama became a suburb of Tokyo, Callao a suburb of Lima, and as Inchon became integrated into the metropolitan area of Seoul. See chapter 10 in volume 2 for a more complete discussion of São Paulo.

In contrast to the unplanned explosion of São Paulo, Belo Horizonte was Brazil's first planned city. Located on the western slopes of a range in the Brazilian highlands, Belo Horizonte's altitude confers on it a moderate climate. The city was planned to be the hub of the large agricultural region north and inland from Rio de Janeiro. The city itself was laid out on a grid modeled after Washington, DC, and La Plata in Argentina. Still relatively isolated, however, the area appears to be undergoing a period of mild growth. Such virtues as a relatively dry climate and uncrowded conditions could, however, be an advantage for future growth.

Three Brazilian cities, Porto Alegre, Recife, and Salvador are located on estuaries: Rio de Janeiro on a fine natural inlet, Fortaleza on a semiprotected bay at the mouth of the Rio Pajeu. The estuarine cities were founded on good harbors and have since grown steadily along with the rest of the country. The physical setting of Rio is a densely built-up enclave rimmed by steep mountains. The bay on which Rio lies has the advantages of deep water and enormous size. Historically, Rio has grown steadily since its founding as a Portuguese port. "Rio does not present the radial distribution of many European and American cities. As its expansion was barred by the mountain and the swamps, it was forced to extend along the seacoast, stopped only by the hills" (Rios, 1972:827).

Comparatively, these tropical cities are not as uncomfortable climatically as their Indian or Indonesian counterparts. Although they are indeed humid, they are not as inundated with water as many Asian cities, because of the moderating effects of the offshore winds.

Buenos Aires is located on the broadest river in the world, and on one of the deepest bays. Its growth has been favored by this excellent location, but historical, economic, cultural, and political factors have also played a vital role. Rosario is the only city of over one million in South America which is a riverine port (Manaus has not yet reached one million).

In contrast to the relative emptiness of the interior, there are large urban concentrations along the Atlantic and Pacific coasts of South America. The wild interior of Brazil, while rich in natural resources, has prevented widespread implantation of European immigrants. Currently the upland areas of the Brazilian highlands are undergoing population expansion. These areas are near enough to the coastal centers to avoid isolation, while at the same time they enjoy a more moderate climate as a result of the altitude. A similar expansion can be observed in Colombia.

To explain this maritime orientation, the historical dimension must be added to the geographical features already discussed. Today's giant South American cities were born as colonial towns for colonial overseas transport, founded by Spaniards and the Portuguese—people whose cultural traditions favored monumental capitals (Violich, 1975:247-48). The foreign trade of these countries and the technological development of maritime transport contributed at a later stage to growth of these maritime gateways.

Today, more than three-quarters of the imports and exports of these countries are channelled through their ports, inland trade among the countries being minimal and air transport being limited to certain light goods. For transporting people, South America has moved from maritime traffic directly to air traffic, even for internal travel, skipping, like Africa and Australia, the railway network. To the giant harbors have been added giant airports.

In 1985 South America had 24 giant cities. According to the most realistic projections, by the year 2000 this continent will have 45 "million" cities. The people who will emigrate to these giant cities—and will enlarge some of them—are already born, and many are already looking to these ports to find a job.

Japan

Japan is a chain of islands. It is today a nation oriented toward the ocean, to an even greater extent than England. Almost all important Japanese cities are situated on the coast, partly because the inland is mountainous. The mountains, covering 70 percent of Japan, rise abruptly. River valleys are steeply cut and the terrain is rugged. The few level areas of any size occur when rivers enter the sea in protected bays, thus enabling their sediment to accumulate without being swept away. These delta plains are highly prized lands in Japan and are the foundation for such large cities as Osaka and Nagoya. Climate also plays a role in favoring Pacific sites, as there is a considerable climatic difference between the Pacific Ocean coast and the Japan Sea coast. Tokyo's average sunshine during July is 175 hours compared with 75 hours for Niigata opposite Tokyo on the Japan Sea coast. Eight of the nine Japanese giant cities are located on coastal alluvial plains while one occupies a flat saddle between two mountain groups.

Demographic pressure has resulted in the urbanization of virtually all delta lands. The three largest coastal plains are located relatively near each other; their cities have now expanded to the point where they meet each other along the mountainous divides which separate them. Because they are tied together by transportation and communication routes they have formed a megalopolis along central Japan's Pacific coast. Within this metropolitan area lie what were originally six autonomous urban centers. Although still administered individually, they are more and more being recognized as merely components of what has been dubbed the Tokaido Megalopolis.

Tokaido means in Japanese "east coast road." The approximate figure for the 1985 population of the megalopolis would be more than 60 million people. The three core areas have stretched "out towards each other, with more and more residences climbing up into the hills as the flat land becomes increasingly scarce" (Nagashima, 1981:283). Within the megalopolis lie the component cities of Tokyo, Kawasaki, Yokohama, Osaka, Kyoto, Nagoya, and Kobe. The three plains on which these cities lie are within 600 kilometers of each other. Each plain meets the sea in a bay, providing the urban center with a good harbor.

Settlements have been located along the Tokaido since feudal times. In fact, with flat land so scarce, it seems inevitable that they were destined to become enormous urban centers. With the advent of industrialization, the Tokaido cities were in the forefront, accelerating their growth even more. Growth in the Tokaido area has been planned to some extent. Unlike the cities of the eastern seaboard of the United States, which haphazardly grew into each other, the Japanese have sought to make the Tokaido component cities complement each other in a productive manner. This concept has been termed the *Pacific Belt*, a plan that calls for concentrated capital investment in the Tokaido cities to create a polynuclear urbanized system. Although it has become common in the United States to associate the notion of megalopolis with the excesses of capitalism, Japanese theorists have taken a different approach. Urban thinkers view the megalopolis as an "alternative growth pattern suitable for a densely populated society, linear and efficient compared with the metropolitan concept, which is centripetal, and therefore inefficient with respect to information links" (Nagashima, 1981:282). Tokyo is one of the case study cities in volume 2.

Fukuoka and Kita Kyushu are physically similar to the plains of Tokyo, Nagoya, and Osaka. Both of these ports have used all their available flat land and are currently expanding gradually into the surrounding hillsides.

The only large Japanese city which is not a seaport, Sapporo, occupies the only sizable expanse of flat land not composed of alluvium. Located on a saddle between two mountain chains, Sapporo is the largest city on the island of Hokkaido. The factors for its growth can be assumed to be the same as for other large Japanese urban centers: availability of sufficient flat land to allow a city to grow.

China

In China in 1982, there were 27 cities of more than one million inhabitants. In relation to the size of its population, China has proportionately fewer giant cities than any other major area in the world. Almost all of the 27 cities have a waterfront. Some are seaports, other riverine ports, and some are reached by canals.

China's coastline forms a large curve stretching 8,700 miles from the Yalu River in the north to the Gulf of Tonkin in the south. Some of China's largest cities are located where the major rivers reach the ocean, situated slightly inland on large estuaries which make fine harbors.

Shanghai (13.5 million inhabitants) is located on the left bank of the Huangpoo River at its junction with Suchon Creek and between the mouth of the Yangtze River to the north and the bays of Hangchow and Yu'pan to the south. The city stands on hardened silt deposited by the Yangtze, forming a flat alluvial plain approximately 15 feet above sea level. The metropolitan area has expanded its total area by setting up state farms around the coastal areas to desalinate the soil and extend the usable land by means of dikes. Shanghai's early population first increased substantially with the influx of refugees during the Mongol period. In more recent times, it was one of the first Chinese cities to be opened to Western commerce after its capture by the British in 1842. The British recognized the advantages of Shanghai's deep water harbor and her strategic location at the mouth of the Yangtze, the natural gateway to the agricultural hinterlands. The area around Shanghai has grown so that structurally the Yangtze delta is a metropolis with Shanghai at the center surrounded by satellite cities (Nanking, Suchou, Wuhsi, Changchou). For a more detailed discussion, see chapter 5 in volume 2.

Similarly, Canton (Kuangchou) is located inland, lying 90 miles from the South China Sea on the Chu Chiang River. The coast around Canton is low and swampy, so the city was located inland on firmer ground, but still within easy navigating distance up the estuary. The Chu Chiang River drains the vast Kwangtung province that provides the products for Canton's commerce. Like Shanghai's, Canton's population grew enormously under European influence and it is now the major industrial center for southern China (2.5 million inhabitants), forming a metropolis with several other moderate-sized cities.

Many Chinese inland cities have harbors on one of the great river systems of China, like the Yangtze, the Hwang Ho, Huai Ho, Chu Chiang, and the 12 other important waterways. Like the large estuarine cities, these cities appear to be located as compromises between being accessible harbors and being located on firm land and high enough to avoid swamps and flooding. They show some interesting similarities.

To the navigable rivers should be added the famous Grand Canal, running north to south through the river basins from Peking to Hangchow. Nowhere in the world is there so much traffic on canals between major cities as in China.

Nanking, Wuhan, Chungking, and Chengtu lie within the Yangtze watershed. All were originally agricultural centers, although they now have diversified economies. The most distinctive feature of the Yangtze valley is its low gradient. Wuhan, 580 miles upstream from Shanghai, is less than 100 feet in elevation. As a result, the Yangtze occasionally shifts its course, despite age-old attempts to prevent it from doing so. Consequently, towns are often located on slightly higher ground when it is available. And in fact, the core areas of the large cities are located on such places, although they may have since expanded beyond them. For example, Wuhan is actually located slightly south of the Yangtze on higher ground, and Chungking occupies a land prominence between the confluence of the Yangtze and the Kialing.

Wuhan is . . . the head of navigation on the major east-west route through the middle of China. It can be reached by ocean-going ships of up to 10,000 tons. At Wuhan the north-south land routes meet the east-west water route and the river must be crossed. The Yangtze is nearly 1.5 kilometres wide at this point. . . . In [1957] the Yangtze Bridge was completed. . . . Ships of 10,000 tons can pass underneath (Speak, 1978:37).

Some Chinese giant cities lack a harbor. With the exception of Kunming all are on the Manchurian or North China plain. Their location appears to have been influenced by such factors as availability of water, overland trade routes, and strategic considerations. Of these cities, the most prominent is Peking.

Located at the northern apex of the triangular North China plain, Peking is the most northerly major settlement of China. From here the ancient land routes left the plains for the high northern plateaus and mountains. The Yen Shan mountain range forms a physical and cultural barrier between China and the Manchurian and Mongolian plains. The two major passes are situated so that roads leading south from them naturally converged where Peking stands today. As China was periodically invaded by hordes from Mongolia and Manchuria, each of these invading groups fortified the area around Peking to serve as an outpost on the southern side of the mountains. As a result of its strategic location, Peking was chosen as the political nexus for China during the last 700 years, despite having had many different ruling dynasties.

As the city grew, the need for a harbor was recognized, and the port of Tientsin was built. Tientsin, at the head of the Gulf of China, developed as a coastal port for goods coming by sea en route to Peking. The city is about 60 miles from Peking, and about 35 miles from the sea. Ships reach Tientsin via the Hai River, a slow-moving, meandering waterway which deposits a good deal of sediment into its riverbed. As a result, the river remains shallow and brackish, permitting ships of only 3,000 tons or less to pass. Constant dredging is necessary.

Tientsin itself is subject to frequent flooding because, although located some distance from the ocean, it is a mere three feet above sea level. Salt water intrusion into the fresh water supply has also been a problem for drinking water and agriculture. The generally unwholesome attributes of Tientsin's location illustrate that the harbor is to a great extent an artificial one, designed to serve Peking. Traffic between Peking and Tientsin by road, railway, or river is the heaviest in China. The peripheries of Peking and Tientsin tend to penetrate each other. Peking (over 11 million inhabitants) and Tientsin (over 5 million) might together become one of the largest megalopolises of the world.

The histories of Fushan, Anshan, Changchun, Shengang (Mukden), and Harbin on the Manchurian plain are more obscure than that of Peking. The Manchurian plain is fertile and productive. Since the Japanese occupation, all the Manchurian cities have grown to be large industrial centers and are well-served by railroads. The one exception to their lack of harbors is Harbin, which possesses a dredged one on the Sungari River.

In terms of structural problems and pathologies, and in spite of the enormous differences between the economic systems of China and Brazil or India or Indonesia, there are many significant ecological analogies among Shanghai, São Paulo, Bombay, and Jakarta.

Southeast Asia

Similar to South America and Africa, most of the giant cities of Southeast Asia were founded by the European colonial powers. The birth and growth of these cities are inscribed in the history of economic colonialism, which explains why they are ports. In this part of the world, as in all others, most seaports are located on estuaries, a deltaic site, or close to the mouth of a navigable river.

Following the coastline we find the following ports with more than one million inhabitants: Dacca and Chittagong in Bangladesh; Rangoon in Burma; Bangkok-Thonburi in Thailand; Singapore; Ho Chi Minh City, Hanoi-Haiphong, and Danang in Vietnam; Hong Kong and Taipei. During this voyage we would leave out only a few inland "million" cities, like Khulna, a river port in Bangladesh, or Kuala Lumpur. (Phnom Penh has been reduced to less than a few hundred thousand inhabitants.)

For each of these ports it would be possible to follow a guide like this one for Rangoon:

> The trend of relief and the flow of the River Irrawaddy make Rangoon the natural outlet of the most developed central region. As the only large port it is the only real link with the outside world, apart from local coastal steamers. The Irrawaddy is navigable to Myitkyina, 1,500 kilometres from the sea. The lower delta region of the Irrawaddy is a maze of shallow, winding distributaries. The city is not on the delta itself but lies to the east, on the left bank of the River Rangoon, about 35 kilometres from its mouth. The channel can be kept clear by dredging (Speak, 1978:40).

Sailing on, we come to two archipelagic countries. First is Indonesia, composed of dozens of islands characterized by rugged volcanic mountains covered with dense, tropical forest. The coastal plains are narrow, often covered with swamps and saltwater marshes. The islands are generally surrounded by reefs extending dangerously near the surface of the shallow seas. The most densely populated island is Java where Indonesia's three largest cities are located.

Indonesia's cities have all experienced rapid population growth. Only Jakarta and Surabaya have the heterogeneity of true cities. The rest are large villages. There has been no industrial growth. Of Java's three large cities, Jakarta-Togjakarta and Surabaya are ports; Bandung is an agricultural center located inland in the mountains. Jakarta's core area is slightly inland on a low, flat alluvial plain, extremely swamp-ridden and easily flooded during the rainy season. As a result, shortages of fresh water occur frequently. Jakarta's port, Togjakarta, is about a mile from the city. The Jakarta plain is one of the most fertile areas along Java's northern coast. Surabaya, also located along the coastal alluvial plain on a delta, is densely populated in its core areas around the harbor. Otherwise it exhibits much of the mixed character of built-up rural Javanese villages interspersed with rice fields. In contrast, Bandung is located at the edge of a plateau at an elevation of 2,400 feet, which gives it a somewhat more moderate climate than the coastal urban centers. Bandung is essentially a large agricultural village with some built-up areas at its core. The farmlands of the central plain support a considerable population, but it remains dispersed. Indonesia's port cities exhibit many of the same traits as the tropical lowland cities of China and India, except that they lack the urban industrial concentration, and their climate is on the whole more humid.

Indonesia is dominated by one of the largest megalopolises in the world: Jabotabeck, centered on Jakarta and including the cities of Bogor, Tangerang, and Beckasi, with a population of over 14 million, expected to reach 22 to 24 million inhabitants by the year 2000. Two other Indonesian cities, not seaports, are more modestly sized, at about one million inhabitants: Palembang and Semarang.

The second archipelagic country is the Philippines, dominated by a single megacity, Manila. Like Jakarta, Manila is located on the most highly developed island, Luzon, and has a deep harbor at the mouth of the Pasig River.

The two island cities of Singapore and Hong Kong are the best examples of the importance and potential of seaports in our world today.

India

Most Indians make their living from agriculture; therefore the bulk of the population lives in the countryside rather than in cities. Except for a few mountainous areas, India is well settled throughout, as evidenced by the central Deccan plateau's being only slightly less densely settled than the coastal areas.

India had, in 1985, 24 metropolitan areas with populations over one million. Of these, only three are port cities: Bombay (11.8 million inhabitants), Calcutta (11.7 million), and Madras (8.7 million). The first two are located on estuaries. Unlike China, India is not penetrated by rivers. Only the Ganges and Brahmaputra systems can be commercially traveled for any great distance.

Bombay occupies a group of islets lying adjacent to the shore. Into the bay formed by the islands run several small rivers that provide water for the city and also increase the usable size of the harbor. In Bombay live more than eight million people, making it one of the world's most densely populated cities. Blessed with an excellent deep water harbor, the best on India's west coast, Bombay's early growth was triggered by the great cotton/textile industry exploited by the British. As labor was needed, the population swelled. Today the cotton industry is still important, but the city's economy is more diversified now. Because of the physical constraints of an island location, Bombay's public transportation is chaotic. To alleviate some of the pressure, Indian authorities have begun constructing a twin city to Bombay on the opposite side of the harbor. The new city is designed to house a population of two million; it will probably be overcrowded even before it is completed.

Calcutta is built on the southeast bank of the River Hooghly, an arm of the Ganges, 80 miles from the Bay of Bengal. The British chose this site largely for its defensibility because it is protected by the Hooghly River on the west and impassable salt lakes on the east. Although commercially well located at the head of the Ganges and Brahmaputra hinterlands, Calcutta is plagued by many common problems of tropical estuarian cities. Because of its low elevation, saltwater intrusion into freshwater sources is a frequent problem. This intrusion is aggravated by the excessive groundwater pumping. Moreover, Calcutta's climate can be extremely uncomfortable, with extremely hot and humid summers. Nevertheless, Calcutta's port handles an estimated one-quarter of India's port cargoes and 40 percent of its exports. The Howrah Bridge on the Hooghly River, with traffic jams at both ends, is, like Shinjuku Metro Station in Tokyo, one of the busiest transit points in the world.

In contrast to Bombay and Calcutta, Madras is located on a straight coastline without an estuary. As a consequence, it handles far less shipping traffic than either of the two larger cities. Madras was the site of a British India Company. Originally lacking a good harbor, the British built an artificial one. Had an artificial harbor not been feasible, the city probably would not have grown as large as it has, especially considering that its most rapid growth followed expansion of the port.

The other 22 largest Indian cities—none as large as these port cities—have no waterfront.

The fourth largest Indian city, with over six million inhabitants, is Delhi, a distinctly different type of city from either Bombay or Calcutta. The city is located near the watershed divide between the Ganges and Indus river valleys. (Between Delhi and the latter lies the Thar desert.) To the north are the Himalayas, to the south are the Avaralli hills. This corridor running east-west in northern India is important as a fertile agricultural area and a strategic zone for control of northern India. Because of its location, Delhi has been a constant target for invaders into the subcontinent. Delhi has been sacked seven times, with each rebuilding using the debris of the prior city. Thanks to its inland location, Delhi's climate is somewhat less humid than Calcutta's.

The cities of the central Deccan Plain, Bangalore (3.1 million) and Hyderabad (3.6 million), are located at sufficient elevation to avoid the worst of the Indian summers. Along with Ahmadabad (3.5 million) to the west of the plain, these cities were important points for overland traffic. Today they are prominent railroad centers. These cities are distinctly native in character rather than British, as they were of lesser importance to the colonial power than the seaports.

Three cities have between 2 and 3 million inhabitants: Poona, Ulhasnagar, and Kanpur; they are followed by Nagpur, Madurai, Jaipur, Coimbatore, each over 1.5 million.

The chapter on Delhi by Hans Nagpaul in volume 2 shows how a megacity without access to the sea manages the transportation of raw material needed by industry, and how a river contributes to the resolution of problems created by the metabolism of a city of six million inhabitants.

The Middle East

From our focus of interest, there are two distinctive kinds of countries in the Middle East region. First are the southern Mediterranean countries with a series of important ports with more than one million inhabitants: Casablanca, Rabat, Algiers, Tunis, Tripoli, Cairo, Alexandria, Tel Aviv, Beirut, Izmir, and Istanbul, as well as many other smaller ports. Second are the non-Mediterranean countries: Iran, Iraq, Pakistan, Afghanistan, Jordan, and Saudi Arabia with several giant cities: Teheran in a site similar to Mexico City's; Baghdad, a riverine port; Lahore, Lyallpur, Mashhad, Basra, and only one important seaport, Karachi (about 7.2 million). We are fortunate to have a chapter in volume 2 by Ahmed Khalifa and Mohamed Mohieddin on Cairo, one of the most famous riverine ports.

Africa

In 1985 Africa had 30 cities with more than a million inhabitants: 22 are seaports, 2 are riverine ports (Cairo and Khartoum), and one is located on the border of a great lake, actually an internal sea (Kampala/Entebbe). Only five giant cities are not ports: Addis Ababa, Johannesburg, Lusaka, Nairobi, and Antananarivo. Africa is a centrifugal continent: the major cities are at the edges. No country without access to the sea has a city of over one million, except Zambia—its capital, Lusaka.

In West Africa, the largest city in each country is located on the coast. As Michael L. McNulty (1976) remarks, this coastal orientation is also a peripheral location and results from colonial economic and administrative structure. The coastal cities were, at the time of the export-oriented colonial system, "the terminal points of transport links which pushed inland to sources of exportable products. . . . The result was a transport system which afforded relatively rapid movements to the coastal towns and ports, but which severely limited lateral movements" (p. 218).

The same phenomenon can be observed in East Africa where "major seaports emerged as the basic points for colonial penetration and as the centers for the export of commercial products from the interior" (Soja and Weaver, 1976:237). Mombasa grew as a major port for Kenya and Uganda. In Tanganyika, Dar es Salaam, chosen by the colonial power because of its harbor, became the terminal point of a railway line. Nairobi is one of the few great African cities which is not a port. Nairobi started as a small railway depot and, because of its climate, became a center for European settlements in the Kenya Highlands.

Nigeria contains about half the population in West Africa and its major port city, Lagos, has been termed the most chaotic city in the world. This city merits special consideration—see chapter 7 in volume 2. In Nigeria the population is polynucleated so that it cannot be said to have a single dominant area.

The Niger River flows through the west central zone, the lifeline for several West African states of the sub-Sahara in addition to Nigeria. Climactically, Nigeria generally experiences hot and wet conditions, with the bulk of the precipitation falling in the southeast and Niger delta regions. To the southwest, around Lagos, temperatures remain hot but rainfall is considerably lighter. The swamps fringing the coastlands and delta plain were havens for parasites and diseases that have been largely controlled.

The most densely populated area is the southwestern territory around Lagos, situated on the Niger delta. Population densities along the coast reach 500 people per square mile and represent the most populated areas of Africa south of the Sahara. The north and the industrializing area around Lagos are experiencing the greatest growth.

The core area of Lagos originally occupied a low sandy island at the mouth of a large estuary. The lagoon is shallow and its water circulates poorly; its widespread use for sewage disposal has created serious health problems. Lagos has since expanded to the western shore of the lagoon, where many businesses have relocated to escape the extreme congestion of downtown Lagos.

As the Niger River is too narrow to permit any but the smallest ships to navigate it, the European influence in Nigeria was largely restricted to the coastal regions and Lagos. Had the Niger been navigable, it is probable that a port would have developed along it, for such a location would have given it favorable status in trade. As the Niger was not penetrable, Lagos, with its large harbor, remained the center of commerce. The Lagos port authority has the reputation of being among the worst in the world. As people from the countryside continue to flood the city in search of work, Lagos and the surrounding area face continued difficulty with slums.

Most of Africa's major seaports and some inland cities are located on sites many geographers and sociologists consider mediocre. Today, problems are generated by this initial site selection. The original location was determined by considerations which several decades later had lost any significance. Many African seaports had as cradles a small island, easy to protect against attacks from the sea or from the continent. Lagos was born on Lagos Island, situated in a lagoon; Dakar on the island of Goree; Konakry on the island of Tombo; Abidjan on the island of Petit-Bassam in the lagoon of Ebrie and on a peninsula between the bays of Banco and Cocody; Port-Harcourt, protected by water on all sides, on the border of the Bonny River; Mombasa on an island four kilometers in diameter; Monrovia between the ocean and an enormous swamp (Vennetier, 1976, chap. 5). Libreville in Gabon, Donala in Cameroon, and Freetown in Sierra Leone are also situated in lagoons.

Zaire has little coastline, so its capital, Kinshasa, is elongated on the river Zaire for 70 kilometers at the point where the river becomes navigable for oceanic ships. It represents a break in transport between water and land. On the other side of the river is Brazzaville, capital of Congo.

South Africa has four important ports: Cape Town, Durban, Port Elizabeth, and East London. A new port, Saldanha, located in the deep water bay of the same name, services the adjacent rich mineral area. Saldanha is the prototype of an important modern harbor which is not an important population center, and performs functions similar to oil ports.

The Soviet Union

Although an enormous country, the Soviet Union has a very limited navigable coastline at the Baltic Sea, the Black Sea, and the Japanese Sea. Its northern frontier is near the pole, so northern waterways are unnavigable much of the year. The Caspian Sea is a great lake without maritime connections. No wonder that among the 25 Soviet cities with more than one million inhabitants only three are seaports: Leningrad, Odessa, and Baku (Vladivostok and Tallinn, two other major Soviet seaports, have only 600 thousand and 420 thousand inhabitants, respectively). Nevertheless, the Soviet Union does have a dozen riverine ports of some importance, such as Volgograd, Rostov, Kiev, Donetsk, Dnepropetrovsk, and even Moscow, which has access to efficient canals for part of the year.

Most of the continental cities on the European side of the Soviet Union were founded early in the country's history. They were distributed fairly evenly since there are no mountains or deserts between the western frontier and the Ural Mountains.

There is an analogy between the transport systems of the United States and the Soviet Union. Most of the rivers in the Soviet Union run north-south. The oldest cities were located on the rivers. Later, the railway system connected the existing cities, crossing the rivers east-west, creating junctions in transportation, and favoring the growth of particular cities.

Geographically, economically, and politically the Soviet Union is not and has never been a maritime power. In no other region of the world has the sea played so modest a role as for the Soviet Union.

The situation is similar in neighboring countries. Czechoslovakia and Hungary have no coastlines. Poland, Romania, and Bulgaria have no single seaport of great importance. Their capitals, Prague, Warsaw, Bucharest, and Sofia have no navigable waterfront. Only Budapest is privileged from this point of view.

Two Empty Countries with Giant Ports: Canada and Australia

On an isotropic map, Canada looks like a serpent attached to the United States' northern frontier. There are three giant cities—all important ports. Toronto and Montreal together represent one quarter of the Canadian population. They are situated on the St. Lawrence-Great Lakes waterway. Vancouver, on Canada's west coast, is the third largest port in North America in cargo tonnage, exceeded only by New York City and Norfolk, VA. Vancouver is a perfect example of a "break" in transportation.

In Australia only a narrow strip of land is suitable for settlement; population of the interior desert is very sparse. The urban centers are located on the coast. Sydney has a magnificent natural setting. Not only does the harbor have an enormous capacity, but rivers and lakes surround the area. The second largest Australian port city is Melbourne, the third is Brisbane. There are no important inland cities. These three cities are Australia's centers of international trade.

In neighboring New Zealand, Auckland is located on an isthmus and Wellington is on a strait between two islands—also an example of a break in transportation.

Why Most Giant Cities Are Ports

Our world is built in historical layers, like geological sediments. We are living at a time when transporting people among the 300 largest cities in the world is done more by air than by sea. Today more goods are transported by train, truck, or pipeline than by ship. Nevertheless, most giant cities are still located at the seacoast or on navigable rivers. They have not moved. On the contrary, many of the most important seaports have meanwhile become the focus of extensive networks of railways, motor highways, and electronic communication.

Historians, geographers, urbanists, and sociologists agree that most of the important cities in West Africa, Southeast Asia, South America, Australia, and many on the east coast of the United States and on the China coast, and other cities like Calcutta, Bombay, or Madras began as trade ports founded by European colonial powers. Traditional Asia, before the 18th century, was composed primarily of inward-facing states and empires. The great cities, and indeed nearly all of the important urban centers, were inland. In present terms, virtually all of the largest coastal or near-coastal cities owe the bulk of their growth and most of their essential nature to Western traders (Rhoads Murphey, chapter 5 in volume 2). The same phenomenon occurred in West Africa where growth of urban systems and transport networks reflects their heavy dependence on foreign markets and sources of imported goods (McNulty, 1976:218). The English, French, Spanish, Portuguese, and to a lesser extent, the Dutch, Belgians, and Germans decided a long time ago, without being conscious of the long-term consequences of their choices, where approximately 800 million metropolitan people would live in the Third World at the start of the 21st century.

The theory of colonial ports amply confirms an older one about the location and growth of cities, formulated in 1894 by Charles Horton Cooley and buried in the recesses of libraries, forgotten by most contemporary scholars of urbanization. To my knowledge, none of the contemporary scholars writing on colonial ports has mentioned the principle of a break in transportation.

According to Cooley (1894:75-76):

> Population and wealth tend to collect wherever there is a break in transportation. By a break is meant an interruption of the movement at least sufficient to cause a transfer of goods and their temporary storage. If this physical interruption of the movement is all that takes place, we have what I may call a mechanical break; but if . . . this physical interruption causes a change in the ownership of the transported goods, we have a commercial break. . . . Where a break of this sort exists on an important line of transportation . . . there must be a commercial city. [The breaks arise] at the junction of land transportation with water transportation, or one kind of water transportation with another, or of one kind of land transportation with another.

The colonial port is the perfect example of a commercial break in transportation, of the intersection between road or river and the sea. Cooley gave few examples outside the United States and none for today's Third World countries, in large part because most seaports in these countries became giant after 1894, when he wrote.

Does the thesis of break in transportation help us to understand better the location of the 285 largest cities of the world? A distinction between three types of countries, based on the age of their urbanization, is useful:

(1) New countries, independent and already advanced economically and relatively urbanized when railway transport expanded during the nineteenth century—typically the United States. In this country the junctions between water transport and railway transport tended to generate large cities. Later the automobile came where the city was, as did the airplane in turn.

(2) Colonial countries, heavily rural and agricultural at the time of the European penetration—particularly Africa, South America, and Southeast Asia. The colonial port represented for them much more than a break in transportation: it was a gateway. The notion of gateway seems for those underdeveloped countries at a certain time more appropriate than the notion of a break in transportation.

(3) The old countries, urbanized long before the technological changes came in land and maritime transport—Western, Central, and Eastern Europe; India, China, Japan, the Middle East. No doubt the development of maritime trade since the first industrial revolution had an impact on the growth of many coastal cities but, except for the Chinese trade ports, for most of the contemporary giant cities in these old countries, the notion of break in transportation helps to explain their growth at a certain moment, rather than their location.

The picture is particularly clear for some continental countries. Among India's 24 largest cities, only three are ports. Most of the 21 cities without waterfronts are very old cities, today connected by one of the densest railway systems in the world. The railway arrived where the great city was, rather than the opposite. In the Soviet Union, most cities have no deep harbor. In this part of the world, too, the train, automobile, and airplane came to the great old cities, despite well-known antimetropolitan policies.

In their useful book, *3000 Years of Urban Growth*, Tertius Chandler and Gerald Fox offer no explanation for the rise and decline of the world's largest cities at various epochs. They seem to have ignored Cooley's theory and have not even observed that a high proportion of the largest cities they mention were ports. Such an analysis is not difficult, but of course the largest cities at the end of the Middle Ages were not "giant" according to present day standards. Of the 25 largest cities Chandler and Fox list for 1750, ten were seaports, eight riverine ports, and three were located on navigable canals. Of the total of 78 cities listed for 1750, two-thirds had access to navigable water.

The proportion of ports remains about the same for the 75 cities listed in 1825, even though some appear to us as tiny ports. For instance, Cairo probably had then about 260 thousand inhabitants; Shanghai, 115 thousand; Calcutta, 250 thousand; Constantinople, 675 thousand; and New York, 170 thousand. In 1925 among the 25 largest cities, only two were not sea or river ports: Manchester and Birmingham, both fed by maritime transport followed by short land transport. Only two of the largest ports were outside the Atlantic *mare nostrum*—Calcutta and Shanghai.

Sixty years later, in 1985, among the 35 largest metropolises in the world, only Mexico City, Milan, Delhi, Teheran, and Madrid are not ports, and only Cairo, Paris, and the Rhine-Ruhr are riverine ports. The other 26 are seaports (considering Peking-Tientsin, São Paulo-Santos, and Seoul-Inchon as ports).

All important seaports or river ports have also become important airports. The inland giant cities, particularly those in the middle of a desert like Mexico City, Teheran, and Tashkent, today have important airports. We live in a world of giant seaports and airports.

Why are some giant cities not seaports or riverine ports? It is relatively easy to find an acceptable explanation for each particular case, but not easy to make generalizations. Sometimes the location was determined primarily by military and security reasons. For instance, Ephesus, near Izmir in Turkey, which is probably the best conserved ancient city, and which was a flourishing mercantile city, had its power based on maritime traffic. Nevertheless, the city, concealed behind a hill so as not to be seen by pirates, was not a port. Military preoccupations played a role in the location of many inland cities which later became giant cities, and this is also true for many cities born on small islands, in estuaries, or in bays.

Some giant cities are not on the coast because climate was a prior criterion and a plateau was preferred. In some countries the coast is inhospitable for human habitation. Many examples could be given. It will suffice to point to Antananarivo in Madagascar and, particularly, Mexico City.

There are few coastal cities on the Pacific side of Latin America. A look at the map instantly offers the explanation: the Cordilleras are too near the coast, depriving the incipient port of a hinterland.

Dozens of pages could be written in the attempt to explain why Milan, Dallas, Addis-Ababa, Madrid, Bogota, Atlanta, Nairobi, Warsaw, or Delhi are where they are, and not on the water.

A basic fact must be stressed. Among the 285 cities with more than one million inhabitants, between 190 and 210 are seaports or riverine ports. The ambiguity in the number comes because it is sometimes difficult to estimate the importance of the harbor to the economy and for the historical growth of some cities.

It is obvious that the harbor has not the same importance for the economy of Los Angeles and Singapore, for Lima and Bombay, for Athens and London, for Barcelona and Manila. In the same city the function of the harbor might have, over the years, declined, as in Antwerp, or might have increased, as in Caracas.

To establish that among the 300 giant cities of the world more than two-thirds are on the water is a significant fact which helps us to understand better how our world is built.

Bibliography

Bagwell, Philip S. 1974. *The Transport Revolution from 1770*. New York: Barnes & Noble.

Berry, Brian J.L., ed. 1976. *Urbanization and Counter-Urbanization*. Newbury Park, CA: Sage.

Bourne, L. S. and M. I. Logan. 1976. "Changing Urbanization Patterns at the Margin: The Examples of Australia and Canada." In B.J.L. Berry, ed., *op cit.*, pp. 111ff.

Chandler, Tertius, and Gerald Fox. 1974. *3000 Years of Urban Growth*. New York: Academic Press.

Cooley, Charles Horton. 1894. "The Theory of Transportation." *Publications of the American Economic Association* 9(3).

Gilmore, Harlan W. 1953. *Transportation and the Growth of Cities*. Glencoe, IL: Free Press.

Hoyle, Brian Stewart, and David Hilling. 1984. *Seaport Systems and Spatial Change: Technology, Industry, and Development Strategies*. Chichester, NY: Wiley.

McKenzie, Roderick Duncan. 1933. *The Metropolitan Community*. Recent Social Trends Monographs. New York: McGraw-Hill.

McNulty, Michael L. 1976. "West African Urbanization." In *Urbanization and Counter-Urbanization*, edited by Brian J.L. Berry. Urban Affairs Annual Reviews, vol. 11. Newbury Park, CA: Sage.

Misra, Rameshwar Prased. 1978. *Million Cities of India*. New Delhi: Vikas.

Murphey, Rhoads. 1969. "Traditionalism and Colonialism: Changing Urban Roles in Asia." *Journal of Asian Studies* 29:67-84.

Nagashima, Catharine. 1981. "The Tokaido Megalopolis." *Ekistics* 48:280-301.

Rios, José Arthur. 1972. "Rio de Janeiro." In *Great Cities of the World: Their Government, Politics and Planning*, vol. 2, edited by William A. Robson and D. E. Regan, 3d ed. Newbury Park, CA: Sage.

Robson, William A., and D. E. Regan, eds. 1972. *Great Cities of the World: Their Government, Politics and Planning*. 3d ed. Newbury Park, CA: Sage.

Sheffield, Charles. 1981. *Earth Watch*. New York: Macmillan.

———. 1983. *Man on Earth, How Civilization and Technology Changed the Face of the World—A Survey from Space*. New York: Macmillan.

Short, M., D. Lowman, C. Freden, and W. A. Finch. 1976. *Mission to Earth: Landsat Views the World*. Washington, DC: National Aeronautics and Space Administration.

Soja, Edward W., and Clyde E. Weaver. 1976. "Urbanization and Underdevelopment in East Africa." In *Urbanization and Counter-Urbanization*, edited by Brian J.L. Berry. Urban Affairs Annual Reviews, vol. 11. Newbury Park, CA: Sage.

Speak, C. M. 1978. *Human Geography: East and South-East Asia and Australasia*. London: Heinemann Educational.

United Nations. Department of International Economic and Social Affairs. 1981. *World Population Prospects as Assessed in 1980*. Population Studies no. 78. ST/ESA/SER.A/78. New York.

Vennetier, Pierre. 1976. *Les villes d'Afrique tropicale*. Paris: Masson.

Violich, Francis. 1975. "Caracas: Focus of the New Venezuela." In *World Capitals: Toward Guided Urbanization*, edited by H. Wentworth Eldredge. Garden City, NY: Anchor Press/Doubleday.

White, Paul. 1984. *The West European City: A Social Geography*. London: Longman.

Wilkie, James W., and Paul Turovsky. 1975. *Statistical Abstracts of Latin America*. Vol. 17. Los Angeles: Latin American Center Publications, University of California, Los Angeles.

Wilkie, Richard W. 1984. *Latin American Population and Urbanization Analysis, Maps and Statistics, 1950-1982*. Statistical Abstract of Latin America Supplement Series, suppl. 8. Los Angeles: Latin American Center Publications, University of California, Los Angeles.

2

Economic Restructuring and America's Urban Dilemma

John D. Kasarda

The oldest and largest cities in the United States traditionally have been at the forefront of the development and restructuring of the nation's economy. These cities spawned North America's industrial revolution, generating massive numbers of blue-collar jobs that served to attract and economically upgrade millions of migrants. More recently, these same cities were instrumental in transforming the U.S. economy from goods processing to basic services (during the 1950s and 1960s) and from a basic service economy to one of information processing and administrative control (during the 1970s and 1980s).

The transformation of major cities from centers of goods processing to centers of information processing was accomplished by corresponding changes in the size and composition of their employment bases. As more efficient transportation and production technologies evolved, manufacturing dispersed to the suburbs, exurbs, nonmetropolitan areas, and abroad. Warehousing activities relocated to more regionally accessible beltways and interstate highways. Retail and consumer service establishments followed their suburbanizing clientele and relocated in peripheral shopping centers and malls.

While most parts of the central city continued to experience an economic base hemorrhage, pockets of employment resurgence emerged. Nowhere is this more visible than in central business district office development where jobs in administration, finance, communications, and the professions have mushroomed since 1970. However, selective job growth in these knowledge-intensive, predominantly white-collar industries has not typically compensated for employment declines in manufacturing and other predominantly blue-collar industries that had once constituted the economic backbone of these cities. As a result, the total number of jobs in most older, larger, U.S. cities has shrunk over the past two decades.

Concurrent with their economic transformation, major U.S. cities have experienced substantial changes in the composition and total size of their residential populations. As predominantly white middle-income groups dispersed from the cities (initially to the suburbs and then increasingly to the exurbs and nonmetropolitan areas), they have been only partially replaced by predominantly lower-income minority groups and relatively small numbers of younger urban professionals. The outcome has been declines in total size and aggregate personal income level of most cities' resident populations since 1970, while concentrations of their economically disadvantaged minorities continue to expand.

These simultaneous, yet conflicting, transformations of the employment and residential bases of the cities have contributed to a number of serious problems, including a widening gap between urban job opportunity structures and skill levels of disadvantaged residents (with corresponding high rates of structural unemployment), spatial isolation of low-income minorities, and rising levels of urban poverty and welfare dependency. Associated with these problems have been a plethora of social and institutional ills further aggravating the predicament of large cities in distress, such as high crime rates, poor public schools, deteriorating public infrastructures, and the decay of once-vibrant residential and commercial subareas of the cities.

The U.S. federal government responded to the hardships confronting its cities and their inhabitants by introducing a variety of urban programs over the past two decades. Unfortunately, these programs have had little effect in stemming blue-collar job decline or improving long-term employment prospects for underprivileged city residents. Indeed, mounting evidence suggests that the plight of economically distressed cities and their underprivileged residents is worse today than before the urban programs began (Bradbury et al., 1982; Fossett and Nathan, 1981; Wilson, 1983).

The inability of federal programs to stem urban blue-collar job decline results primarily from the overarching technological and economic dynamics influencing the locational choices of various industries and the changing roles of major cities in advanced service economies. This chapter elaborates on these dynamics and changes, especially as they have altered the capacity of North America's oldest and largest cities to offer entry-level jobs and socially upgrade disadvantaged resident groups.

The guiding thesis is that large cities always have and always will perform essential economic and social functions, but changing technological and industrial conditions (both nationally and internationally) alter these functions over time. In this regard, it must be remembered that blue-collar job opportunities expanded most rapidly in the largest and oldest U.S. cities during a transportation and industrial era that no longer exists. During this era, cities had comparative advantages over other locations that substantially reduced costs and raised efficiency of firms concentrating in them. These advantages included superior long-distance transportation and terminal storage facilities; an abundance of immigrant labor willing to work for extremely low wages; many nearby complementary businesses; and availability of private and public municipal services such as police and fire protection, sewage systems, and running water.

Modern advances in transportation and communication technologies, the spread of population and public services, and changing modes of production organization have virtually wiped out the comparative advantages of major U.S. cities as locations for large-scale manufacturing and warehousing facilities. The precipitous decline of blue-collar jobs in cities during recent decades reflects, in part, their loss of comparative locational advantages in these industrial sectors.

Nevertheless, just as older, larger U.S. cities have lost their competitive strength to hold or attract jobs in their traditional blue-collar industries, many are exhibiting new competitive strengths as locations for a variety of postindustrial growth industries. These emerging strengths are in the administrative office, communication, financial, professional, and business service sectors and in cultural, leisure, and tourist industries. I will argue that the exploitation of these competitive strengths and a reversal of the buildup of growing concentrations of disadvantaged persons in U.S. cities are essential to urban economic revitalization and renewed minority opportunity. This argument is made because no policy that would be either politically or economically feasible could overcome the cost disadvantages of central city locations for manufacturing and most other blue-collar industries that once served to support and socially upgrade massive numbers of low-skill residents.

To explicate the latter contention and provide an interpretive foundation for analyzing contemporary urban problems, the following section provides a broad brush sweep of the spatial-economic development of North America's oldest cities. Next, recent changes in the employment bases and demographic compositions of these cities are documented, with illustration of the widening mismatches between educational requirements of new urban growth industries and educational attainment of urban minorities. Implications of these mismatches for post-1970 rises in rates of urban minority unemployment are examined. The chapter concludes with policy suggestions to revitalize U.S. cities and facilitate the mobility of economically displaced minorities from distressed inner city areas to reduce their social isolation and high rates of unemployment and welfare dependency.

Evolving Structure and Functions of America's Cities

The oldest cities in the U.S. originated during its colonial era as mercantile centers of commerce and shipping. Since their chief function was funneling raw materials from their immediate hinterlands to the nascent metropolitan economy of 18th century Europe, these cities evolved at deep-water sites on long-distance riverine and oceanic routes (Berry and Kasarda, 1977:272-74). By today's urban standards, the cities of colonial North America were miniscule—not a single place would have met the current metropolitan criterion of 50 thousand residents.

As America entered the 19th century, it remained a rural-agrarian nation. Fewer than 4 percent of the U.S. population resided in places of eight thousand or more. Urban expansion during the first two decades of the 19th century progressed at a sluggish rate, as externally oriented commerce and shipping continued to dominate the economic bases of fledgling cities. Regional and national markets were constrained by limited transport systems, while the bulk of urban manufacturing and goods production was of a cottage industry and handicraft form.

The next 50 years witnessed a dramatic development of railroad, highway, and canal systems that substantially broadened urban access to the country's rich raw material resources, spun off new inland towns and cities, and created a nationally integrated exchange network among cities, smaller trade centers, and hinterland villages. At the same time, a revolution was occurring in food production technology and organization that generated the agricultural surplus that supported a growing nonagricultural population and displaced large numbers of subsistence farmers. Steam-powered machinery and mass production technologies were introduced that, together with entrepreneurial innovations in credit, banking, and finance capitalism, gave rise to large corporate business organizations and their primary units of production—factories. With high domestic fertility rates and increased immigration from abroad providing both an expanded domestic market for manufactured goods and abundant, cheap labor to staff the factories, all the requisites for the rise of America's industrial cities were in place.

Viewed spatially, the industrial cities of North America developed as highly concentrated forms around rail and waterway terminal points within access of northern coal fields. The reasons for such concentration rested in transport costs and the mechanized application of steam power. Steam cools quickly and dissipates easily, so it must be used where it is produced. Since coal—the fuel used to produce steam—had high transport and terminal costs, business expenses could be minimized by factories clustering around the terminal and sharing bulk carriage costs (Hawley, 1981). Because the terminal was also the point where other raw materials used in the production process were received and finished products distributed, factories concentrating there accrued further cost advantages.

Moreover, during the latter half of the 19th century, raw materials used in the manufacturing process and finished products tended to be far heavier and bulkier than they are today, and short-distance transport technologies were, at best, primitive. Under such conditions, agglomeration became cumulative and mutually reinforcing. Complementary units that serviced factories or used their by-products located as close to the factories as possible, as did storage warehouses and wholesalers distributing finished products. Workers employed by the factories, service shops, and warehouses, constrained by pedestrian movement, clustered tightly within walking distance of their place of employment. As late as 1899, the average commuting distance of workers in New York City from their home to their place of employment was roughly two blocks (Palen, 1975:53).

Thus U.S. industrial cities evolving in the late 19th century were compact agglomerations of production and distribution facilities, centers of corporate capital accumulation, and both residence and work place of millions of unskilled and semiskilled migrants. Industrial expansion and concentrative migration converged in space, generating explosive city growth. Chicago, for example, incorporated in 1833 with a population of 4,100, grew to over two million residents by 1910, the vast majority of whom resided and worked within a three-mile radius of the city's center.

Spurring the dramatic growth of U.S. industrial cities were a rapidly advancing western resource frontier and strengthening commercial markets interacting with a powerful entrepreneurial spirit which held that individualism, competition, the pursuit of profit, and economic growth were uniformly positive and beneficial. In this political-economic climate, urbanization and industrialization surged together, catapulting the entire country into a period of enormous economic expansion. By the end of the 19th century, the output of America's industrial cities surpassed the industrial output of Britain, France, and Germany, the world's leaders in 1860 (Mohl, 1976).

A point that merits emphasis is that the employment bases of America's early industrial cities were characterized by entry-level job abundance, compared with their current entry-level job scarcity (Downs, 1973; Hawley, 1981; Kovaleff, 1974). This job abundance, with few requisites for entry, attracted the waves of migrants and offered them a foothold in the urban economy. The rapidly expanding job base accompanying national economic growth in turn provided ladders of opportunity and social mobility for the migrants, most of whom were fleeing nations or regions of substantial labor surplus (Bodner et al., 1982; Farley, 1968; Lieberson, 1980; Vinyard, 1976; Zunz, 1982).

Migration thus served an important equilibrating function, bringing into better balance local population numbers with local labor needs, while also improving immigrants' life chances. The latter was particularly true for the millions who were economically disadvantaged—initially arriving from Europe followed by southern whites and southern blacks (as well as rural northerners) displaced by technological and organizational transformations in 20th-century American agriculture (Fligstein, 1981).

The access to opportunity and social mobility that America's industrial cities provided migrants was obtained at significant human cost, however. Migrants frequently had to pull up deep community roots, permanently leave close friends and relatives behind, and adjust to a totally different way of life. On arriving at their urban destinations, migrants were usually greeted with scorn and prejudice by other groups who had preceded them to the city (Hauser, 1960). Lacking financial resources, unaccustomed to city ways, and often without English language skills, immigrants were ascribed the lowest status and were residentially segregated in overcrowded ghettos in the least desirable sections of the city. A polluted, unsanitary physical environment contributed to high morbidity and mortality rates, as did hazardous factory working conditions. Political corruption and exploitation were common, working hours were long, and there was no such thing as a minimum wage. By today's dual labor market theory classification scheme, virtually all immigrants held "dead end" jobs (cf. Berg, 1981).

Nonetheless, there was an abundance of jobs for which the only requisites were a person's willingness and physical ability to work. This surplus of low-skill jobs and overall economic growth provided North America's older industrial cities with a unique role in the nation's history as developers of manpower and springboards for social mobility (Norton, 1979).

Technological Advance and Blue-Collar Job Deconcentration

During the first half of the 20th century, a number of innovations occurred in transportation, communication, and production technologies that markedly reduced the locational advantages the older, compactly structured cities had previously held for manufacturing and distributing activities. Manufacturers and wholesalers soon found that older, urban street patterns were not conducive to automobile and truck movement. Traffic congestion, lack of employee parking space, and problems of freight transfer greatly increased direct and indirect costs, particularly for manufacturers and wholesalers located in the old, densely settled sections of the cities. On the other hand, development of suburban highway systems, widespread automobile ownership, and increased dependence by manufacturers and wholesalers on trucking for receiving and shipping made uncongested suburban sites more attractive. Manufacturers and wholesalers recognized that by locating on or near the suburban expressway they could reduce transport costs, tap an adequate automobile-owning labor supply, and solve problems of employee parking and freight transfer.

The changing mode of manufacturing technology from unit processing to mass production and assembly lines methods also hastened the urban exodus of blue-collar industries. Early central-city manufacturing facilities had been constructed as multistory loft-type structures not adaptable to much of today's mass production technology. The assembly line, in particular, has large horizontal space requirements, more difficult and costly to obtain in the central city than in the suburban rings. Likewise, large-scale wholesalers shifting to automated freight-transfer systems found it more practical and less costly to build newly designed facilities on relatively inexpensive suburban ring land (often *more* accessible to regional and national markets) than to convert their obsolete inner-city structures. Owners of newer manufacturing and warehousing facilities that had large space requirements would rarely even consider a central-city site, most locating their facilities in the suburban rings and beyond.

A third, yet no less important, factor stimulating the deconcentration of blue-collar industry was the widespread development of suburban public services and external economies, previously restricted to central cities and their immediate built-up areas. The spread of electric and gas power lines, running water, sanitary waste systems, police and fire protection, and highway services throughout the suburban rings released manufacturing and related industries from their previous dependency on central cities. Moreover, rapid suburban development after World War II brought housing, local suppliers, subcontractors, and other complementary services to nearby areas.

By 1960 further advances in transportation and communication technologies and growing industrial competition from nonmetropolitan areas and abroad made the larger, older cities almost obsolete as locations for manufacturing and warehousing facilities (Kasarda, 1980; Sternlieb and Hughes, 1975; Suttles, 1978). A massive exodus of blue-collar jobs from the cities began and has continued ever since.

Retail and Service Sector Shifts

Retail trade and consumer services followed their traditional middle- and upper-income patrons to the suburbs, exacerbating blue-collar job declines in the cities. Between 1954 and 1978 more than 15 thousand shopping centers and malls were constructed to serve expanding suburban and exurban populations. By 1975 these shopping areas were the locus of over half the annual retail sales in the United States (Muller, 1976).

For the current generation of suburbanites, enclosed regional malls have become their Main Street, Fifth Avenue, and community social and entertainment center—all wrapped up into one. The malls not only contain retail establishments of every size, price range, and variety, but also offer professional offices, restaurants, movie theaters, public service outlets, and common space used for free or inexpensive patron attractions. The broad spectrum of goods and services offered by the malls along with special customer amenities (such as controlled climate; pedestrian arcades insulated from automobile traffic, noise, and polluted air; good security and lighting for safer and more relaxed nighttime shopping; free parking) and their accessibility to the automobile-oriented suburban population have combined to give regional malls a distinct competitive edge in capturing the metropolitan retail dollar. Indeed, large enclosed malls have become so successful that many major cities are now trying to simulate them in their downtowns in an attempt to recoup some of their lost share of metropolitan retail sales and employment.

Urban Growth Industries

There are, nonetheless, significant countertrends occurring in certain retail and service sectors as businesses and institutions offering highly specialized goods and services continue to be attracted to downtown areas. The specialized and complementary nature of these establishments still makes it advantageous for many to choose central locations that maximize their accessibility to certain consumers and businesses. Such establishments as advertising agencies, brokerage houses, consulting firms, financial institutions, small luxury goods shops, and legal, accounting, and professional complexes have been accumulating in the central business districts, replacing traditional department stores and many other standardized retail goods and consumer services establishments unable to compete effectively or afford rents at a central location.

The past two decades have also witnessed a remarkable growth of high-rise administrative office buildings in the largest cities' central business districts. Even with major advances in telecommunications technology, many administrative headquarters rely on a complement of legal, financial, public relations, and other specialized business services most readily available in the large cities' central business districts. Moreover, unlike manufacturing, wholesale trade, and retail trade—which typically have large space requirements and whose products cannot be efficiently moved in a vertial direction—most managerial, clerical, professional, and business service functions are highly space intensive and their basic product (information) can be transferred vertically as efficiently as horizontally. Thus people who process information can be stacked, layer after layer, in downtown high-rise office buildings with no loss in productivity. Indeed, such stacking and the resulting proximity often enhance the productivity of those requiring extensive, nonroutine personal interaction. The outcome has been an office building boom in central business districts.

Two other growth industries in major cities have been increasingly important to downtown vitality. The first is convention and tourism business, fostered by an increasing number of professional and trade associations and rising incomes of a sizable, mobile segment of the national and international population. The second is industry associated with entertainment, cultural, and leisure services, catering to the above groups and to increasing numbers of city dwellers who are single professionals or childless dual-income couples.

The convention-tourism and cultural-entertainment industries may be referred to as "gathering-service" industries because both operate via the convergence of large numbers of individuals from widely dispersed locations to a single site where service is provided. Gathering-service industries typically have high agglomeration and mutual-scale economies; concentrating together enhances their functional complementarity, ambiance, and market appeal. Many large cities are recognizing that their dense clusters of hotels, restaurants, nightclubs, sport arenas, museums, and theaters, if situated in clean, pedestrian-safe areas, are effective magnets for attracting round-the-clock growth industry resources back to their cores. In addition, these clusters serve as important social and aesthetic complements to the increasing number of nine-to-five high-rise office towers beginning to dominate their downtowns.

Effects on Job Opportunities

The transformation of older U.S. cities from centers of production and distribution of material goods to centers of information exchange and service consumption has profoundly altered their capacity to offer employment opportunities for disadvantaged residents. The increments in urban blue-collar jobs generated by newer service industries employing and catering primarily to higher-income persons have been overwhelmed by job losses in their traditional goods-producing and distributing industries. Aggravating these blue-collar job losses has been the exodus of middle-income population and general retail trade and service establishments from much of the city beyond the central business districts. These movements have combined to further weaken secondary labor markets and isolate disadvantaged groups in economically distressed subareas where opportunities for employment are minimal.

Major cities in the northern industrial belt of the U.S. have been particularly hard hit by changes in their economic base and loss of blue-collar jobs. Unfortunately, many of these same cities experienced the largest post-World War II migration flows of people whose educational backgrounds do not equip them for employment in new urban growth industries. As a result of this skill mismatch, inner-city unemployment rates are well above the national average and inordinately high among their educationally disadvantaged urban minorities caught in the web of change.

The next section illustrates the scope of the problem by documenting the spatially conflicting industrial and demographic transformations that have occurred in the largest northern cities since World War II, first in terms of changing compositions and educational requisites of their employment bases and then in terms of the changing composition and educational attainment of their residents. An assessment of the consequences of these transformations for minority unemployment follows.

Changing Urban Employment Bases

An overview of urban industrial restructuring is provided by table 2.1, which describes employment changes in 12 northern U.S. cities between 1972 and 1982. The two largest cities, New York and Chicago, each lost over 200 thousand manufacturing jobs during this ten-year period. For Chicago, this constituted nearly 50 percent of its manufacturing jobs. On a percentage basis, Baltimore, Detroit, and Philadelphia also lost more than one-third of their

TABLE 2.1
Central City Employment Changes in Manufacturing, Wholesale Trade, Retail Trade, and Selected Services (in thousands)—United States, 1972-1982

	Manufacturing		*Wholesale*		*Retail*		*Services*		*Total*	
City	*N*	*%*	*N*	*%*	*N*	*%*	*N*	*%*	*N*	*%*
Baltimore	−32	−35	−6	−25	−11	−20	2	6	−47	−19
Boston	−12	−20	−7	−28	−5	−9	32	57	7	4
Chicago	−203	−47	−30	−29	−53	−26	58	37	−228	−25
Cleveland	−39	−29	−4	−15	−13	−29	1	2	−55	−22
Detroit	−75	−41	−16	−45	−29	−41	−10	−18	−129	−38
Milwaukee	−28	−27	−1	−4	8	21	8	29	−13	−7
Minneapolis/ St. Paul	−29	−27	−2	−5	−5	−9	19	42	−17	−7
Newark	−13	−28	−4	−35	−10	−49	<1	1	−27	−28
New York	−228	−30	−39	−15	−75	−18	113	26	−229	−12
Philadelphia	−78	−38	−14	−28	−21	−20	9	12	−104	−24
Pittsburgh	−10	−16	−2	−11	1	3	10	33	−1	−1
St. Louis	−29	−29	−7	−26	−7	−19	−5	−14	−48	−24

SOURCE: U.S. Bureau of the Census, Economic Censuses (1972, 1982).

manufacturing jobs between 1972 and 1982.

Wholesale employment also declined in all cities, particularly Baltimore, Boston, Chicago, Detroit, Newark, Philadelphia, and St. Louis. The retail employment picture in these cities is not much brighter with all cities but Pittsburgh losing retail employment between 1972 and 1982. Cities with major losses of retail employment, on a proportional basis, include Chicago, Cleveland, Detroit, and Newark, all declining 25 percent or more.

In contrast to sharp declines in the retail, wholesale, and manufacturing sectors of these cities, service sector employment has expanded in all but two—Detroit and St. Louis. Boston more than doubled its employment in the service sector between 1972 and 1982, whereas Chicago, Minneapolis/St. Paul, and Pittsburgh increased their service sector employment by at least one-third. Boston's economy, in fact, added more service sector jobs than it lost employment in other sectors, making this city the only net job gainer between 1972 and 1982.

Note the dire straights facing Detroit. During the same ten-year period, this city lost 41 percent of its manufacturing jobs, 45 percent of its wholesale jobs, 41 percent of its retail jobs, *and* 18 percent of its service sector jobs. These figures will become particularly telling (given our mismatch thesis) when we describe demographic changes occurring in Detroit and other large northern cities between 1970 and 1980.

The Rise of Information-Processing Industries

Examination of industrial employment classifications within the service sector (U.S. Management and Budget Office, 1972) shows that selected service employment encompasses a wide variety of functions (e.g., bootblacks, parking lot attendants, economic researchers, and medical doctors). Such diversity in aggregate selected service employment clouds more than it clarifies significant transformations taking place in the employment bases of cities (cf. Stanback and Noyelle, 1981). Adequate appraisal of transforming city economies therefore requires decomposition of their selected service employment changes into more detailed categories.

Such decomposition was done for five major northern cities whose boundaries match or approximate county boundaries (New York City, Philadelphia, Boston, Baltimore, and St. Louis). Table 2.2 shows that between 1953 and 1984, New York City lost approximately 600,000 jobs in manufacturing. During this same period, white-collar service industries (defined as those service industries where executives, managers, professionals, and clerical employees exceed more than 50 percent of the industry work force) grew by nearly 700,000. Corresponding to industrial redistributional patterns discussed above, retail and wholesale employment in New York City declined by 200,000 and blue-collar service employment by 135,000 since 1970.[1]

Philadelphia, Baltimore, Boston, and St. Louis likewise experienced substantial employment declines in manufacturing, retail, and wholesale trade, as well as blue-collar services. Between 1953 and 1984, Philadelphia lost over two-thirds of its manufacturing jobs (from 359,000 to 115,000), dropping employment in this industry from over 45 percent of the city total in 1953 to under 20 percent in 1984. During the same period, manufacturing employment declined in Boston from 114,000 to 49,000; in Baltimore from 130,000 to 55,000; and in St. Louis from 194,000 to 67,000. Employment declines in retail and wholesale trade and in blue-collar services followed suit, though the absolute and proportional losses were not as steep as for manufacturing employment.

As in New York City, white-collar service employment expanded substantially in the four cities between 1953 and 1984. Boston's white-collar service employment increased from 87,000 to 279,000; Philadelphia's from 98,000 to 276,000; Baltimore's from 44,000 to 128,000; and St. Louis's from 50,000 to 101,000. St. Louis is the only northern city where white-collar service employment did not exceed 40 percent of the city total in 1984. By the same token, 58 percent of all Boston's jobs were in white-collar service industries in 1984, compared to 22 percent in 1953. Such increases in white-collar service employment across these northern cities clearly manifest their emerging information-processing roles in the computer age.

Table 2.3 further highlights the overriding significance of predominantly information-processing industries for contemporary urban employment growth. This table divides total employment change between 1970 and 1984 for each city into that accounted for by (1) its service sector industries where more than 60 percent of the employees in 1978 were classified in executive, managerial, professional, and clerical occupations and (2) all other industries combined (U.S. Bureau of Labor Statistics, 1981). In addition to the five northern cities discussed above, four southern and western cities with closely

TABLE 2.2
Central-City Employment (in thousands) by Sector, 1953, 1970, 1984

Central City and Sector	1953 N	1953 %	1970 N	1970 %	1984 N	1984 %
New York						
Total employment[a]	2977	100.0	3350	100.0	2926	100.0
Manufacturing	1070	35.9	864	25.8	486	16.6
Retail/wholesale	805	27.1	779	23.3	579	19.8
White-collar services[b]	646	21.7	1172	35.0	1445	49.4
Blue-collar services[c]	344	11.6	424	12.6	289	9.9
Other	111	3.7	112	3.3	127	4.3
Philadelphia						
Total employment[a]	788	100.0	772	100.0	593	100.0
Manufacturing	359	45.5	257	33.3	115	19.4
Retail/wholesale	206	26.1	180	23.3	130	21.9
White-collar services[b]	98	12.5	220	28.5	276	46.5
Blue-collar services[c]	85	10.8	81	10.5	50	8.4
Other	40	5.1	35	4.5	22	3.7
Boston						
Total employment[a]	402	100.0	465	100.0	481	100.0
Manufacturing	114	28.4	84	18.1	49	10.2
Retail/wholesale	132	32.8	111	23.9	84	17.5
White-collar services[b]	87	21.7	194	41.6	279	58.0
Blue-collar services[c]	51	12.6	55	11.7	47	9.7
Other	18	4.4	21	4.6	22	4.6
Baltimore						
Total employment[a]	342	100.0	367	100.0	302	100.0
Manufacturing	130	38.1	105	28.6	55	18.2
Retail/wholesale	89	26.1	94	25.5	68	22.4
White-collar services[b]	44	12.8	108	29.5	128	42.4
Blue-collar services[c]	51	14.9	44	12.1	32	10.5
Other	28	8.1	16	4.2	20	6.5
St. Louis						
Total employment[a]	431	100.0	376	100.0	259	100.0
Manufacturing	194	44.9	133	35.3	67	25.9
Retail/wholesale	103	23.9	89	23.6	54	20.8
White-collar services[b]	50	11.5	96	25.5	101	39.0
Blue-collar services[c]	46	10.6	44	11.8	26	10.0
Other	39	9.1	14	3.8	11	4.2

SOURCE: U.S. Bureau of the Census (1953, 1970, 1984), *County Business Patterns*; U.S. Bureau of Labor Statistics (1981), used to compute service statistics.

a. Total classified employment and industry subcategories exclude government employees and sole proprietors.

b. Service industries (excluding government, retail, and wholesale) in which more than one-half the employees hold executive, managerial, professional or clerical positions.

c. Service industries (excluding government, retail, and wholesale) in which fewer than one-half the employees hold executive, managerial, professional or clerical positions.

TABLE 2.3
Central City Industry Employment by Percentage Employment in Industry Classified as Information Processors, 1970-1984
(in thousands)

City and Industry Type[a]	*1970* N	*1970* %	*1984* N	*1984* %	*Change 1970-84* N	*Change 1970-84* %
New York						
Industries with over 60% I.P.	946	28.0	1340	45.4	394	41.6
All other industries	2404	71.8	1613	54.6	–791	–32.9
Philadelphia						
Industries with over 60% I.P.	208	26.9	254	42.5	46	22.1
All other industries	564	73.1	343	57.5	–221	–39.2
Boston						
Industries with over 60% I.P.	189	40.7	256	53.2	67	35.4
All other industries	276	59.3	225	46.8	–51	–18.5
Baltimore						
Industries with over 60% I.P.	95	25.5	120	39.5	25	26.3
All other industries	272	74.5	184	60.5	–88	–32.4
St. Louis						
Industries with over 60% I.P.	92	24.5	96	36.9	4	4.3
All other industries	284	75.5	164	63.0	–120	–42.3
Atlanta						
Industries	92	24.7	136	31.7	44	47.8
All other industries	280	75.1	293	68.3	13	4.6
Houston						
Industries with over 60% I.P.	130	20.6	317	26.0	187	143.5
All other industries	500	79.4	904	74.0	404	80.8
Denver						
Industries with over 60% I.P.	70	27.6	122	35.5	52	73.6
All other industries	183	72.0	222	64.5	39	21.0
San Francisco						
Industries with over 60% I.P.	149	37.1	209	40.9	60	40.6
All other industries	253	62.9	301	59.1	48	19.0

SOURCE: U.S. Bureau of the Census (1970, 1984), *County Business Patterns;* U.S. Bureau of the Census (1981) matrix.
a. Information processors include executive, managerial, professional, and clerical occupations.

corresponding city-county boundaries (Atlanta, Houston, Denver, and San Francisco) are presented for comparative purposes.

Observe that all five northern cities experienced substantial employment growth in their predominantly information-processing industries and marked employment declines in their other combined industries. New York City added nearly 400,000 jobs between 1970 and 1984 in its predominantly information-processing industries (a 41% increase), while losing nearly 800,000 jobs in other industries (a 32.9% decrease). By 1984, 41.6 percent of all jobs in New York City were in service industries in which executives, managers, professionals, and clerical workers composed more than 60 percent of the industry employment total.

Boston's information-processing industries expanded by 35.4 percent between 1970 and 1984, while its other industries declined by 18.5 percent. For the other three northern cities—Philadelphia, Baltimore, and St. Louis—job increases in their predominantly information-processing industries were overwhelmed by job losses in their more traditional industries. This is especially the case for St. Louis, which lost one-half of its manufacturing jobs between 1970 and 1984.

In contrast to larger, older cities in the North, Atlanta, Houston, Denver, and San Francisco experienced employment gains both in their predominantly information-processing industries and other industries combined. Like larger cities in the North, though, the older, major cities in the South and West (Atlanta, San Francisco, and Denver) exhibited substantially greater absolute and proportional gains in their information-processing industries than in their other combined industries. Indeed, all three of these cities lost manufacturing employment between 1970 and 1984. Houston, on the other hand, added employment across all industries between 1970 and 1984, no doubt reflecting the city's economic surge during much of this period.

A major difference, then, between large cities in the Frostbelt (Northeast and Midwest) and Sunbelt (South and West) is that, since 1970, Sunbelt cities have added jobs in many other industries besides information-processing which have contributed to their overall employment growth. Conversely, many Frostbelt cities have experienced overall employment decline since 1970 as growth in their predominantly information-processing industries did not compensate for substantial losses in their more traditional industrial sectors, especially manufacturing. In this regard, one finds a strong negative correlation between the percentage of the city's employment in manufacturing in prior decades and total job change since 1970. Those cities where manufacturing constituted at least 35 percent of their employment bases in 1953 (Philadelphia, Baltimore, St. Louis, and New York) all experienced significant overall job declines between 1970 and 1984, whereas the others (Boston, Atlanta, Houston, Denver, and San Francisco) all added jobs during this period.

Changing Educational Requirements for Urban Work

Corresponding to the functional transformation of major northern cities from centers of goods processing to centers of information processing during the past three decades, there has been an important change in the education required for employment in these cities. Job losses have been greatest in northern urban industries where educational requirements for employment tend to be low (e.g., a high school degree is typically not required). Job growth has been primarily concentrated in urban industries where education beyond a high school degree is the norm.

To illustrate this phenomenon, table 2.4 presents employment change from 1959 to 1984 in industries classified by the mean years of schooling

TABLE 2.4
Central City Jobs in Industries, by Mean Education of Employees, 1959, 1970, 1984 (in thousands)

City and Educational Mean of Industry	*Number of Jobs 1959*	*1970*	*1984*	*Change 1959-70*	*1970-84*
New York					
Less than high school	1449	1445	953	–4	–492
Some higher education	682	1002	1241	320	239
Philadelphia					
Less than high school	434	396	224	–38	–172
Some higher education	135	205	244	70	39
Boston					
Less than high school	176	168	124	–8	–44
Some higher education	117	185	252	68	67
Baltimore					
Less than high school	215	187	114	–28	–73
Some higher education	59	90	105	31	15
St. Louis					
Less than high school	207	197	108	–10	–89
Some higher education	61	98	96	37	–2
Atlanta					
Less than high school	117	157	148	40	–9
Some higher education	42	92	129	50	37
Houston					
Less than high school	162	280	468	118	188
Some higher education	59	144	361	85	217
Denver					
Less than high school	79	106	111	27	5
Some higher education	42	72	131	30	59
San Francisco					
Less than high school	125	132	135	7	3
Some higher education	82	135	206	53	71

SOURCE: U.S. Bureau of the Census (1959, 1970, 1984), *County Business Patterns* and (1982) *Current Population Survey*, machine-readable files.

completed by their jobholders. Two categories of industries were selected: (1) industries whose jobholder education levels in 1982 averaged less than 12 years (high school not completed), and (2) industries whose jobholders averaged more than 13 years of schooling (i.e., employees, on average, acquired some higher education).[2]

The figures reveal that all major northern cities had consistent employment losses in industries with lower education requisites, with the heaviest job losses occurring in these industries after 1970. New York City, for instance, lost 492,000 jobs between 1970 and 1984 in those industries where mean jobholder education levels were less than high school completed, while adding 239,000 jobs in those industries where mean employee education levels exceeded 13 years of schooling. Philadelphia, Baltimore, and St. Louis also lost substantial numbers of low education requisite jobs since 1970, with St. Louis experiencing a small loss of jobs in its industries with high mean jobholder education levels, as well.

Boston, on the other hand, added more jobs in its highest education requisite industries than it lost in its lowest education requisite industries, contributing to overall city job growth since 1970. By 1984, Boston had twice as many jobs in industries with high mean employment education levels than in industries with low mean levels of employee education. This would indicate that Boston's economy has adapted especially well to the emerging post-industrial order which should sustain that city's employment growth into the 1990s. New York City likewise has a much higher percentage of its labor employed in knowledge-intensive service industries, implying that this city, too, should fare well in employment growth during the remainder of the 1980s.

Employment growth in industries whose jobholders' education averaged more than 13 years was also marked in major cities in the South and West. Yet, in contrast to major cities in the North, each of the four cities in the South and West gained substantial numbers of jobs in low education requisite industries between 1959 and 1984. Even after 1970, all except Atlanta added jobs in their low education requisite industries, though Houston (Harris County) is the only city to experience a boom in low education requisite jobs during the 1970s and early 1980s.

The oil price decline of the mid-1980s cooled Houston's economy considerably and also depressed employment growth in Denver. At the same time, the transformed service economies of Boston, New York City, and a number of other northeastern cities have led to a recent resurgence in their employment growth. While job losses are continuing in manufacturing and other blue-collar sectors of these cities, their vibrant information-processing sectors are more than compensating for blue-collar job losses, reversing decades of net employment decline.

Changing Demographic Compositions

A rather clear picture emerges from the analysis above. There have been precipitous declines since 1960 in blue-collar employment in northern central-cities that traditionally sustained large numbers of lesser-educated persons. Job losses have been partially replaced by growth in white-collar service industries with substantially higher educational requisites. This situation has implications for the expansion in large northern cities of population groups whose typical educational backgrounds place them at a serious disadvantage as the economies of these cities transform.

Table 2.5 summarizes the racial and ethnic compositional changes between 1970 and 1980 within the four largest northern cities (New York City, Chicago, Philadelphia, and Detroit). New York City, which experienced an overall population decline of over 823 thousand during the decade, actually lost 1.4 million non-Hispanic whites. Thus, in just ten years, New York's non-Hispanic white population (its nonminority population) dropped by an amount larger than the total population of any other U.S. city except Los

TABLE 2.5
Decomposition of Total Population and Population Changes by Race and Ethnicity—United States, Four Largest Northern Cities 1970-1980 (in thousands)

Central City	Total Population	Non-Hispanic Whites	Non-Hispanic Blacks	Non-Hispanic Other	Hispanic Population	Percentage Minority
New York						
1980	7072	3669	1694	303	1406	48
1970	7895	5062	1518	113	1202	36
Change	–823	–1393	+176	+190	+204	
Chicago						
1980	3005	1300	1188	96	422	57
1970	3363	1999	1076	40	248	41
Change	–358	–699	+111	+56	+174	
Philadelphia						
1980	1688	963	633	28	64	43
1970	1949	1247	646	11	45	36
Change	–260	–283	–13	+17	+19	
Detroit						
1980	1203	402	754	18	29	67
1970	1511	820	652	9	30	46
Change	–308	–418	+102	+9	–1	
Total change four cities	–1749	–2794	+377	+271	+396	

SOURCE: U.S. Census of Population (1970, 1980), machine-readable files.

Angeles, Chicago, Philadelphia, and Houston. Approximately 40 percent of the loss of non-Hispanic whites in New York City was replaced by an infusion of over 570 thousand Hispanics, blacks, and other minorities during the 1970s. By 1983, New York City's minority population had climbed to 53 percent of the city's total (Kasarda, 1984).

Chicago's demographic experience during the 1970s was similar to New York City's, but at about half the scale. More than 50 percent of Chicago's minority population increase during the decade consisted of Hispanics. However, it is important to point out that Chicago also experienced the third largest increase of non-Hispanic blacks of any U.S. city. By 1980, 57 percent of Chicago's resident population was composed of minorities with projections showing the minority proportion climbing to nearly 70 percent during the 1990s (Kasarda, 1984).

Philadelphia had the smallest aggregate population decline of the four cities. Both the number of non-Hispanic whites and non-Hispanic blacks declined in Philadelphia between 1970 and 1980, while Hispanics and non-Hispanic other minorities increased. Philadelphia's substantial decline in non-Hispanic whites, together with its net increase of 23 thousand minority residents during the 1970s, raised its minority percentage to 43 percent in 1980.

Detroit experienced the highest rate of non-Hispanic white residential decline of any major city in the country. Between 1970 and 1980 Detroit lost more than half its non-Hispanic white residents. Concurrently, Detroit had the fourth largest absolute increase of non-Hispanic blacks of any city in the country, falling just behind Chicago in black population increase. Combined with modest increases in "non-Hispanic others," Detroit's large increase in black residents and sharp drop in non-Hispanic white residents transformed the city's residential base from 46 percent minority in 1970 to 67 percent minority in 1980.

The bottom row of table 2.5 provides a summary of demographic transformation and residential decline in the four cities. Between 1970 and 1980 they experienced an aggregate loss of 2.8 million non-Hispanic whites, while their Hispanic residential bases increased by nearly 400 thousand. Added to the Hispanic increase during 1970 were substantial cumulative increases of non-Hispanic blacks and other non-Hispanic minorities, resulting in a total increase of more than 1 million minority residents in the four cities during the decade. Aggregated data from the March 1985 Current Population Survey indicate that major cities in the Northeast and Midwest continued to grow in minority population during the first half of the 1980s, while their non-Hispanic white residents declined further.

The New Structure of Minority Unemployment

Growing numbers of minorities are at a structural disadvantage in cities losing blue-collar and other entry-level jobs because substantially larger proportions of these minorities lack the necessary schooling to participate in new urban growth industries. To illustrate this structural disadvantage, table 2.6 presents data on years of schooling completed for white and black males residing in the central cities of the Northeast and Midwest.

Bearing in mind the selective employment changes in these cities during the past 15 years, note that the modal category of education completed by white males is "exceed one year of higher education." The smallest representative category for white males is "did not complete high school." The education completed distribution of white males residing in central cities in the Northeast and Midwest is therefore consistent with the distribution of job changes classified by education (table 2.4).

The opposite educational distributions hold for black males. Despite substantial gains in educational attainment during the past two decades, black males (over age 16) in northern cities are still most concentrated in the education completed category where employment opportunities declined the fastest since 1970 and least represented in that category where northern central city employment has most expanded (table 2.4).

An additional structural impediment inner-city minorities face is their increased distance from current sources of blue-collar and other entry-level jobs. As industries providing these jobs have dispersed to the suburbs, exurbs, and nonmetropolitan peripheries, racial discrimination and inadequate incomes of inner-city minorities have prevented many from moving with their

TABLE 2.6
Number of Central City Residents in Northeast and Midwest
Ages 16-64, by Race and Years of School Completed, 1985

Race and Schooling	*Northeast*	*Midwest*
White males		
Did not complete high school	944,964	743,105
Completed high school only	1,096,986	1,136,702
Attended college one year or more	1,205,944	1,291,168
Black males		
Did not complete high school	455,349	479,141
Completed high school only	366,932	404,121
Attended college one year or more	234,723	352,993

SOURCE: U.S. Bureau of the Census (March 1985), *Current Population Survey*, machine-readable file.

traditional sources of employment. Moreover, the dispersed nature of job growth sites makes public transportation from inner-city neighborhoods impractical, requiring virtually all city residents who work in peripheral areas to commute by automobiles.

Combined costs of maintaining, operating, and insuring an automobile in major cities are substantially higher than elsewhere, particularly in older, larger, densely settled cities. In fact, automobile ownership in the core areas of these cities is so expensive relative to the actual or potential incomes of their disadvantaged residents that most cannot afford this increasingly essential means of securing and maintaining blue-collar employment.

The severity of the problem of access to deconcentrating blue-collar job opportunities is documented in table 2.7 by the percentages of black and Hispanic households in New York City, Philadelphia, and Boston who do not have an automobile or truck available. More than half the minority households in Philadelphia and Boston are without a means of personal transportation. New York City's proportions are even higher, with only three of ten black or Hispanic households having a vehicle available. More detailed breakdowns of data show that in New York City's core boroughs (Brooklyn, Bronx, and Manhattan) fewer than three of ten black or Hispanic households had an automobile or truck available in 1980.

The figures in table 2.7 refer to *all* black and Hispanic households in each city or borough. It stands to reason that black and Hispanic households whose members have lower income and educational attainment have even less access to personal means of transportation. Thus, with automobile or truck ownership becoming increasingly necessary for blue-collar job access by poorly educated inner-city minorities, many find themselves confined to areas of declining opportunity.

Consequences of Minority Confinement

The confinement of large numbers of poorly educated minorities in economically transforming northern cities is one reason why their unem-

TABLE 2.7
Central City Minority Households with No Automobile or Truck Available—United States, 1980

Central City	Households	
	Percentage Black	*Percentage Hispanic*
Boston	51.3	50.8
New York	69.3	71.3
Bronx	71.1	72.9
Brooklyn	72.2	73.9
Manhattan	84.7	83.5
Queens	43.4	49.4
Staten Island	54.2	38.6
Philadelphia	50.9	50.8

SOURCE: U.S. Census of Housing (1980), *Detailed Housing Characteristics*

ployment rates and labor force dropout rates are much higher than those of central city white residents, and why black unemployment rates, in particular, have not responded well to the recent economic recovery of many northern cities. Let us assess these concrete manifestations of mismatch.

The Current Population Survey, which is the basis of the nation's unemployment statistics, does not contain a sufficiently large sample to provide reliable estimates of unemployment rates by both race and education level for individual cities. It is possible, however, to aggregate central city samples by region and calculate central city unemployment rates, by race and education completed, within each region. These computations, which are presented in table 2.8, document the increasingly important role education has played in urban employment prospects. Consistent with the mismatch thesis, one finds a sharp rise in unemployment rates between 1969 and 1985 for those who have not completed high school. Within central cities in the Northeast and Midwest, unemployment rates of black males without a high school degree exceeded 30 percent in 1985. Indeed, for blacks in northeastern cities who lack a high school degree, the national economic recovery since 1982 has had no effect, with black male unemployment rates actually increasing from 26.2 percent in 1982 to 30.4 percent in 1985.

Another statistic revealing the growing importance of education for urban employment prospects is the substantial increase in the absolute gap in unemployment rates between the poorly educated and the better-educated in 1969 versus 1985, regardless of race. For central city white residents in all regions combined, the education level gap in unemployment rates in 1969 was 2.7 (4.3 – 1.6). By 1985, it was 11.9 (15.5 – 3.6). For central city blacks, the gap was 2.9 in 1969 (6.6 – 3.7) and 14.2 in 1985 (27.3 – 13.1). This widening gap over time in unemployment rates between less and more highly educated labor is slightly larger in cities in the Northeast and Midwest.

In examining central city unemployment changes since 1982, note that while white male unemployment rate declines were generally consistent for all education categories, declines in black male unemployment were consistent only for those who had completed education beyond high school. The

TABLE 2.8
Unemployment Rates of Central-City Males Ages 16-64
by Race, Region, and Years of School Completed,
1969, 1977, 1982, and 1985

	White				*Black*			
Region and Schooling	*1969*	*1977*	*1982*	*1985*	*1969*	*1977*	*1982*	*1985*
All regions								
Did not complete high school	4.3	12.2	17.7	15.5	6.6	19.8	29.7	27.3
Completed high school only	1.7	8.0	11.0	8.3	4.1	16.2	23.5	18.4
Exceed one year of higher education	1.6	4.7	4.4	3.6	3.7	10.7	16.1	13.1
Northeast								
Did not complete high school	3.7	13.9	17.2	16.0	7.6	20.9	26.2	30.4
Completed high school only	1.7	9.4	10.3	9.7	3.4	18.2	21.9	13.6
Exceed one year of higher education	1.4	6.0	4.8	3.9	7.1	13.9	18.6	11.7
Midwest								
Did not complete high school	4.9	12.8	24.3	23.2	8.3	26.2	34.8	32.8
Completed high school only	1.1	8.0	14.5	10.8	3.3	18.0	35.8	24.9
Exceed one year of higher education	1.3	3.5	3.8	3.7	1.4	12.3	22.2	18.4

SOURCE: Computed from U.S. Bureau of the Census (1969, 1977, 1982, 1985), *Current Population Survey,* machine-readable files.

relatively large 1985 unemployment rates among better educated black urban males (particularly in the Midwest) is troublesome and difficult to interpret. Quality of education may play a role here as could racial ceilings in hiring practices of firms in major midwestern cities with large minority percentages.

Unemployment rates reveal only part of the picture of mismatch and the corresponding displacement of many minorities from the urban economic mainstream. These rates do not include persons who have given up looking for work because they believe no jobs are available (discouraged workers) and those who want work but cannot hold employment for a variety of physical or personal reasons. Such individuals are not considered to be in the labor force and, therefore, they are not counted among the unemployed. Computing proportions of black males (aged 16-64) residing in northeastern and midwestern central cities who are neither in school nor in the labor force, one again finds a sharp rise in these proportions since 1969. By 1985, the not in school, not in labor force component among this demographic group had exceeded 20 percent in cities in the Northeast and Midwest. Moreover, as with unemployment rates, labor-force dropout rates for blacks were, by a wide margin, highest for those with the least education. If one combines black unemployment rates with these labor force nonparticipation rates, the dire straights of black males residing in central cities in the Northeast and Midwest is quickly apparent.

Skill Mismatch and Demographic Disequilibria

A key policy construct developed above is "mismatch" which is defined as a discordant distribution of labor qualifications vis-à-vis qualifications required for available jobs at a point in time. Mismatch has both nonspatial (nationwide) and spatially-specific (community) aspects. The nonspatial aspect results from transformations in the overall economy from industrial to post-industrial and corresponding shrinking demands for traditional blue-collar labor (Bell, 1973; Singelmann, 1978). A tacit assumption in much of the literature on post-industrial society is that, through the interplay of market forces, displaced labor will adapt to the transforming economy by "shifting" from one sector to another (e.g., from manufacturing to services). Appropriate skills will eventually be acquired or sufficient numbers of service sector jobs (both low skill and high skill) will be created, absorbing the displaced and relieving the mismatch. This, of course, has been slow to happen in the United States, giving credence to those who argue that structural unemployment will remain a permanent feature of the national economy.

Spatially-specific mismatches emerge where transformations in local employment bases occur at a faster pace than their local labor can adapt, either through retraining or relocation. These mismatches are most apparent in larger, older cities in the North where declines in traditional blue-collar industries and growth of information-processing industries have been rapid and substantial. So different are the skills used and education required in these growing versus declining urban industrial sectors that adaptation by the displaced is exceedingly difficult. This difficulty is concretely represented in the exceptionally high unemployment rates of those central city residents who have not completed high school, regardless of race, and the widening gap over time in central city unemployment rates between the poorly and better educated.

It follows from the above that unemployment rates and labor force dropout rates will be higher for those resident groups whose educational distributions are inconsistent with the changing job opportunity structures of their localities. Such is particularly the case for black males (over age 16) in major cities in the North who are most concentrated in the education-completed category (less than high school degree) where matching local jobs are contracting and least represented in the education-completed category where local jobs are expanding (some higher education typically required).

Exacerbating resident skill level-job opportunity mismatches have been recent demographic trends in these cities. During the past two decades, northern cities that lost the largest numbers of blue-collar and other low-education requisite jobs simultaneously added large numbers of poorly educated minorities to their working age population. This demographic phenomenon, which contrasts sharply with that anticipated on the basis of market equilibrium models, leads to the thorny question: What is continuing to attract and/or hold large numbers of lesser skilled minorities in urban centers while employment opportunities appropriate to their skills are disappearing? For sure, such factors as racial discrimination, lack of sufficient lower income housing in outlying areas, and dependence of low income minorities on public transportation account for a significant part of the explanation. There is also the emergence of a vast urban underground economy that enables many of those displaced from the economic mainstream to survive. Indeed, for many who lack the education, technical or interpersonal skills for employment in mainstream institutions, the inner city may provide the only environment where they can economically stay afloat.

There is also the real possibility that certain public policies are anchoring disadvantaged persons in areas of rapid blue-collar job decline. These policies are based on the seemingly reasonable principle of spatially targeting public assistance, whereby areas of greatest economic distress (measured by such factors as poverty rates and persistent unemployment) receive the largest allocations of funds for public housing, community nutritional and health care, and other locationally-focused government aid. Formula-based community assistance programs have also been introduced such that the greater a locality's employment loss or other indicator of economic distress, the more federal aid it could receive (U.S. Department of Housing and Urban Development, 1978, 1980). Thus, as the blue-collar employment bases of cities withered, additional public assistance was provided, serving as a partial subsistence surrogate for many of those displaced from the economic mainstream.

While these policies have definitely helped relieve certain problems associated with declining urban blue-collar job bases (such as the inability of the unemployed to afford private sector housing or adequate nutrition and health care), they did nothing to reduce the growing skill mismatch between the resident labor force and available jobs. In fact, such spatially targeted assistance may have inadvertently increased the mismatch and the plight of the poor by bonding distressed people to places of entry-level job decline.

For those with some resources and the lucky portion whose efforts break the bonds of poverty, spatially concentrated public assistance may not impede mobility. But for many inner-city poor without skills, local concentrations of public assistance and community services can be sticking forces. With a low perceived marginal utility of migration relative to opportunity costs of giving up their in-place assistance, they see themselves better off by staying where they are. The upshot is that growing numbers of economically displaced minorities find themselves socially, economically, and spatially isolated in segregated areas of urban decline where they subsist on a combination of public assistance and their own informal economies. Such isolation, dependency, and blocked mobility breed hopelessness, despair, and alienation which, in turn, foster drug abuse, family dissolution, and other social ills that disproportionately afflict the urban disadvantaged.

It may be argued that this demographic disequilibrium works against the long-term economic health of distressed cities and their structurally unemployed residents. Imagine, for instance, what would have happened in the first half of this century if the great numbers of structurally displaced U.S. southerners who migrated to economically expanding northern cities in search of jobs and a better life had remained in their distressed localities. It is possible that the significant advances in income levels and living standards that both the South and its outmigrants experienced would not have occurred.

Circumstances today are analogous, but regionally the reverse. Many economically changing northern cities are now characterized by excesses of structurally displaced labor as their blue-collar job bases wither. For example, the March 1984 unemployment rates in Gary, Indiana, and Pontiac, Michigan, were both over 20 percent, whereas in the Raleigh-Durham-Chapel Hill area of North Carolina and in Austin, Texas, rates were under 4 percent (U.S. Department of Labor, 1984).

A more equal and, it may be contended, socially equitable territorial distribution of unemployment (say, no place continuously carrying more than 10 percent unemployment) could prevent the downward economic spiral that persists in some distressed cities. It is proposed that this downward spiral continues because the social overhead burdens and problems associated with exceptionally high unemployment rates create negative externalities that dissuade new businesses from locating in such cities. If this is so, under conditions of national economic growth a locality with excessively high unemployment has much poorer prospects of attracting new employers and economically rebounding than a locality with a moderate proportion of its labor force unemployed.

The conclusion to be drawn is that greater spatial equity in sharing the nation's unemployment burden may be essential to greater equality of opportunity for distressed cities and their disadvantaged residents. As long as disproportionately large numbers of unemployed persons remain concentrated in these cities, the economic future of both will be bleak.

New Urban Policies for New Urban Realities

To address the problem of inequities resulting from insufficient private-sector investment in areas of severe blue-collar employment decline, some have suggested a national development bank, a reconstruction finance corporation, enterprise zones, or government-business-labor partnerships, which might "reindustrialize" these areas or otherwise rebuild their historic employment bases. Such suggestions are as unrealistic as they are nostalgic. Government subsidies, tax incentives, and regulatory relief contained in existing and proposed U.S. urban policies are not nearly sufficient to overcome technological and market-driven forces redistributing blue-collar jobs and shaping the economies of the major cities. Nor would reversing these redistributional trends and inhibiting urban economic transformation necessarily be in the longer term interest of either the cities or the national economy.

On the contrary, encouraging the return of older blue-collar industries to urban core areas may additionally saddle urban economies with stagnating or declining industries and further weaken the competitive economic position of cities. Indeed, efforts to assist distressed urban areas through policies that direct older industries to the inner cities where these industries could experience greater costs or lower productivity may well conflict with efforts to strengthen the national economy on which the health of the cities is inextricably dependent. Economic advancement of cities and maximum job creation can best be accomplished through government programs and private-sector initiatives that promote urban service industries whose functions are consistent with the roles computer-age cities most effectively perform. The functional transformation of cities from centers of goods processing to centers of information processing must be recognized as inexorable.

This new urban reality also calls for fresh thinking and more future-oriented public policies regarding infrastructure development. Just as canals, railways, paved streets, running water, and electric power lines once provided cities with comparative advantages for processing and transporting goods, successful cities of the future will develop computer-age infrastructures that will provide them with comparative advantages for processing and transmitting information. As a start, concerted efforts must be made to "wire" cities so that businesses locating in them quickly and efficiently receive, process, store, and transmit immense amounts of data and information. Cities should also take advantage of their economies of scale (size) and provide municipally owned supercomputer facilities to service their growing information-processing industries on a cost-share basis.

Wiring a city and linking local businesses to municipally owned supercomputer facilities will be as essential to urban economic prosperity in the computer age as were water, sewer, and power lines linking businesses to public works facilities in the industrial age. The products are different; the concept is the same. Cities that take the lead in providing advanced information processing infrastructures for computer-age industries will hold a competitive edge over those places where infrastructure development remains geared to an industrial age winding down.

The competitive edge of major cities in cultural-leisure service provision offers additional opportunities for urban economic and demographic revitalization. Whereas the outflow of middle-income residents from cities still surpasses those returning, research indicates that important selection mechanisms based on consumer tastes and household composition are operating. Middle- and upper-income persons moving back to large cities or choosing to remain in them are predominantly single professionals, childless couples with dual incomes, older families whose last son or daughter has left the household, those leading unconventional lifestyles, and others who cherish and can afford the rich cultural amenities, specialized services, and myriad stimuli that downtown living offers (see Laska and Spain, 1980; Kasarda, 1980). These groups, no doubt, will serve as the primary demographic component of inner-city residential revitalization in the years ahead. For them, the emerging consumption-oriented city provides attractive advantages compared to dispersed, low density, home-centered communities.

A related issue new urban policies must address in their revitalization policies is the growing importance people and businesses are placing on "quality of life" in residential and commercial location decision making. In the earlier industrial age, technological and product constraints provided cities with a near monopoly in capturing national economic and demographic growth. Today, most of those constraints are gone. People and businesses are increasingly footloose—free to locate where they perceive the social and physical environment as more desirable. If large cities are to compete effectively for higher income residents and better jobs, they must improve their image as being safe, clean, and well managed. Cities that continue to be viewed as crime-ridden, dirty, and politically profligate will consistently lose out, regardless of other comparative advantages they offer.

Helping Those Caught in the Web of Change

Cities that improve their social and physical environments and adapt to their emerging service-sector roles should experience renewed demographic and economic vitality. However, many urban residents who lack appropriate skills for advanced service-sector industries are likely to remain on the bottom rungs of the socioeconomic ladder. Indeed, their economic plight could further deteriorate. For example, New York City, capitalizing on its strength as an international financial and administrative center, experienced a net increase of over 200 thousand jobs between 1980 and 1986. Yet, while the city's total employment base was expanding, its black unemployment rates remained intractably high. The latter persisted because most of New York's employment expansion was concentrated in white-collar service industries whereas blue-collar jobs continued to drop. Manufacturing employment, for example, declined by 100 thousand jobs during the first half of the 1980s. These figures, together with the demographic and employment data presented for New York City earlier in this chapter, testify that the mismatch between urban resident skill levels and new job opportunities can worsen even under conditions of overall central city employment gains.

The stark reality of rising urban minority unemployment under transforming (and possibly even growing) city economies, together with the improbability of government programs stimulating sufficient numbers of blue-collar jobs in transforming cities, calls for a shift in policy emphasis. Politically popular (but ineffective) jobs-to-people strategies and essential urban welfare programs must be better balanced with serious efforts to upgrade the education and skills of disadvantaged city residents *and* with people-to-jobs strategies that would facilitate the migration of the structurally unemployed to places where job opportunities appropriate to their skills are still expanding. Contrary to conventional wisdom, there have been massive increases in entry-level jobs nationwide during the past decade. Nearly 2.1 million nonadministrative jobs were *added* in the food and drink industry alone between 1975 and 1985 (U.S. Department of Labor, 1975, 1985). This employment growth, the vast majority at the entry level in fast food and drink establishments, is more than the total number of production jobs that *exist* in America's automobile, textile, and steel industries combined. Unfortunately, entry-level job expansion in such rapidly multiplying service establishments has occurred almost exclusively in the suburbs, exurbs, and nonmetropolitan areas, far removed from concentrations of low-skill urban minorities.

The inability of most inner-city minorities to follow decentralizing entry-level jobs (either because of racial discrimination, inadequate knowledge and resources, or subsidized anchoring) is among the chief reasons for the widening gap in black-white rates of unemployment and labor force nonparticipation. It is also a major contribution to rising demographic disequilibria in the transforming cities and their correspondingly high social overhead burdens. In this regard, a thinner central city, composed largely of those who actually desire to live in the city (minorities as well as whites) and who have appropriate employment opportunities nearby, would create a more vibrant city than results from present circumstances in which millions of disadvantaged persons are confined in cities without jobs or hope for them.

To increase mobility options of the urban disadvantaged, revised policies should be considered that would partially underwrite their more distant job searches and relocation expenses. Additional policies must be aimed at further reducing housing and employment discrimination and other institutional impediments to the mobility of minorities who wish to leave distressed urban areas. Finally, existing public assistance programs should be reviewed to ensure that they are not inadvertently attracting or bonding large numbers of disadvantaged persons to inner-city areas offering limited opportunities for employment.

All of the above, of course, must be complemented by broader economic development policies fostering sustained private-sector job generation. Programs assisting the retraining or mobility of disadvantaged urban minorities will prove fruitless unless new and enduring jobs are available at the end of the training programs or moves.

The vexing dilemma of the urban underclass in the U.S. will certainly not be entirely solved by these and other strategies proposed here. However, without an expanding national economy, improved education and technical training programs for the urban disadvantaged, stricter enforcement of civil rights legislation, and the mobility of the underclass from socially isolated, economically distressed ghettos, the permanence *and* growth of the underclass will be assured.

Bibliography

Bell, Daniel. 1973. *The Coming of Post-Industrial Society*. New York: Basic Books.

Berg, Ivar E., ed. 1981. *Sociological Perspectives on Labor Markets*. Quantitative Studies in Social Relations. New York: Academic Books.

Berry, Brian J.L., and John D. Kasarda, 1977. *Contemporary Urban Ecology*. New York: Macmillan.

Bodnar, John E., Roger Simon, and Michael P. Weber. 1982. *Lives of Their Own: Blacks, Italians, and Poles in Pittsburgh, 1900-1960. The Working Class in American History*. Urbana: University of Illinois Press.

Bradbury, Katharine L., Anthony Downs, and Kenneth A. Small. 1982. *Urban Decline and the Future of American Cities*. Washington, DC: Brookings Institution.

Downs, Anthony. 1973. *Opening Up the Suburbs: An Urban Strategy for America*. New Haven, CT: Yale University Press.

Farley, Reynolds. 1968. "The Urbanization of Negroes in the United States." *Journal of Social History* 1:241-58.

Fligstein, Neil. 1981. *Going North: Migration of Blacks and Whites from the South, 1900-1950*. Quantitative Studies in Social Relations. New York: Academic Press.

Fossett, James W., and Richard P. Nathan. 1981. "The Prospects for Urban Revival." In *Urban Government Finances Emerging Trends*, Urban Affairs Annual Reviews, vol. 20, edited by Roy Bahl. Newbury Park, CA: Sage.

Hauser, Phillip M. 1960. *Population Perspectives*. New Brunswick, NJ: Rutgers University Press.

Hawley, Amos H. 1971. *Urban Society: An Ecological Approach*. New York: Ronald Press.

———. 1981. *Urban Society: An Ecological Approach*. 2d ed. New York: Wiley.

Kasarda, John D. 1980. "The Implications of Contemporary Redistribution Trends for National Urban Policy." *Social Science Quarterly* 61:373-400.

———. 1982. "Urban Structural Transformation and Minority Opportunity." Report prepared for the Office of Policy Development and Research, U.S. Department of Housing and Urban Development, Washington, DC.

———. 1984. "Hispanics and City Change." *American Demographics* 6(11):25-29.

Kovaleff, Theodore P. 1974. "Industrialization and the American City." In *Cities in Transition: From the Ancient World to Urban America*, edited by Frank J. Coopa and Philip C. Dolce. Chicago: Nelson Hall.

Laska, Shirley Bradway, and Daphne Spain, eds. 1980. *Back to the City: Issues in Neighborhood Renovation*. Pergamon Policy Studies on Urban Affairs. New York: Pergamon.

Lieberson, Stanley. 1980. *A Piece of the Pie: Black and White Immigrants since 1880*. Berkeley: University of California Press.

Mohl, Raymond A. 1976. "The Industrial City." *Environment* 18(5):28-38.

Muller, Peter O. 1976. "The Outer City: Geographical Consequences of the Urbanization of Suburbs." Resource Paper for College Geography no. 75-2. Washington, DC: Association of American Geographers.

Norton, R. D. 1979. *City Life-Cycles and American Urban Policy*. Studies in Urban Economics. New York: Academic Press.

Palen, J. John. 1975. *The Urban World*. New York: McGraw-Hill.

Singelmann, Joachim. 1978. *From Agriculture to Services—The Transformation of Industrial Employment*. Newbury Park, CA: Sage.

Stanback, Thomas M., Jr., and Thierry J. Noyelle. 1981. *Metropolitan Labor Markets in Transition: A Study of Seven SMSAs—Atlanta, Denver, Buffalo, Phoenix, Columbus, Ohio, Nashville, Charlotte*. Final report, Contract no. USDL 21-36-78-33, U.S. Department of Commerce, Washington, DC.

Sternlieb, George, and James W. Hughes, eds. 1975. *Post-Industrial America: Metropolitan Decline and Inter-Regional Job Shifts*. New Brunswick, NJ: The Center for Urban Policy Research, Rutgers-The State University of New Jersey.

Suttles, Gerald D. 1978. "Changing Priorities for the Urban Heartland." In *Handbook of Contemporary Urban Life: An Examination of Urbanization, Social Organization, and Metropolitan Politics*, edited by David Street and Associates. The Jossey-Bass Social and Behavioral Science Series. San Francisco: Jossey-Bass.

U.S. Bureau of the Census. 1972. *1970 Census of Population*. Fourth Count Summary Tapes from 1970 Census of Population and Housing for Places. Suitland, MD.

U.S. Bureau of Labor Statistics. 1981. *The National Industry-Occupation Employment Matrix, 1970, 1978, and Projected 1990*. Washington, DC: U.S. Government Printing Office.

U.S. Department of Commerce. (selected years). *County Business Patterns.* Washington, DC.
U.S. Department of Housing and Urban Development. 1978, 1980. *The President's National Urban Policy Report.* Washington, DC: Department of Housing and Urban Development.
U.S. Department of Labor, Bureau of Labor Statistics. 1975 and 1985. *Employment and Earnings.*
———. 1984. "Employment and Unemployment in States and Local Areas." (Mar.). Microfiche. Washington, DC.
U.S. Management and Budget Office. Statistical Policy Division. 1972. *Standard Industrial Classification Manual, 1972.* Washington, DC.
Vaughan, Roger J. 1977. *The Urban Impacts of Federal Policies: Vol. 2. Economic Development.* R-2028-KF/RC. Santa Monica, CA: The Rand Corporation.
Vinyard, Jo Ellen. 1976. *The Irish on the Urban Frontier: Nineteenth Century Detroit, 1850-1880.* The Irish Americans. New York: Arno Press.
Wilson, William J. 1983. "Inner-City Dislocations." *Society* 21(1):80-86.
Zunz, Olivier. 1982. *The Changing Face of Inequality: Urbanization, Industrial Development, and Immigrants in Detroit, 1880-1920.* Chicago: University of Chicago Press.

3

High Technology and Urban Dynamics in the United States

Manuel Castells

The spatial structure of the United States has been undergoing substantial changes in recent years, not the least being consolidation of the large metropolitan areas as the dominant pattern of settlement (National Decision Systems, 1982). In 1970, 24 metropolitan areas had populations over one million. In 1980 this number had grown to 29. The largest conurbations in the country (based on 1980 census figures) include the New York-New Jersey-Southern Connecticut area (17.6 million); Los Angeles-Long Beach-San Bernardino-Riverside (10.2 million); Chicago-Northwest Indiana-Aurora-Elgin-Joliet (7.2 million); Philadelphia-Wilmington-Trenton (4.8 million); and San Francisco-Oakland-San Jose (4.5 million).

Nevertheless, for the first time, some of these giant cities, such as New York, have lost population in absolute numbers—between 1970 and 1980, the demographic decline was particularly severe in large metropolitan areas, such as Detroit, Cleveland, Pittsburgh, Milwaukee, and Buffalo. Furthermore, urbanization as a whole has stagnated or declined, also for the first time. Percentages of the U.S. population living in urban areas (over 2,500 inhabitants) were 64 in 1950, 70 in 1960, 73.5 in 1970, and 73.3 in 1980 (Long and DeAre, 1983).

In the 1970-1980 decade a *process of ruralization* took place as the rural population grew by 12 percent, as fast as growth of metropolitan areas. The urban population would have declined even more as a percentage of the total population, except that some rural areas grew so fast that they became urban.

The interesting point is that this rural increase was mainly related to *industrial* growth in rural areas. The process of growth and decline varied tremendously across regions, with the West, Southwest, and Florida growing fast, and the Northeast and North Central regions experiencing decline. Yet, within the Northeast some regions, like New England, have shown considerable economic vitality in certain years (1980-85), holding their demographic share, while upgrading the social and income status of their residents. Central cities of large metropolitan areas show a general tendency to lose population to the benefit of the suburbs, with middle-class whites leading the exodus.

Yet, here also, some significant new trends, reversing the secular pattern of urban evolution, pinpoint a new spatial logic at work. Thus, after losing population for 30 years, the city of Boston increased its population from 561 thousand in 1982 to 571 thousand in 1985. New York City also *gained* population between 1980 (7.072 million) and 1984 (7.165 million). San Francisco, while slightly losing population, has increased the number of households and upgraded the economic and social status of its dwellers, because the flight to the suburbs is led by low-income families unable to afford housing prices in the city, leading San Francisco toward the predominant pattern of residential segregation in European cities.

Yet, the so-called "back to the city" movement is still insignificant on a national scale, though it reveals new trends in *some* areas of *some* central cities. Even at this level, data on New York City show that the movement has been largely overestimated: in the period 1970-80, while New York's total population fell by 23 thousand, the white non-Hispanic population shrank by 1.4 million (Kasarda, 1984); the share of families with income over $50,000 declined from 62 to 54 percent; and an analysis of census tracts showed "few dramatic changes in neighborhood distribution" (Koretz, 1984). In fact, as a general trend, metropolitan suburbs have continued to grow in absolute and relative terms. The proportion of urban population living in the suburbs in 1960 was 30 percent; by 1980, it was 41 percent, occupying 42 percent of the urban land. Interestingly, the increase in the suburbs has been faster in the older, metropolitan areas than in newer areas. Density of settlement is decreasing in all areas: between 1970 and 1980 it decreased by 15 percent in the suburbs, by 17 percent in central cities, and by 18 percent in small cities and towns (Long and DeAre, 1983).

At a more synthetic level, one can summarize current spatial trends in six major developments and one overarching process. Cities and regions seem to have evolved toward a series of specific sociospatial forms, each with its own dynamics, yet structurally interconnected.

(1) The process of the new *interregional division of labor* defines the relative weight of each process in a given metropolitan area.
(2) Most central business districts (CBDs) in the giant cities increasingly concentrate directional activities and continue to engage in a process of economic growth, political dominance, and cultural centrality. Around this dynamism, *gentrified neighborhoods* preserve the space of the new urban elite.

(3) Around the CBDs, large areas of central cities continue their decline and become more and more "*urban reservations*" for destitute workers and minorities, increasingly accounted for by female-headed families.
(4) Yet, some physically declining areas in the central cities, as well as some inner suburbs, are revitalized economically and socially by the new *immigrant city* connected to the growth pole of the metropolitan economy, yet suffering from exploitation and discrimination.
(5) Beyond city borders, *suburbs* continue to spread in a pattern of living and working increasingly individualized and diversified.
(6) Suburban sprawl is also becoming *territorial sprawl*, with diffusion of activities and residence across rural and semirural areas, sometimes in autonomous settlements, but often in some form of functional relationship with a nearby giant city.

The whole structure appears to follow a logic that combines *spatial diffusion, territorial hierarchy (including an urban concentration of the highest level),* and *functional interconnectedness.*

Nevertheless, while there is, to a large extent, general agreement on the profile and significance of the changes underway in the spatial structure of the United States, few coherent, comprehensive interpretations have been proposed concerning causes of such changes. This chapter aims at exploring the hypothesis according to which the nucleus of the process affecting urban-regional change in the United States relies on the combination of some secular economic trends with recent policies of techno-economic restructuring adopted to supersede the structural crisis of the 1970s. The two major secular trends are the process of uneven development and the movement toward a postindustrial, service economy. Policies of techno-economic restructuring include, mainly, establishment of new relationships between capital and labor, transformation of the public sector, and furthering the internationalization of the economy. The current technological revolution has played a decisive role in the feasibility of such economic policies for business and government. Therefore, I subsequently analyze the two major secular trends and three main dimensions of the process of techno-economic restructuring, pinpointing their specific impact on reshaping the spatial structure of the United States.

High Technology, Uneven Development, and Reshaping the Spatial Structure

Uneven development is a functional mechanism of capitalist production. Uneven development refers to the differential rate of investment (and therefore of allocation of resources) between economic sectors and spatial areas, according to their changing potential of profitability. Uneven development follows a differential rate of growth that, by the different level of wealth it generates in various places and activities, tends to have cumulative effects: areas that first rose to economic prominence have a clear advantage over the laggard ones as far, *and only as far,* as there is no major altering of the historical pattern of economic growth. When such an alteration occurs, we then assist a fundamental transformation of the spatial economy, which we call *urban and regional restructuring.*

The entire world is currently undergoing such a process, and my hypothesis is that this spatial transformation is the result of a fundamental economic transformation of which the rise of high technology is a major element. In this sense, the impact of high tech on the secular trend of uneven development gives new meaning to such a trend and results in historically distinct spatial dynamics.

In recent years, the debate in the field of regional development has been dominated by the "Sunbelt-Snowbelt" issue (Sawers and Tabb, 1984). Starting with David Perry and Alfred Watkin's seminal volume, *The Rise of Sunbelt Cities* (1977), and based on research presented by Sternlieb and coauthors (1975) or Weinstein and Firestine (1978), the stage was set for a reassessment of the spatial structure of American economic and demographic trends, and indeed, of its political system. Since then, the trend has continued and is expected to go on in the near future. Investment, jobs, people, and income are leaving the traditional Northeastern and North Central centers of U.S. capitalism in growing numbers to start a new process of growth in the, at first sight, more promising lands of the Southwest, the West, and, to some extent, the South (particularly in Florida).

But while the phenomenon is clear and widely recognized, its causes and consequences are not. The standard interpretation generally presented by both neoliberal economists and leftist critics finds the source for the move to the Sunbelt in the tendency for capital to search for lower wages, non-unionized labor, fewer taxes, and less government control. Indeed, these are important factors in the process of relocating activities. But it is by no means clear that those features are the overarching consideration for companies to move from their present locations.

Perry and Watkins (1977) compared the South's performance between 1940 and 1960 for low-wage and high-wage manufacturing sectors. They found that while both sectors expanded in the South, growth was particularly rapid in *high-wage* industries, multiplying by two the national average of growth in these sectors. Of course, one can say that high-wage industries in the Southwest will still pay lower wages than in the Northeast, but it is crucial to observe that the Sunbelt's growth is not concentrated in low-wage sectors but, if anything, in "underpaid" high-wage, high-growth sectors, which is different and very important in drawing conclusions for the overall dynamics of the economy.

Along the same line of reasoning, John Mollenkopf (1983) has contributed the most systematic critique of the Sternlieb-Hughes interpretation of the movement, by comparing 10 northern to 11 southwestern cities on the basis of a series of indicators of factors supposed to account for the move. He writes:

> It is fallacious to conclude that the manufacturing jobs which have been disappearing from the Northeast are reappearing in southwestern cities. Manufacturing actually accounts for *less* growth in southwestern urban employment than services, and these manufacturing jobs differ qualitatively from those the Northeast has lost. Finally, those new manufacturing establishments

which have located in the Southwest in preference to the Northeast have not necessarily done so to find low-wage workers, cheaper land or closer proximity to their markets. Indeed, many firms have had to import trained labor for plants being established in the Southwest (1983:200).

When controlled by occupational categories, wages in the Southwest are rapidly closing the gap with the North, and while some infrastructural advantages (communications) seem to be offset by others (serious water shortage problems in many cities, for instance), Mollenkopf finds that a probusiness political environment, including weaker labor unions, community groups, and elected officials, is the most important element characteristic of the Southwest. But he introduces a novel hypothesis that relates Sunbelt development to the structural transformation of the economy. Given the growth of services and the emergence of high technology industries as the new source of economic dynamism, those areas that will grow with the new sectors will be the fastest growing ones, simply because they are new and their economic growth is met simultaneously with decline of preexisting activities. Accordingly, Mollenkopf proposes a typology of cities in three different situations:

(1) *Cities based on old manufacturing activities, such as Cleveland, Detroit, or Buffalo.* Because of the decline in these activities, uncompensated by growth of new sectors, the entire urban economy declines.
(2) *Cities representing a mixture of old manufacturing and new activities, either in services or high tech, such as New York or Boston.* In these metropolitan areas, we observe a process of *simultaneous growth and decline.*
(3) *Finally, newer cities whose growth is associated with the rise of the new economy, particularly in high-tech and corporate services.* This pattern characterizes much of the Sunbelt.

The general trend is less one of interregional shift of activities (in fact, runaway shops are more likely to go abroad) than one of increasing sectoral differentiation of the economy, with a specific spatial expression for location of new activities. And while advanced business services have a tendency to remain in the oldest metropolitan areas, restructuring internally their social and spatial fabric, high-tech manufacturing has shown a clear tendency for the newest areas in the South, and, particularly, the West. In this sense, high-technology industry seems to be, *along with energy resources,* and *some* advanced services, the real engine for the rise of Sunbelt cities.

In fact, the U.S. economy (as well as capitalist economy in general) is undergoing a dramatic process of intersectoral differentiation. Some industries literally die while others explode. In this sense we have uneven development at its peak.

Within this perspective, the "deindustrialization of America" thesis Bluestone and Harrison forcefully argue takes its full meaning.

By *deindustrialization* is meant a widespread, systematic disinvestment in the nation's productive capacity. Controversial as it may be, the essential problem with the U.S. economy can be traced to the way capitalism—in terms of financial resources and of real plant and equipment—has been diverted from productive investment in our basic natural industries into unproductive speculation, mergers and acquisitions, and foreign investment. Left behind are shuttered factories, displaced workers, and a newly emerging group of ghost towns (1982:6).

The result is a pattern of increasing disparity between those sectors (and others) where disinvestment occurs and those (speculative or otherwise) where investment goes. Again, this situation is the updated version of the process of uneven development, but we should not consider it as being something "natural." Quite the contrary, these are *decisions* taken by corporations and governments—but decisions that follow the social logic of a capitalist economy: investment for the sake of profit. In this sense, this situation is a secular trend within capitalism.

So, it is the new wave of uneven development, with its spearheads mainly in advanced services and high-technology manufacturing that represents the driving force behind the current process of urban and regional restructuring. But why so? Why do high-technology activities choose new locations, developing new regions, instead of renewing the traditional manufacturing centers? In other words, why the Sunbelt? There are specific arguments relating the characteristics of the new high-tech industries to the expansion of the southwestern and western regions of the United States.

Steven J. Pinkerton (1984) completed a straightforward statistical study based on a series of step-wise regressions among the increase in high-tech jobs in 1975-81 in each state of the U.S., the state's total share of high-tech jobs in 1981, and a number of variables calculated for each state.

The results, even if too general to be definitive, are very revealing of some basic trends. First, in terms of the absolute share of high-tech jobs, the study shows that the old industrial states are still dominant, given their sheer size, particularly if we include California in the group (as Pinkerton's statistical model did). As he writes:

The typical high-tech state is highly populated, highly taxed, has numerous defense-related industries and has had a stagnant population growth over the past ten years. In other words, the declining industrial sector of the U.S. still holds a significant share of total U.S. high-tech jobs (Pinkerton, 1984:9).

But, interestingly enough, the results completely change when the dependent variable is the increase in high-tech jobs in 1975-81, instead of their absolute share, thus, showing the new spatial trends. This variable is negatively correlated with the size of the state population, and positively with population increase 1970-80; so there is evidence of a trend toward decentralization of new industries in new areas. The high *negative* correlation with percentage of unionization, as Pinkerton observes, could reflect not only "the movement from the old industrial corridors but also the reduction in growth in California and Massachusetts" (p. 43). And, as in the study by Glasmeier et al. (1984), there is a strong correlation between increase in high-tech jobs and government transfers received, and defense spending, underlining anew the government and military inducement of the new spatial structure of the United States.

Thus, the rise of the Sunbelt is the result of an increasing sectoral differentiation of the economy, in which high-tech manufacturing and energy production associate themselves with a new regional environment. Together with continuing location of advanced corporate services in CBDs of the top metropolitan areas, these leading trends characteristic of the new process of uneven development seem to be largely responsible for the new geography of America.

Nonetheless, the economic effect of high-tech manufacturing activities is by no means homogeneous, and their growth rate is extremely diversified among industries. Of the 100 high-tech industries Glasmeier et al. (1983) studied for the period 1972-77, 34 actually lost employment. Yet, the group as a whole increased its employment by 8 percent nationally, well above the 3 percent performance for overall manufacturing. But again, this growth was extremely differentiated among cities and regions. Eighty-six of the 264 SMSAs Glasmeier and colleagues analyzed lost high-tech employment, including 28 southern metropolitan areas. Generally speaking, "losing metro areas consisted of older big-city SMSAs; winners, on the other hand, reflected newer, adjacent suburban metropolitan areas" (p. 11). So, along with New York, Philadelphia, Cleveland, and other northern cities, we also find among the list of top ten losers, Miami and Los Angeles. On the other hand, attracting high-tech industries is not the sure way to stable economic development. High-tech employment tends to be highly concentrated in one industry in many of the areas studied, particularly in the South and the Southwest. For instance, in Austin and Tucson more than 50 percent of high-tech employment is concentrated in *one* industry—radio-TV transmitting devices and equipment in Austin, guided missiles in Tucson (p. 11). This concentration lends credit to the argument Noyelle and Stanback (1984) put forward about the extreme vulnerability of current high-tech areas based on rapid development of a limited number of industries without much connection to the regional economy.

In sum, high-tech growth accelerates the process of uneven development through intersectorial differentiation of the economy and profound restructuring of the spatial organization of production largely determined by the new industries' characteristics and functional requirement.

New Technologies, the Service Economy, and Internal Structure of Metropolitan Areas

Advanced capitalist societies have often been characterized as entering the *postindustrial* age (Bell, 1973; Touraine, 1969). As long as we recognize that societies remain capitalist—a fundamental feature of the entire social and economic organization—we have no major argument with such a statement, with only one reservation—the term and underlying theory are ambiguous, and the exact implications embodied in the term are unspecified. But the postindustrialist perspective has a major merit: to point out the fundamental transformation of our process of production within the overall capitalist logic.

The most immediate indicator of some fundamental transformation of the way we produce, work, and live is the expansion of the proportion of the U.S. population employed in the so-called service sector from 48.1 percent in 1940 to 66.7 percent in 1980. Yet, a number of remarks should be recalled to assess fully this transformation before introducing into the process the variables in which we are interested here, namely new technologies and the urban structure.

First of all, rise of services does not equal demise of manufacturing. The proportion of the labor force in transformative activities (including construction) was still 29.8 percent in 1980. Also many service jobs depend on their linkage to manufacturing activities.

Second, it is crucial to distinguish different activities within the service sector, given the broad range of activities it generally includes. Joachim Singelmann (1978) has provided a useful typology that differentiates, in terms compatible with government statistics, among distributive, producer, social, and personal services. Between 1960 and 1980, distributive services grew moderately, while still employing 23.7 percent of the total labor force. Personal services declined, with the major exception of the eating and drinking category. Producer services almost doubled, while still employing, overall, fewer workers than personal services. And social services literally exploded, jumping from 10 to 24.3 percent in 1980. Even during the critical 1970s, the decline in education has been more than compensated by employment growth in hospitals and medical services, reaching 7.8 percent of the labor force in 1980. That figure is more than twice the total labor force in agriculture: a small indication of what secular transformation means in practical terms.

Nevertheless, while the importance of the shift of employment toward service activities can hardly be overstated, it must be put into perspective. Because, as Fuchs (1968) and Stanback (1979) have reminded us, when *measured in constant dollars* there is very little shift in the share of GNP for the output of goods-producing or service-producing sectors between 1969 and 1976, since the small gains of service outputs here are at the expense of agriculture. Thus, while industry represented in 1980 less than 30 percent of employment, it accounted for about 41 percent of the GNP. What are the implications in terms of the secular evolution of the economy? The main one is, it seems to me, the growing disjunction between activities people engage in for a living and those that generate growth in our society. This situation comes about because, as I proposed years ago (Castells, 1976), the service sector has expanded under a variety of pressures, most attributable to social mobilization and political concession. A substantial part of service employment, particularly in social services, is, in fact, a form of absorbing surplus population generated by increased productivity in agriculture and industry, in a society that still requires full-time salaried work to survive, even if we could afford less collective work with more collective production.

Yet, that most services show a lower labor productivity does not mean that they are "parasite" activities. In fact, most producer services are crucial for the development of productivity in the economy as a whole: personal services structure the consumption process, distributive services make possible exchange, and social services, to a large degree, are crucial for collective consumption and social stability. So, one can hypothesize that for the entire economy to keep growing, we would need more and better service activities. In fact, the fastest growth of office-type occupations, in industry and services, has been in highly skilled occupations, particularly in 1970-80. Such an expansion clearly indicates the fundamental evolution occurring within the service activities themselves: the handling of all types of information is increasingly at the core of the entire economic activity. Here is where the contradiction appears: on the one hand, information and knowledge are key productive forces in increasing productivity; on the other hand, much of the process of information management is in itself labor-intensive and prone to low productivity. Thus increased productivity in the economy as a whole is countervailed by the increase of low-productive, yet generally indispensable, activities. Thus, between 1948 and 1982, labor output increased substantially as did, to a lesser extent, "multifactor productivity" (that accounts for a number of elements outside capital and labor).

Yet, what generally failed was the productivity of capital: overhead costs of production skyrocketed (Mark and Waldorf, 1983). Part of this situation has to do with government defense and social expenditures. But a substantial part goes to management costs—handling information—for a long time more difficult to rationalize and automate than goods production. Since new technologies are basically aimed at information processing, we can see the possibility emerging of the historical supersession of the barrier that exists to extend the productivity drive from industry to services. Office automation and telecommunications are leading the change. The office of the future is already in our lives (Baran, 1982), and with it, a fundamental reorganization of the production process, with manufacturing and services being increasingly complementary and feeding each other in demand and in enhancement of their productive capacities. We will see below the tremendous potential impact of this trend on the labor market and occupational structure. But, for the moment, let us say, concerning the service sector, that improved productivity generated by the new technologies will trigger an explosion of producer services, particularly in finance, insurance, and real estate (FIRE) activities, eventually the fastest growing subsector. As a whole, services will take off, this time not only in terms of employment, but in output. It is now that we are entering into the service economy, with the additional caveat that the traditional distinction of the production of goods and services will soon be obsolete.

On the basis of these trends, what can we say of urban processes and spatial forms that express the interaction between high technology, services, and corporate capitalism? How does the service economy affect the urban structure and how do the new technologies intervene in this relationship?

Thomas M. Stanback (1979), Thierry Noyelle, and their coworkers (1981; Noyelle and Stanback, 1984) have been systematically studying the question for a number of years. On the basis of these findings we can observe a number of trends:

- Increasing concentration of export-oriented services in nodal metropolitan areas.
- Increasing concentration of these services within the CBDs of large metropolitan areas.
- Both of these tendencies are particularly accentuated for "advanced corporate services," as has been established by Robert Cohen's pioneer research (1978). Cohen also shows that even the booming Sunbelt remains largely dependent on the old established basis of headquarters and corporate services in New York, Chicago, and, for banking, San Francisco. In particular, the predominance of New York in all advanced services is overwhelming (Mollenkopf, 1984).
- There has been a process of suburbanization of secondary offices of major companies, as well as of small businesses unable to afford competition for land in the highly priced CBDs. Nevertheless, the bulk of service activities remains in the downtown areas (Armstrong, 1979).
- Consumer services have, by and large, followed the suburbanization of middle-class residents.
- Local sector services, particularly in health and education, tend to follow the spatial distribution of the population they serve. Yet, some services (like big hospitals) are also export-oriented. In this case they tend to dominate a given space, generally in their former location in the inner city (Shonick, 1979).

Current technological change, by accelerating the expansion of services because of the removal of the productivity barrier, is likely to reinforce these *observed trends.* CBDs of nodal metropolitan areas are (and will be) experiencing major growth. While some decentralization of secondary offices to the suburbs and nonmetropolitan areas is occurring (given the possibilities offered by new communication systems), the overall *economic impact of high technology*, by enhancing the role of the large corporation and expanding the service sector, *works actually for the development of downtowns,* with the major exception of metropolitan areas based in old-line manufacturing (such as Buffalo). Even so, the revival of downtown Baltimore (in spite of the crises of steel and automobiles) is one example of the possible successful transition to a service economy on the basis of renewed productivity in the sector.

So, there are three major spatial consequences of the economic effects of new technologies on the service sector:

- Increasing mixture of consumer services and residential areas in the suburban landscape
- Decentralization of second-level service activities to suburbs and nonmetropolitan areas
- Rapid growth of CBDs, particularly in metropolitan areas highly placed in the urban hierarchy of corporate networks.

The last development, by far the most significant, is confronted by two major, and potentially explosive contradictions:

(1) On one hand, because of the lower level of automation and rationalization of public sector services, these services are not matching the increase in their demand resulting from the new dynamism of downtown areas. As I discuss below, the situation is being made worse by the simultaneous fiscal retrenchment that accompanies, in most cities, the current economic restructuring. Thus, a booming corporate service economy is revitalizing the CBDs, but declining, low-productive public sector services could slow down such a revival or create very serious functional difficulties.

(2) On the other hand, the service economy labor force is characterized by polarization between a white, male-dominated sector of professionals and a large pool of low-paid labor, overwhelmingly female and minority. As Stanback and Noyelle write (1982:141):

> The developments adumbrated above suggest that our society may be moving toward an unstable dual economy, in the sense that the transformation of employment systems is pushing toward an increasing separation between two socio-economic strata with increasingly restricted bridges between them: a stratum of managers, professionals, technicians, teachers, and other highly skilled employees living in a relatively well protected and well paying economic world, and a large structure of assemblyworkers, clericals, and service workers, who find it increasingly difficult to make ends meet and to deal with the stress associated with unrewarding and somewhat insecure jobs.

To the extent that expansion of services is being fueled by the new technologies, there will be a growing polarization of the urban labor market, particularly in the central cities of large metropolitan areas. And this situation is regardless of the impact of high technology on service jobs themselves, a different question that we examine in detail in the following section. Thus, the current trend shows a growing corporate service sector, located in high-level metropolitan CBDs in full development, yet daily populated by a growing number of low-paid female and minority workers, and coming under the pressure of functional and social demands that cannot be met by a public sector in deep crisis.

The economic and urban evolution of New York City since 1977 exemplifies these trends. Studies by John Mollenkopf (1984) and Saskia Sassen-Koob (1984a) converge toward the same conclusion: there is a process of simultaneous growth and decline, related to rapid expansion of services, particularly producer services, and to closing or downgrading manufacturing activities. After its notorious 1975 fiscal crisis, New York City engaged in a major process of restructuring that led to a concentration of resources and public incentives in the leading sector of corporate services (1982). As a consequence, there was also a *building boom for office space in Manhattan's CBD*. Mollenkopf, in another study (1982), calculated that employment in the construction industry, which had *declined* by 39 percent between 1969 and 1977, experienced the highest percentage of growth (along with business services) in 1977-82: it grew by 31.3 percent. Evidence cited in these studies confirms that expansion of services in New York resulted in a much more polarized social structure, in terms of jobs and income.

Besides the occupational structure characteristic of large corporations in the service sector, Sassen-Koob cites an additional reason for such a process of social polarization: the simultaneous development of entertainment and personal consumption services, attracted by a larger market of high-income professionals. As she writes,

> high-income gentrification is labor-intensive, in contrast to the typical middle class suburb that represents a capital intensive process. . . . In sum, the existence of major growth sectors, notably the producer services, generates low-wage jobs directly through the structure of the work process, and indirectly through the structure of the high-income lifestyles of those therein employed, and the consumption needs of the low-wage work force (1984a:7).

Economic and social polarization *within the central city* of the largest metropolitan areas, and within the metropolitan areas themselves, is increasing dramatically as the service economy takes a new qualitative leap forward, under the combined impact of economic restructuring and technological change. New York is the showcase of the process we are analyzing: a booming CBD, a growing gentrified Manhattan, and yet the persistence of rundown, devastated areas are the spatial signs of a new process of growth that deepens its contradictions, while showing an extraordinary vitality as the economic and informational capital of the world.

Nevertheless, the objection can be raised that this particular pattern of urban development is related to expansion of the service economy and has only a remote connection with high technology. In fact, I have no real argument with such a statement. But my general perspective is precisely that we cannot isolate high technology from the overall process of the economic and organizational change of which it is an indispensable part. In other words, without the current technological change, large corporations would not flourish, nor could the services they require be performed. What characterizes an advanced economy is that the core of its activity is *information processing*; that is the object of the major technological innovations under way. Thus, the new dynamism of central cities of the high-ranking metropolitan areas is basically the result of the rise of advanced services, with their specific pattern of polarized growth. But the type of services that are expanding, and their organizational logic, are the function *both* of the current capitalist interests *and* of the new technological environment in which they operate. High technology is a *process*, rather than a *product*. And as such, it is inextricably linked to the expression of the advanced service economy and to the new urban process they jointly determine.

Techno-Economic Restructuring and the Urban-Regional Process

Secular trends of economic structural transformation exert a decisive influence as spatial patterns. Yet their pace has been dramatically accelerated, and their direction considerably altered by the emergence of a new socio-economic model of growth, in recent years in the U.S. The term *techno-economic restructuring* refers to the process through which this new model comes into being based on a series of business decisions and government policies that take advantage of the current technological revolution.

As I have argued elsewhere (Castells, 1980), the capitalist system went through a major crisis in the 1973-82 decade (Bowles et al., 1983; Carnoy et al., 1983; O'Connor, 1984) and is emerging from it by means of setting up a new model of economic accumulation, social organization, and political legitimization that, while still being capitalist, will be as different from the Keynesian model of 1945-73 as the latter was different from pre-Great Depression capitalism (Carnoy and Castells, 1984). Such a model relies on three main processes.

(1) A fundamental *transformation of the capital-labor relationship in the work process,* with capital gaining the initiative again over the wages and regulations conquered by the labor movement after decades of class struggle (Carnoy and Shearer, 1980; Gordon et al., 1982; Portes and Walton, 1981; Sabel, 1982).

(2) *A new role of the state and the public sector,* not so much reducing government intervention in the economy, but shifting its emphasis from collective consumption to capital accumulation, and from legitimation to domination—something that can schematically be called the transition from the Welfare State to the Warfare State (Crouch, 1979; Dumas, 1982; Gough, 1979; Leontieff and Duchin, 1983; O'Connor, 1984; OECD, 1981; Wilensky, 1974).

(3) *A new international and interregional division of labor,* with capital, labor, production, markets, and management shifting locations in a continuing variable geometry to take advantage of the best possible conditions for the strategy of the large corporations, regardless of the social and political consequences for specific territorial units (Bluestone and Harrison, 1982; Frobel et al., 1980; Palloix, 1975; Sawers and Tabb, 1984).

I propose as a hypothesis that the new technologies are playing a major role in these three fundamental processes and that such a technological-economic impact is crucial to understanding the emerging spatial structure and new dynamics of giant cities in the United States.

Rising Social Dualism in Large Metropolitan Areas

The first process of economic restructuring concerns deterioration of wages and working conditions obtained by labor over capital after hard social struggles for many decades. Given the substantial share of labor costs in total production costs, the new strategy of cheapening labor and weakening union power is a paramount goal for corporations. The new technologies play a major role in the reorganization of the labor process to implement this strategy along four major lines:

(1) Automation in factories and offices allows labor-saving procedures that dramatically increase productivity while eliminating jobs on a massive scale, particularly, in manufacturing and heavily unionized sectors (Leontieff and Duchin, 1984).
(2) By creating a permanent threat of substituting machines for workers, management puts a tremendous pressure on labor, which will be forced to accept, in many cases, whatever conditions capital considers profitable.
(3) Automation of information-processing functions allows capital to supersede one of the major barriers for the growth of economic productivity, as I have argued in the preceding section. The possibility of office automation allows corporations to grow in size without losing flexibility and efficiency and makes it possible to boost productivity in the management process, potentially overcoming obstacles to labor productivity in the service sector (Chandler, 1977; Cohen, 1978). Yet, as services were the refuge for the growing labor surplus from other sectors, as well as a response to political pressures on the public sector to provide jobs and services, automation of services will be uneven, provoking bitter social conflicts (Castells, 1976).
(4) Scattered observations on the impact of automation on the occupational structure seem to indicate that it will result in a bifurcated labor market, with the upgrading of a minority of workers and rapid growth of professional sectors, while a majority of workers are deskilled and reduced to low-paying jobs, either in labor-intensive services or down-graded manufacturing (Glasmeier, 1985; Hirschhorn, 1984; Markusen, 1983; Storper, 1982). This bifurcated occupational structure reflects itself in the income distributions, with a tendency to what Lester Thurow (1984) calls "the disappearance of the middle class."

Nevertheless, although many experts recognize the massive elimination of jobs in traditional activities, through the introduction of new technologies, the potential for growth and jobs generated in the new high-tech sectors has become a question. In fact, new technologies do generate jobs, but not enough to compensate for the ones eliminated by technological change. Ann Markusen (1983) estimates that during the 1980s high-tech activities will generate 3 million new jobs, but their use will eliminate 25 million jobs.

In this sense, it is important to distinguish between growth rates and actual aggregate growth in employment. For instance, between 1982 and 1990 the Bureau of Labor Statistics projects a 27 percent growth in computers and peripherals, but total employment in the sector will be about 600 thousand jobs in 1990, while the moderate 15 percent growth in restaurants and other retailing will take the number of jobs in this sector to 20 million in 1990. Thus, high-tech will eliminate many more jobs than it will create by itself.

Yet it could still be argued that the overall dynamics of the economy, helped by high technology, will generate new jobs outside high-technology activities in new sectors different from those being automated. And such seems to be the current trend in the United States (Wayne, 1984). As the point of comparison, during the crisis of the 1970s, while Western Europe lost almost 3 million jobs, the American economy created *20 million new jobs*. Given the slowing of the growth rate of the U.S. work force in the 1980s (a 1.5 percent annual rate, against 2.7 in the 1970s), political acceptance of 7-8 percent unemployment as "normal," and expansion of the "underground economy," the impact of technology on labor is likely to be socially assimilated in terms of numbers of available jobs.

Nonetheless, the fundamental feature is the mismatching between characteristics of jobs that disappear and those of newly created jobs. On one hand, an upgrading of professional and technical jobs in advanced services and high-technology manufacturing requires a fundamental retraining of labor, something the educational system is hardly able to assure, particularly in the secondary public school. On the other hand, the new unskilled jobs are found in low-paying services and down-graded manufacturing, where cheap labor is cheaper than automation, where companies cannot afford capital investment, or where the nature of the activity makes it difficult to standardize work (Bluestone and Harrison, 1982; Serrin, 1983). So routine service jobs, new "sweatshop manufacturing," and nonunion labor are substituting for the traditional unionized automobile or steel worker, the government employee or stable clerk. Bluestone and Harrison (1982) have documented the mismatching between old and new jobs. Other studies find similar trends in the new labor market (Hirshhorn, 1984; Storper, 1982; Teitz, 1984; Tomaskovic-Devey and Miller, 1982; Walker, 1983).

Restructuring the labor process, and thus labor markets, by new social relationships and high technology, is having a major effect on the urban-regional structure at four levels.

(1) Traditional manufacturing regions and cities where old-line factories were located are being abandoned or automated, with considerable decline and outmigration of jobs (Hicks and Glickman, 1983; Mollenkopf, 1983; Noyelle and Stanback, 1984).

(2) New regions emerge as manufacturing centers: those where high-tech nests are located; those where labor costs are low and the sociopolitical environment is "good for business"; and some isolated rural areas where new high-tech industries decentralize assembly-line operations.

(3) Expansion of advanced corporate services, supported by office automation, is concentrating companies, headquarters, and their constellation of auxiliary services (producer and consumer services) in the downtowns of top metropolitan areas, many of which are literally booming (Cohen, 1978; Mollenkopf, 1983; Stanback et al., 1981; Strassman, 1979-1980), thus eliminating a surge in construction of new high-rise buildings as well as rehabilitation and conversion of old structures (Mollenkopf, 1984). Paradoxically, the decline in public urban services is hampering the perspective of further downtown development.

(4) An even more fundamental spatial restructuring is underway, directly connected with the new capital-labor relationship facilitated by high technology: *the new economic and social dualism within the largest metropolitan areas*, as has been shown by empirical studies of the recent evolution of New York (Mollenkopf, 1984; Sassen-Koob, 1984a) and Los Angeles (Soja et al., 1983; Sassen-Koob, 1984b). Several processes are taking place simultaneously within the same metropolitan area: rapid growth of advanced corporate services and high-tech manufacturing; decline of traditional activities (in services and manufacturing); and development of the new downgraded, yet booming, economic sectors ("sweatshops" with undocumented workers, eating and drinking places, luxury consumption, and so on).

This process of *polarized growth* creates distinct social spheres, yet it has to link these spheres within the same functional unit. This trend differs from the old phenomenon of social inequality and spatial segregation in the big city.

There is something other than the distinction between rich and poor or white and nonwhite; it is the formation of different systems of production and social organization, equally dynamic and equally new, yet profoundly different in the wealth, power, and prestige they accumulate. The new immigrant neighborhoods of New York or Los Angeles are lively areas, not run-down ghettoes, in spite of the misery and oppression experienced by many of their dwellers. Their role in the new metropolitan economy is crucial for production and consumption, and their cultural autonomy transforms the life and politics of the city. Yet these different worlds (the high-tech world, the advanced services world, the auxiliary services world, various immigrant worlds, traditional black ghettoes, protected middle-class suburbs, etc.) develop along separate lines in terms of their own dynamics, while still contributing altogether to the complex picture of the new supercity. The new labor market is at the basis of this newly polarized sociospatial structure. We witness the rise of dualized giant cities that segregate internally their activities, social groups, and cultures while reconnecting them in terms of their structural interdependency. These metropolises are magnets on a world level, attracting people, capital, minds, information, materials, and energy while keeping separate the channels of operation for all these elements in the actual fabric of the metropolis. We are not in a situation of urban-regional crisis (as it could be in Detroit or Buffalo), but in a process of interactive growth between elements that ignore one another while being part of the same system.

Impact of the Transformation of the Public Sector on the Spatial Structure

The most far-reaching element of the current economic restructuring is the transformation of the state's role in the U.S. economy for the 50 years since the New Deal (Carnoy, 1984; Janowitz, 1980; O'Connor, 1973; Wolfe, 1977). The state is not withdrawing from economic interventionism—quite the opposite. But the form and content of government intervention are drastically changing, shifting their focus from accumulation and redistribution to *selective* accumulation and military reinforcement. It would be a mistake to consider this trend a product of the Reagan Administration, although its policy represents a qualitative step in this direction (Lekachman, 1982; Palmer and Sawhill, 1982; Wilmoth, 1983). The crisis of the Welfare State results from the method chosen to fight inflation: to cut back social expenditures, thus limiting the budget deficit and the need for public indebtment and excessive money supply (Adams and Freeman, 1982; Gough, 1979; OECD, 1981). Yet, as everybody knows, the budget deficit has skyrocketed in recent years because of the military build-up and interest payments required by massive borrowing. So we witness *simultaneously* the partial dismantlement of the Welfare State, particularly in the sphere of urban programs and community services (Hirschhorn et al., 1983), and the rise of the Warfare State: a gigantic investment in military technology and warfare systems, to prepare for all kinds of wars on a "galactic" scale (Dellums, 1983; Dumas, 1982; Gansler, 1980).

While the reasons for such trends are primarily political (as are all processes related to the state) their impact on the economy is decisive (Carnoy and Castells, 1984). Not only has the budget deficit gone out of control, but the economy as a whole is becoming increasingly militarized, reversing the trend of the 1960s and 1970s, when dependency from military expenditures was decreasing. The current economic dynamism is being restructured around a core of highly profitable industries related to military production for domestic consumption and weapons export (Markusen, 1985).

High technology is directly connected to the militarization of the economy in the United States—again, not because technologies by themselves are military-prone (as development of Japanese high tech oriented toward consumer electronics shows). In fact, even in the United States, after an initial period (particularly during the 1950s) when military markets overwhelmingly dominated the industry, particularly in electronics, consumer products and industrial applications became the most important part of high-tech activities during the 1970s, as evidenced by the declining share of government in the semiconductors market. Nevertheless, during the 1980s the trend has been reversed and high-tech manufacturing, particularly microelectronics, has been greatly stimulated by the military and space programs.

A study by Markusen (1984a:27) based on several data sources, including her own, concludes that "heavily defense-dependent manufacturing sectors are dominated by those in the high tech category, and a substantial proportion of high-tech industries are military suppliers."

She estimated that in 1977 military-related high-tech production accounted for 47 percent of all high-tech manufacturing employment. Given the Reagan military build-up, it is likely that this proportion increased considerably in recent years. Other empirical studies also show a close link between high-tech production and military procurement, in terms of employment (Dempsey and Schmude, 1971; Henry, 1983; Markusen, 1986; Rutziek, 1970) and location of high-tech industries (Clayton, 1962; Glasmeier, 1985; Pinkerton, 1984). In any case, what experts agree on is the decisive role played by the military in the origins of high-technology developments (Carlson and Lyman, 1984; Mutlu, 1979). The emphasis on performance, regardless of cost, generous funding for long periods of time, willingness to support innovation outside the large companies, and the requirement of miniaturization of electronic devices (to be used in airplanes and mobile supports) were all factors that facilitated the electronics revolution and shaped its products, even if many were developed afterward for all kinds of civilian applications.

These origins of military-related high-tech production, along with current trends of renewed militarization, have had decisive effects in the urban-regional process of the United States. While dismantlement of the Welfare State has led to the urban fiscal crisis of most inner cities, the rise of the new Warfare State has spurred growth in new regions: in the suburbs of some metropolitan areas and among professional-technical sectors of a few cities.

Norman Glickman (1984) has attempted to measure this impact by calculating rates of growth and decline of federal outlays for selected programs between 1981 and 1984 for different cities and regions, typologized by a number of key variables. All welfare-related programs decline substantially, while defense spending increases sharply, particularly in procurement. Interurban and interregional variations of social and defense expenditures follow almost systematically opposite patterns. Defense-spending increase is more pronounced than average in high-income, nondeclining, low-unemployment, and "low-hardship" areas; in medium and small, rather than large, cities; and is overwhelmingly concentrated in suburbs, although this last result (interestingly enough) is exclusively because of the high suburban concentration of defense procurement. In terms of regions, while concentration ratios vary across the country depending on programs, defense expenditures are heavily concentrated in the West, South, Central, South Atlantic, Mountain, and especially in the Pacific regions.

Glasmeier et al. (1983) found similar trends with high-technology defense-related activities: the most spatially concentrated of all high-tech sectors show a preference for the Southwest and West, medium-sized cities, and suburban locations. The reasons for this regional-urban location pattern seem to be functional, historical, and cultural (Markusen, 1984b): (1) the military needs large extensions of isolated, undeveloped land for construction of huge facilities and testing sites; (2) it needs good weather all year round for aircraft and missile testing; (3) it was located in the Pacific rim during the build-up against Japan in World War II; (4) it prefers self-contained residential communities close to the production facility, so that the political environment will not jeopardize production continuity and shared promilitary values, given the considerable cross-over between the army and management positions in the military industry. Also, there is an anti-big city, anti-urban culture that characterizes many military personnel, eager to live within their own world, far away from the social and political complexities of large inner cities. Thus, semirural regions and isolated suburbs in metropolitan areas in the western United States seem to fit the cultural pattern of this "frontier spirit" of the military and related industries.

Internationalization of the Economy and the New Spatial Division of Labor

The new spatial division of labor in the United States also results from the growing internationalization of the economy, accelerated after the 1974-75 crisis on several levels (Glickman, 1980). Take, for example, the increasing role of international trade, with the United States importing more and more manufactured goods while exporting services, agricultural products, and high-tech manufacturing. And capital itself has become more internationalized. Now, along with the growing outflow of U.S. capital abroad, has come a massive increase of direct foreign investment in the United States, which grew 600 percent between 1973 and 1983, reaching $111.3 billion. Much of this foreign investment is in manufacturing, in old and new sectors (Schoenberger, 1984).

The trend is accelerating. While in 1983 there were 280 new foreign manufacturing investments, accounting for $48 billion, in 1984 the corresponding figures were 325 investments for $51 billion, and in 1985 the estimates foresee about 375 new investments. Geographically, New York continues to be the most attractive location, with 69 foreign investments in 1984, followed by California (33), North Carolina (20), Georgia (17), and Texas (16) (*New York Times*, 1985).

Another trend concerns renewed internationalization of labor in the United States (Glickman and Petras, 1981), in which we witness an immigration flow unprecedented since the 1920s, fueling, among other things, the informal economy in the largest cities (Portes and Bach, 1984; Sassen-Koob, 1984b). There is also an internationalization of the production process, with the location of different facilities in different countries, and reintegration of the process through improved communications (Henderson and Scott, 1984; Mutlu, 1979). Along with electronics, the automobile industry is the most significant example of this global factory, with production of the "world car," whose different components are produced in various countries and assembled in the final markets: by 1980 37.2 percent of the total motor vehicle production of the four leading U.S. automobile companies was located abroad (Trachte and Ross, 1983). Another expression of the new internationalization of production is formation of the so-called border regions, particularly in the United States. Mexico's Border Region has profoundly transformed the socioeconomic structure of large areas in Mexico and the United States as well (Hansen, 1981).

Finally, the process of internationalization in the United States has a specific historical actor: the multinational corporation. Emergence of this particular type of organization—transcending national boundaries, cultural specificity, and political controls within a global strategy, whose logic cannot be fully understood in any given spatial unit—allows for the expression of the four processes of internationalization we have described (Barnet and Muller, 1974; Palloix, 1975; United Nations, 1978; Vernon, 1971). So we find a tendency toward delocalization of the logic of the economic process and concentration of decision-making units in a few commanding heights of the international economy.

High technology plays a major role in this process of internationalization on different levels. First, it allows communication and decentralized unified management between spatially scattered units, through new telecommunication technologies. Second, high-technology manufacturing epitomizes the new spatial division of labor, with the locationally distinct hierarchy between research and design and assembly-line operations, therefore spearheading the new space of global production, facilitated by the light weight of many electronic components, whose value is basically their informational content.

Third, the process of automation and increasing precision of machines (through robots and numerical controlled machine tools) makes possible a large-scale standardization of the components of most manufacturing activities and their recombination wherever the location appears to be convenient.

Finally, we should remember that high technology is, above all, a new process of production and management. Therefore, the more the economy becomes open, internationalized, and competitive, the more the appropriation of high tech and its implementation in factories, offices, and communication systems of a company or of a country become crucial elements in winning a competitive edge. In this sense, internationalization of the economy greatly reinforces growth and importance of high-tech activities. Because of this importance, investment in high-tech production and conquest of markets for high-tech products in growing demand become key elements in the strategy to dominate the international economy (Zysman and Cohen, 1982).

We can now summarize the current trends of spatial restructuring in the United States in the context of the new international economy and in interaction with high-technology activities along the following lines:

(1) The internationalization of the economy actually reinforces the spatial polarization between sectors in different regions and within their metropolitan areas. Particularly because manufacturing exports are decreasing and manufacturing imports are on the rise, the old-line industries are increasingly hurt by international competition. Furthermore, outflow of capital investment (for instance, in the automobile) and shift of capital toward the promising high-tech sectors will augment intersectoral differentiation and, therefore, the distance between the spatial areas we have been able to associate with each economic sector.

(2) The concentration of economic power on a world level in a few hundred major corporations is spurring the growth of advanced corporate services and headquarters in a few major metropolises, consolidating the formation of what Friedmann and Wolff (1982) have called *the world city*.

(3) The new process of international migration concentrates a new labor force and, therefore, the informal economy precisely in these world cities, whose "underground" component also has to be underlined as a part of the same system.

(4) Finally, perhaps the most striking effect of the new international economy on cities and regions is the loss of their autonomy vis-à-vis the worldwide economic actors that control their activities in terms of a global logic largely ignored and uncontrolled by local and regional societies. A rapidly changing economic space determined by economic units whose size and transnationality places them above social pressures and political controls is a *tendency* that, favored by the internationalization process *and* by high tech, attempts to impose the abstraction of a space of strategic decisions over the experience of place-based activities, cultures, and politics. Yet the reactions of *developmental states* (such as Singapore, Japan, or Brazil) in the international economy creates a new dynamics, equally abstract in its horizon, but more directly rooted in political pressures and social values (Zysman and Cohen, 1982). The process of internationalization gives rise to the space of flows, but the political dynamics of the decision-making process in the world economy partially restore the space of historical meaning. In the middle of this dialectic cities and regions live, die, struggle, and change.

Conclusion

Changes that are deeply modifying the spatial structure of the United States can be traced back to a large extent to processes of secular economic change as they are shaped and accelerated by recent policies of techno-economic restructuring. High-tech-led economic development and the search for less astringent conditions for capital have fostered a new interregional division of labor. Advanced corporate services required by the informational mode of development and global needs of multinational capitals have provoked the revitalization of high-level CBDs and created conditions for a limited, yet significant, gentrification of their adjacent neighborhoods. The transition from Welfare State to Warfare State, along with the planned shrinkage of urban services, has precipitated decline of large central cities and expansion of suburbs, particularly in regions where the "new economy" is booming. Interpenetration of economies and societies at the international level, facilitated by the new communication technologies, has laid the ground for the "immigrant city," as well as for the exindustrial "ghost towns" that often become the "back door" for newcomers into the city. The polarized occupational structure favored by high tech in production and applications has considerably contributed to the deepening of cleavages between cities, between inner cities and suburbs, and within central cities. And, finally, new communication technologies, the coming of the "electronic home," and the "electronic office" have stimulated territorial sprawl, suburbanization, and individualization of sociospatial relations.

Altogether, two major processes appear to express, at the spatial level, the dynamics of the current technological and economic restructuring:

(1) New technologies allow the emergence of a *space of flows,* substituting for a space of places, whose meaning is largely determined by their position in a network of exchanges. The logic of large-scale organizations fits perfectly into a spatial form that abstracts from historical reality and cultural specificity to accommodate new information and instructions.

(2) The new model of capitalist growth, supported by the Warfare State and the informational mode of production, induces a *new territorial division of labor,* based on *polarized growth* and *selective development,* which reflects itself in the interregional cleavages, intrametropolitan dualism, and *simultaneous* life and death of our great cities.

Thus, although giant cities are still the predominant form of settlement patterns in the U.S., their shape, functions, and dynamics present some new, historically original characteristics that make them depart from the classical model of large metropolitan areas. Gottman's "megalopolis" is obsolete as an ideal type for understanding the new macrourban systems, because what characterizes U.S. giant cities today is less their size and the spatial vicinity of diverse functions than their complexity and interdependence within a nonlocalized network of flows of variable geometry; locational constraints and geographical rigidities play a much lesser role in the current process of urbanization than worldwide business decisions and governments' political strategies. Instead of a natural economic evolution toward the megalopolis, we observe a constant reshaping of spatial structures by conflictive processes of interaction among social, economic, and political actors.

Bibliography

Adams, Paul, and Gary Freeman. 1982. "Social Services under Reagan and Thatcher." In *Urban Policy under Capitalism,* edited by Norman I. Fainstein and Susan S. Fainstein. Urban Affairs Annual Review Series, vol. 22. Newbury Park, CA: Sage.

Armstrong, Regina B. 1979. "National Trends in Office Construction, Employment and Headquarters Location in Metropolitan Areas." In *Spatial Patterns of Office Growth and Location.* See Daniels, 1979.

Baran, Barbara. 1982. "The Transformation of the Office Industry: Impacts on the Workforce." Masters thesis, University of California, Berkeley.

Baran, Barbara, and Suzanne Teegarden. 1983. "Women's Labor in the Office of the Future." Research paper. Berkeley: Institute of Urban and Regional Development, University of California.

Barnet, Richard J., and Ronald E. Muller. 1974. *Global Reach: The Power of the Multinational Corporations.* New York: Simon & Schuster.

Bell, Daniel. 1973. *The Coming of Post-Industrial Society.* New York: Basic Books.

Bluestone, Barry, and Bennett Harrison. 1982. *The Deindustrialization of America: Plant Closings, Community Abandonment, and the Dismantling of Basic Industries.* New York: Basic Books.

Borrus, Michael, and James Millstein. 1982. "Technological Innovations and Industrial Growth: A Comparative Assessment of Biotechnology and Semiconductors." Research report prepared for the U.S. Congress Office of Technology Assessment. Berkeley: Berkeley Roundtable on the International Economy, University of California.

Bowles, Samuel, David M. Gordon, and Thomas E. Weisskopf. 1983. *Beyond the Wasteland. A Democratic Alternative to Economic Decline.* Garden City, NY: Anchor Press/Doubleday.

Carlson, Roger, and Terence Lyman. 1984. "U.S. Government Programs and Their Influence on Silicon Valley." Research report. Menlo Park, CA: SRI International.

Carnoy, Martin. 1984. *The State and Political Theory.* Princeton, NJ: Princeton University Press.

Carnoy, Martin, and Manuel Castells. 1984. "After the Crisis?" *World Policy Journal* 1:495-515.

Carnoy, Martin, and Derek Shearer. 1980. *Economic Democracy: The Challenge of the 1980s.* White Plains, NY: Sharpe.

Carnoy, Martin, Derek Shearer, and Russell Rumberger. 1983. *A New Social Contract: The Economy and Government after Reagan.* New York: Harper & Row.

Castells, Manuel. 1976. "The Service Economy and the Postindustrial Society:" A Sociological Critique." *International Journal of Health Services* 6:595-607.

———. 1980. *The Economic Crisis and American Society.* Princeton, NJ: Princeton University Press.

Chandler, Alfred D. 1977. *The Visible Hand: The Managerial Revolution in American Business.* Cambridge, MA: Belknap Press.

Clayton, James L. 1962. "Defense Spending: Key to California's Growth." *Western Political Quarterly* 15:280-93.

Cohen, Robert. 1978. *The Corporation and the City.* New York: Conservation of Human Resources Project, Columbia University.

Crouch, Colin, ed. 1979. *State and Economy in Contemporary Capitalism.* London: Croom Helm.

Daniels, P. W., ed. 1979. *Spatial Patterns of Office Growth and Location.* Chichester, England: Wiley.

DeGrasse, Robert W., Jr. 1983. *Military Expansion, Economic Decline: The Impact of Military Spending on U.S. Economic Performance.* Armonk, NY: Sharpe.

Dellums, Ronald V., with R. H. (Max) Miller and H. Lee Halterman. 1983. *Defense Sense: The Search for a National Military Policy*. Cambridge, MA: Ballinger.

Dempsey, Richard, and Douglas Schmude. 1971. "Occupational Impact of Defense Expenditures." *Monthly Labor Review* 94(12):12-15.

Dumas, Lloyd J., ed. 1982. *The Political Economy of Arms Reduction: Reversing Economic Decay*. AAAS Selected Symposium no. 80. Boulder, CO: Westview for the American Association for the Advancement of Science.

Friedmann, John F., and Goetz Wolff. 1982. "World City Formation: An Agenda for Research and Action." *International Journal of Urban and Regional Research* 6:309-44.

Frobel, Folker, Jurgen Heinrichs, and Otto Kreye. 1980. *The New International Division of Labor*. Studies in Modern Capitalism. Cambridge: Cambridge University Press.

Fuchs, Victor R. 1968. *The Service Economy*. New York: National Bureau of Economic Research.

Gansler, Jacques S. 1980. *The Defense Industry*. Cambridge, MA: MIT Press.

Glasmeier, Amy K. 1985. "Spatial Differentiation of High Technology Industries: Implications for Planning." Ph.D. dissertation, University of California, Berkeley.

Glasmeier, Amy K., Peter Hall, and Ann Markusen. 1983. "Recent Evidence on High Technology Industries' Spatial Tendencies." Institute of Urban and Regional Development Working Paper no. 417. Berkeley: Institute of Urban and Regional Development, University of California.

———. 1984. "Can Everyone Have a Slice of the High Tech Pie?" *Berkeley Planning Journal* 1(Summer).

Glickman, Norman J. 1980. "International Trade, Capital Mobility, and Economic Growth: Some Implications for American Cities and Regions." Report to the President's Commission for a National Agenda for the 1980s. Washington, DC.

———. 1984. "Economic Policy and the Cities. In Search of Reagan's Real Urban Policy." *Journal of the American Planning Association* 50:471-78.

Glickman, Norman J., and E. M. Petras. 1981. "International Capital and International Labor Flows: Implications for Public Policy." Working Papers in Regional Science. Philadelphia: University of Pennsylvania.

Gordon, David M., Richard Edwards, and Michael Reich. 1982. *Segmented Work, Divided Workers: The Historical Transformation of Labor in the United States*. Cambridge: Cambridge University Press.

Gough, Ian. 1979. *The Political Economy of the Welfare State*. Critical Texts in Social Work and the Welfare State. London: Macmillan.

Hall, Peter G., Ann R. Markusen, R. Osborn, and B. Wachsman. 1983. "The Computer Software Industry: Prospects and Policy Issues." Working Paper No. 410. Berkeley: Institute of Urban and Regional Development, University of California.

Hansen, Niles. 1981. "Mexico's Border Industry and the International Divison of Labor." *Annals of Regional Science* 15(2):1-12.

Henderson, Jeff, and Allen Scott. 1984. "The American Semiconductors Industry and the New International Division of Labor." Center of Urban Studies and Planning Working Paper. Hong Kong: University of Hong Kong.

Henry, David K. 1983. "Defense Spending: A Growth Market for Industry." In *1983 U.S. Industrial Outlook for 250 Industries with Projections for 1987*. Washington, DC: Bureau of Industrial Economics, U.S. Department of Commerce.

Hicks, Donald A., and Norman J. Glickman, eds. 1983. *Transition to the 21st Century: Prospect and Policies for Economic and Urban-Regional Transformation*. Greenwich, CT: JAI Press.

Hirschhorn, Larry. 1984. *Beyond Mechanization: Work and Technology in a Post-Industrial Age*. Cambridge, MA: MIT Press.

Hirschhorn, Larry, and Associates. 1983. *Cutting Back: Retrenchment and Redevelopment in Human and Community Services*. Jossey-Bass Social and Behavioral Science Series; Jossey-Bass Management Series. San Francisco: Jossey-Bass.

Janowitz, Morris. 1980. *The Last Half-Century: Societal Change and Politics in America*. Chicago: University of Chicago Press.

Kasarda, John D. 1984. "Hispanics and City Change." *American Demographics* 6(Nov.):25-29.

Koretz, Gene, ed. 1984. "Economic Diary: Are the 'Gentry' Really Moving Back to the City?" *Business Week* no. 2845 (4 June):12+.

Lekachman, Robert. 1982. *Greed Is Not Enough: Reaganomics*. New York: Pantheon.

Leontief, Wassily, and Faye Duchin. 1983. *Military Spending: Facts and Figures, Worldwide Implications, and Future Outlook*. New York: Oxford University Press.

Long, Larry, and Diana DeAre. 1983. "The Slowing of Urbanization in U.S." *Scientific American* 249(1):33-41.

Mahony, Sheila, Nick DeMartino, and Robert Stengel. 1980. *Keeping PACE with the New Television: Public Television and Changing Technology*. New York: Carnegie Corporation of New York; VNU Books International.

Mark, Jerome A., and William H. Waldorf. 1983. "Multifactor Productivity: A New BLS Measure." *Monthly Labor Review* 106(12):3-15.

Markusen, Ann R. 1983. "High Tech Jobs, Markets, and Economic Development Prospects." Working Paper no. 403. Berkeley: Institute of Urban and Regional Development, University of California, Berkeley.

———. 1984a. "Defense Spending and the Geography of High Tech Industries." Working Paper no. 423. Berkeley: Institute of Urban and Regional Development, University of California, Berkeley.

———. 1984b. *Profit Cycles, Oligopoly, and Regional Development*.Cambridge, MA: MIT Press.

———. 1985. "The Economic and Regional Consequences of Military Innovation." Institute of Urban and Regional Development Working paper no. 442. Berkeley: Institute of Urban and Regional Development, University of California, Berkeley.

———. 1986. "Defense Spending: A Successful Industrial Policy?" *International Journal of Urban and Regional Research* 10:105-22.

Mollenkopf, John H. 1982. "Economic Development." In *Setting Municipal Priorities, 1984*, edited by Charles Brecher and Raymond D. Horton. New York: New York University Press.

———. 1983. *The Contested City*. Princeton, NJ: Princeton University Press.

———. 1984. "The Post-Industrial Transformation of the Political Order in New York City." Paper presented at the Social Science Research Council Conference on New York City, New York.

Mutlu, Servet. 1979. "Interregional and International Mobility of Industrial Capital: The Case of the American Automobile and Electronics Companies." Ph.D. dissertation, University of California, Berkeley.

National Decision Systems. 1982. *1980 U.S. Census Population and Housing Characteristics*. 5 vols. San Diego.

New York Times. 1985. "Foreign Investment in U.S. Up Sharply." (16 Sept.):D-9.

Noyelle, Theirry J., and Thomas M. Stanback. 1984. *The Economic Transformation of American Cities*. Totowa, NJ: Rowman & Allanheld.

O'Connor, James R. 1973. *The Fiscal Crisis of the State*. New York: St. Martin's Press.

———. 1984. *Accumulation Crisis*. Oxford, England: Blackwell.

OECD. See Organisation for Economic Cooperation and Development.

Organisation for Economic Cooperation and Development. 1981. *The Welfare State in Crisis. An Account of the Conference on Social Policies in the 1980s*. Paris.

Palloix, Christian. 1975. *L'économie mondiale capitaliste et les firmes multinationales*. 2d ed. Economie et socialisme, no. 24-25. Paris: Maspero.

Palmer, John L., and Isabel V. Sawhill, eds. 1982. *The Reagan Experiment: An Examination of Economic and Social Policies under the Reagan Administration*. Washington, DC: Urban Institute Press.

Perry, David C., and Alfred J. Watkins, eds. 1977. *The Rise of Sunbelt Cities*. Urban Affairs Annual Reviews, vol. 14. Newbury Park, CA: Sage.

Pinkerton, Steven J. 1984. "High Technology Growth and Regional Structure." Seminar paper for PLUS 508. School of Urban Planning, University of Southern California, Los Angeles.

Portes, Alejandro, and Robert L. Bach. 1984. *Latin Journey: Cuban and Mexican Immigrants in the United States*. Berkeley: University of California Press.

Portes, Alejandro, and John Walton. 1981. *Labor, Class, and the International System*. Studies in Social Discontinuity. New York: Academic Press.

Rutziek, Max A. 1970. "Skills and Location of Defense-Related Workers." *Monthly Labor Review* 93(2):11-16.

Sabel, Charles F. 1982. *Work and Politics: The Division of Labor in Industry*. Cambridge: Cambridge University Press.

Sassen-Koob, Saskia. 1984a. "Growth and Informalization at the Core: The Case of New York City." Paper presented at the Seminar on the Urban Informal Economy in Core and Periphery, Department of Sociology, Johns Hopkins University, Baltimore, 8-10 June.

———. 1984b. "The New Labor Demand in Global Cities." In *Cities in Transformation. Class, Capital, and the State,* edited by Michael Peter Smith. Urban Affairs Annual Reviews, vol. 26. Newbury Park, CA: Sage.

Sawers, Larry, and William K. Tabb, eds. 1984. *Sunbelt/Snowbelt, Urban Development, and Regional Restructuring*. New York: Oxford University Press.

Schoenberger, E. 1984. "The Regional Impact of Foreign Investment in U.S. Manufacturing." Ph.D. dissertation, University of California, Berkeley.

Serrin, William. 1983. " 'High Tech' Is No Jobs Panacea, Experts Say." *New York Times* (18 Sept.):I-1+.

Shonick, William. 1979. "The Public Hospital and Its Local Ecology in the United States: Some Relationships between the 'Plight of the Public Hospital' and the 'Plight of the Cities'." *International Journal of Health Services* 9:359-96.

Singelmann, Joachim. 1978. *Agriculture to Services: The Transformation of Industrial Employment*. Sage Library of Social Research vol. 69. Newbury Park, CA: Sage.

Soja, Edward, Rebecca Morales, and Goetz Wolff. 1983. "Urban Restructuring: An Analysis of Social and Spatial Changes in Los Angeles." *Economic Geography* 59:195-230.

Stanback, Thomas M., Jr. 1979. *Understanding the Service Economy: Employment Productivity, Location*. Policy Studies in Environment and Welfare no. 35. Baltimore: Johns Hopkins University Press.

Stanback, Thomas M., Jr., and Thierry J. Noyelle. 1982. *Cities in Transition: Changing Job Structures in Atlanta, Denver, Buffalo, Phoenix, Columbus, Ohio, Nashville and Charlotte*. Conservation of Human Resources Series no. 15. Totowa, NJ: Allanheld, Osmun.

Stanback, Thomas M., Peter J. Bearse, Thierry J. Noyelle, and R. A. Karasek. 1981. *Services: The New Economy*. Conservation of Human Resources Series no. 20. Montclair, NJ: Allanheld, Osmun.

Stack, Peter. 1984. "Special Report: VCR Revolution. The Big Changes in Entertainment." *San Francisco Chronicle* (27 Feb.):1.

Sternlieb, George, and James W. Hughes, eds. 1975. *Post-Industrial America: Metropolitan Decline and Inter-Regional Job Shifts*. New Brunswick, NJ: Center for Urban Policy Research, Rutgers-The State University of New Jersey.

Storper, Michael. 1982. "The Spatial Division of Labor: Technology, the Labor Process, and the Location of Industries." Ph.D. dissertation, University of California, Berkeley.

Strassman, Paul A. 1979-80. "The Office of the Future: Information Management for the New Age." *Technology Review* 82(3):54-65.

Tabb, William K. 1982. *The Long Default: New York City and the Urban Fiscal Crisis*. New York: Monthly Review Press.

Teitz, M. 1984. "The California Economy, Changing Structure and Policy Responses." In *California Policy Choices, 1984,* edited by John J. Kirlin and Donald R. Winkler. 2 vols. Los Angeles: University of Southern California Press.

Thurow, Lester C. 1984. "The Disappearance of the Middle Class." *New York Times* (5 Feb.):3-3.

Tomaskovic-Devey, D., and S. M. Miller. 1982. "Recapitalization: The Basic U.S. Urban Policy of the 1980s." In *Urban Policy under Capitalism,* edited by Norman I. Fainstein and Susan S. Fainstein. Urban Affairs Annual Reviews Series, vol. 22. Newbury Park, CA: Sage.

Touraine, Alain. 1969. *La société post-industrielle*. Bibliothèque Méditations no. 61. Paris: Denoël.

Trachte, Kent C., and R. Ross. 1983. "The Crisis of Detroit and the Emergence of Global Capitalism." Paper presented at the annual meeting of the American Sociological Association, Detroit, 31 August-4 September.

United Nations. 1978. *Transnational Corporations in World Development: A Reexamination.* E/C.10/38. New York.

Vernon, Raymond. 1971. *Sovereignty at Bay: The Multinational Spread of U.S. Enterprises.* The Harvard Multinational Enterprise Series. New York: Basic Books.

Walker, Patrick C. 1983. "The Distribution of Skill and the Division of Labor, 1950-1978." Ph.D. dissertation, University of Massachusetts.

Wayne, Leslie. 1984. "America's Astounding Job Machine." *New York Times* (17 June):III-1+.

Weinstein, Bernard L., and Robert E. Firestine. 1978. *Regional Growth and Decline in the United States: The Rise of the Sunbelt and the Decline of the Northeast.* New York: Praeger.

Wilensky, Harold L. 1974. *The Welfare State and Equality: Structural and Ideological Roots of Public Expenditures.* Berkeley: University of California Press.

Wilmoth, David. 1983. "The Evolution of National Urban Policy in the U.S." Ph.D. dissertation, University of California, Berkeley.

Wolfe, Alan. 1977. *The Limits of Legitimacy: Political Contradictions of Contemporary Capitalism.* New York: Free Press.

Zysman, John, and Stephen Cohen. 1982. *The Mercantilist Challenge to the Liberal International Trade Order.* A Study Prepared for the Joint Economic Committee, Congress of the United States. Washington, DC: U.S. Government Printing Office.

4

Urban Growth and Decline in Western Europe

Peter Hall

The central question to ask in any study of Western European cities is whether, and if so to what extent, they are suffering from the phenomenon of deurbanization so well chronicled for cities of the United States (Berry, 1976; Hauser, 1981; Kasarda, 1980; Sternlieb and Hughes, 1975). More specifically, the objective must be to ask whether Western European countries are beginning to exhibit the same kinds of geographical change as the United States, summarized as follows:

(1) *Suburbanization.* Growth passes from the central city to adjacent areas, leading first to a slowing-down of growth in the city, then to its actual decline.
(2) *Deurbanization.* This process causes the whole metropolitan area to experience lower growth rates and then eventually to decline; nonmetropolitan areas begin to grow faster than metropolitan ones, the phenomenon Daniel Vining has labelled the clean break with the past (Vining and Kontuly, 1977; Vining and Strauss, 1977).
(3) *Negative returns to urban scale.* This process occurs earliest and to the greatest extent in the largest metropolitan areas.
(4) *Older to newer regions.* Deindustrialization also occurs earliest, and most acutely, in older-settled and especially in older-industrialized regions (New England, the Middle Atlantic seaboard, the Midwest); it may not yet be apparent in newly industrializing regions (much of the Sunbelt), where even large metropolitan areas may continue to grow.

In the United States, these processes are associated and difficult to disentangle. Many of the largest metropolitan areas (and the largest cities within them) are in the older-industrialized regions, and appear to be suffering from stagnation and decline because of a recent massive contraction of their traditional employment base in manufacturing and the associated goods-handling service sectors. Newer metropolitan areas, many of them small but some of them large, do not experience the same degree of contraction and are in addition growing as regional (or even, in the case of Los Angeles and San Francisco, national) service sectors (Noyelle and Stanback, 1984).

To answer the question for Western Europe is inevitably more complex, conceptually and technically. Though the European Common Market is now more than a quarter century old, it is not yet by any means a single continental economy like the United States. Even multinational corporations find it necessary to adapt to national and linguistic barriers. The urban hierarchy reflects these barriers, but is also profoundly affected by the complex political changes of two millennia of history. Thus the French system of cities is relatively very primate, reflecting the long existence of a centralized nation-state; the German and Italian systems have no dominant city, reflecting the late achievement of national unity in those two countries (and, in Germany, the effective loss of the national capital in 1945-49). City size and distribution also in part reflect the existence of powerful historic forces, which have operated over many centuries despite profound changes in economic organization; thus a line of major cities still follows the historic *Hellweg*, the great east-west Central European trade route of medieval times, while the concentration of cities in Flanders and the Netherlands recalls their trading role in the early modern period.

Data

National barriers have another consequence in that the urban researcher in Western Europe faces a more difficult technical task than do American counterparts. Despite attempts to standardize data collection and analysis, European statistical offices retain a strong tradition of national autonomy and individuality. It proved impossible, despite efforts by the statistical office of the Commission of the European Communities, to persuade all nine European Economic Community (EEC) countries to conduct a census in the same year; indeed, because of internal political protests, the Federal Republic of Germany and the Netherlands failed to conduct a census at all in 1980-81—the first time, apart from World War Two, that their regular sequence has been broken in nearly two centuries. Even where information is available, it may not prove readily comparable. The administrative divisions, for which most published data are available, may differ profoundly from country to country; classification systems, for industrial composition of the work force for instance, may similarly differ. Therefore, the European researcher faces a major task of data reorganization and standardization before any systematic comparison is possible.

Nevertheless, the attempt has been made in three major studies during the last five years (Berg et al., 1982; Cheshire et al., 1984; Hall and Hay, 1980). All have used very similar modes of analysis directly derived from American work. The basic geographical building block in all three is the *Functional Urban Region* (FUR), a unit based on the Standard Metropolitan Statistical Area (SMSA) of the United States Census, though in practice somewhat more generously defined. The FUR has a central urban *core* corresponding to the American *central city*, defined in terms of an absolute level of density threshold of employment, together with a suburban *ring* of administrative units having close daily ties (generally, expressed in commuting flows) with it.

The study by Hall and Hay, conducted in conjunction with the International Institute for Applied Statistical Analysis (IIASA), reported data for 539 such units in 13 countries of western and central Europe; later, some eastern European data were added (Kawashima and Korcelli, 1982). The data referred to the census dates 1950-51, 1960-61, 1970-71 wherever possible; where data were not available for these dates (as for instance in France, where censuses were taken in 1954, 1962, 1968, and 1975), data were interpolated. Additionally, official population estimates (and, in the case of France and Sweden, censuses) were used to produce comparable data for 1975. The study by Cheshire et al. builds on the Hall-Hay study by using basically the same functional regions and producing comparable data for 1980; these refer, however, only to the ten countries of the EEC as constituted in 1983.

In both studies, the bulk of the data refer only to population. Hall and Hay produced some limited data on industrial structure, which Cheshire et al. extended for 23 cities which seemed to present special problems of economic adjustment (Athens, Belfast, Bochum, Catania, Charleroi, Copenhagen, Cosenza, Dortmund, Dublin, Essen, Genoa, Glasgow, Liège, Liverpool, Manchester, Nancy, Naples, Saarbrücken, St. Etienne, Turin, Valenciennes, West Berlin, Wuppertal). In addition, Cheshire et al. present new data on migration and (for the selected 23 cities) unemployment. The most important limitation of the studies so far, then, is that no systematic across-the-board analysis is possible of employment structure and changes in it.

A Framework for Comparison

All three studies used a similar spatial framework of analysis, first developed by Hall et al. (1973) in their study of urban change in Britain. The hypothesis was that over the course of time, FURs within national urban systems tended to go fairly regularly through six stages of evolution (Hall, 1984, chap. 10). Each is described in terms of the relationship between change in the urban core and in the suburban ring.

(1) *Centralization during Loss (LC)*. In the initial stage, represented by early industrialization and agricultural rationalization, a surplus rural population pours off the land looking for industrial work in the nearest city. The city therefore grows in population, but this growth cannot make up for the loss from the immediately-surrounding ring, which at this stage is still agrarian

(core and ring definitions are assumed fixed over time). Because of this, the whole functional region is in decline. The resulting state, with growing city, declining ring, and loss overall, may fairly be described as centralizing during loss. Surplus agrarian populations migrate to the leading (primate) national city, where the most vigorous industrial growth occurs (the familiar case of uneven development, now witnessed in developing countries). Thus this city reaches the second stage (below) before the others.

(2) *Absolute Centralization (AC).* As economic development proceeds, advanced industrialization spreads from the primate city to second- and third-order cities, which in consequence are now able to absorb their local surplus agrarian populations. Meanwhile, as in stage 1, natural increase of these local populations remains high. Within functional urban regions, therefore, the gain in core city population now exceeds loss in the still-agrarian ring; the overall population change becomes positive, but with increasing centralization in the city. This pattern is characteristic of many cities in countries in the early stages of industrialization.

(3) *Relative Centralization (RC).* As industrialization proceeds even further, cities grow so that they overrun their boundaries: suburban invasion begins to occur in the hitherto-rural ring. (This process occurs first in the primate city, only subsequently in second- and third-order cities). At this point, both core city and ring are growing in population; however, the city's rate of growth is faster, so that an increasing portion of the population of the FUR is in the city.

(4) *Relative Decentralization (RD).* Now, a critical turning point occurs. Suburban growth around the city becomes so rapid that the rate of growth of the ring exceeds that of the city itself; the city now begins to account for a diminishing portion of total FUR growth. This stage, as with all others, is reached first in the primate city and is characteristic of mature industrial cities with well-developed commuter transport systems; in earlier decades it could not occur until such systems appeared.

(5) *Absolute Decentralization (AD).* At least three separate forces now begin to affect the central city population. First, with growth of the entire Functional Urban Region, commercial activities begin to displace large numbers of people in and around the Central Business District. Second, increased affluence allows people to live at lower densities of occupation within the existing built stock. Third, slum clearance and redevelopment at lower densities displace some low-income people to peripheral locations. All these factors work to decant the population of the core city, which—since by now there is no available land within the city limits—for the first time records a loss of population. Meanwhile, the suburban ring continues to witness rapid expansion. This stage, again first reached by the leading city, began shortly after 1900 in some of the largest European cities.

(6) *Decentralization during Loss (LD).* In the original formulation of the model this stage was regarded as the final step in urban evolution. In some ways this stage is the most complex and difficult to interpret. In it, the

outward wave of population movement rolls out from the central city to the suburbs. Suburban growth can no longer compensate for the continuing loss in the city; in some cases, suburbs too, may begin to decline. Thus the pattern—the mirror-image of stage 1—is one of decline of the entire Functional Urban Region, accompanied by a relative outward shift.

Stage 6 is especially significant since a number of larger SMSAs in the United States (and, as will be seen, also their European equivalents) have now entered it. The difficulty about stage 6 is to understand precisely what is the dominant cause. As I emphasize later, in such FURs several forces seem to be operating simultaneously. Population continues to move outward for the same reasons as were described for stage 5; some of these may indeed operate with additional force, such as central commercial redevelopment, inner-ring "gentrification," and slum clearance schemes. All these can be described as agents of *decentralization.* If these trends go far enough, they may reach the outer limits of the Functional Urban Region, thus leading to a transfer of growth to neighboring FURs. But superimposed on them, in the last decade, have been quite different forces of *structural economic decline* of the manufacturing and associated goods-handling service base of the city. These declines operate via multiplier effects to reduce the entire economic base and thus to encourage out-migration from the region altogether.

Disentangling these forces is difficult because of an almost accidental contingency. The decentralization effect is likely to express itself most strongly in FURs closely bounded by the existence of neighboring FURs—FURs in large and complex urban agglomerations (a *megalopolis,* in the language of Jean Gottmann [1961]; a *Standard Consolidated Statistical Area* [SCSA], in the language of the U.S. Bureau of the Census). But it so happens that many such agglomerations were developed in the 19th century on the basis of concentrated manufacturing industry at either coalfield or port locations—Northern England, Wallonia, the Ruhrgebiet, the U.S. Northeastern Seaboard, Pennsylvania and Ohio—which have recently experienced structural economic decline. In such regions, the decentralization effect and structural effect thus tend to coincide in time and space. Indeed one explanation of the structural effect is that it is compounded by a lack of space for the necessary reorganization of manufacturing processes (Fothergill and Gudgin, 1982).

Not merely in this final stage of the model, but also throughout it, two processes are intertwined. One is the spatial impact first of industrialization, then of deindustrialization. Product cycle theory, first developed by Raymond Vernon (1966), would suggest that an industrial-urban region would develop on the basis of innovations which produce what are in effect new industries (cotton textiles in 18th-century Lancashire, steel and engineering in mid-19th-century Ruhrgebiet, cars in early 20th-century Oxford). As the industry matures through the product cycle, process innovations result in substitution of capital for labor, leading first to "jobless growth" and finally to contraction of employment and stagnation of output (Rothwell and

Zegveld, 1981); newly industrializing regions become steadily more capable of borrowing the necessary production techniques, so as to produce at lower cost than the original innovators. The theory thus provides a convincing account of the progress of an urban region from innovatory youth to industrial senescence, through one or more long waves of economic development (Duijn, 1983; Freeman, 1984; Kondratieff, 1935).

At the same time, the very processes set in motion by the original innovation-led growth would tend to lead to progressive decentralization, first of people, then of jobs. Affluence resulting from high profits and wages in the early stages of development (when the region enjoyed an effective monopoly or at least oligopoly of production knowledge) would lead to rapid suburbanization as well as to development of central commercial complexes (both in producer and consumer services) which would displace people. Thus an innovative, growing region would also be a decentralizing region. If such a region consisted of a series of contiguous Functional Urban Regions, that would provide an explanation of why such decentralization would before long run its course through to the sixth stage; decentralization would become decline. And this might well arise at just the stage when the product cycle was leading to an erosion of the region's competitive ability.

Testing the Model

The data produced in the three major comparative studies can be used, to a limited degree, to test the validity of this model of development for Western Europe as a whole and for its constituent parts. Limitations of the data base should, however, again be stressed: first, in its present form it overwhelmingly measures changes in residential population rather than in the underlying employment base; second, even for population the data are not fully comparable, in particular as between the fuller Hall-Hay study and the more limited Cheshire-Hay-Carbonaro sample. These limitations recognized, tables 3.1 to 3.3 reproduce the most important statistical evidence.

Tables 4.1 and 4.2 both measure the six stages of population shift as outlined above: table 4.1 for the three decades 1950-60, 1960-70, 1970-80; table 4.2 by population size classes for the most recent decade, 1970-80 (and, additionally, for 1975-80). Both tables include data for France, Italy, Benelux (combined), the Federal Republic of Germany, and the United Kingdom. (Data for Denmark, Greece, and Ireland, where only a few FURs exist, are not tabulated.) Taken together, the tables confirm the conclusion of Hall and Hay (1980) in their original analysis: no good generalization is possible for Western Europe as a whole. The different countries' urban systems, on the contrary, display marked differences from one to another.

Table 4.1 shows consistent results in one respect: there is indeed a general tendency over time for all countries to move progressively through the stages of the model. The dates at which they do so, however, differ sharply from one

TABLE 4.1
Shift Patterns—Selected EEC Countries, 1950-1980

Country, Dates and Total Number of FURs[a]	Numbers of Functional Urban Regions in Shift Pattern					
	1	2	3	4	5	6
	Centralization			Decentralization		
	During Loss (LC)	*Absolute (AC)*	*Relative (RC)*	*Relative (RD)*	*Absolute (AD)*	*During Loss (LD)*
France						
1950-60 (86)	5	47	28	5	1	0
1960-70 (86)	3	21	55	5	1	1
1970-80 (81)	10	5	14	39	11	2
1975-80 (81)	10	3	4	27	30	7
Italy						
1950-60 (84)	23	25	30	6	0	0
1960-70 (84)	15	21	34	14	0	0
1970-80 (84)	7	0	12	23	27	15
1975-80 (84)	8	2	10	16	25	23
Benelux (Belgium, Luxembourg, The Netherlands)						
1950-60 (33)	0	0	12	17	3	1
1960-70 (33)	0	0	4	19	9	1
1970-80 (34)	0	0	0	12	20	2
1975-80 (34)	0	0	1	12	17	4
Germany (Federal Republic)						
1950-60 (138)	24	11	31	0	1	2
1960-70 (138)	1	2	18	37	19	1
1970-80 (78)	2	0	4	6	27	14
1975-80 (78)	2	2	3	8	11	27
United Kingdom						
1950-60	8	8	55	39	14	14
1960-70	4	1	30	57	34	12
1970-80	10	1	4	15	27	21
1975-80	12	4	1	9	22	30

SOURCE: Cheshire et al. (1983); Hall and Hay (1980).
a. Functional urban regions, in parentheses.

TABLE 4.2
Shift Patterns by Size of Functional Urban Regions (FUR)—Selected EEC Countries, 1970-1980

		Numbers of FURs in Shift Pattern					
		1	*2*	*3*	*4*	*5*	*6*
		Centralization			*Decentralization*		
Country, Size and Total Number of FURs		*During Loss (LC)*	*Absolute (AC)*	*Relative (RC)*	*Relative (RD)*	*Absolute (AD))*	*During Loss (LD)*
France							
Million +	(6)				3	3	
500k-million	(11)	1		2	7	1	
250k-500k	(38)	2	3	8	21	3	1
100k-250k	(26)	7	2	4	8	4	1
<100k	(0)						
Italy							
Million+	(7)			1		6	
500k-million	(15)				5	6	4
250k-500k	(23)	2		5	7	4	5
100k-250k	(36)	5		5	10	11	5
<100k	(3)			1	1		1
Germany (Federal Republic)							
Million +	(21)				1	14	6
500k-million	(17)			3	2	9	4
250k-500k	(14)	2		1	3	4	4
100k-250k	(1)					1	
<100k	(0)						
Benelux (Belgium, Luxembourg, The Netherlands)							
Million +	(5)					5	
500k-million	(12)				4	6	2
250k-500k	(10)				8	3	
100k-250k	(7)					7	
<100k	(0)						
United Kingdom							
Million +	(8)					1	7
500k-million	(12)				1	8	3
250k-500k	(21)	2			3	10	6
100k-250k	(25)	5	1	3	8	6	2
<100k	(0)	3		1	3	2	3

SOURCE: Cheshire et al. (1983).
NOTE: k = thousand.

country to another; in table 4.1 I have attempted to order them in this regard. In France, the biggest single category of FURs in the 1950s was in stage 2; by the 1960s this had become stage 3, and in the 1970s stage 4. In Italy corresponding categories were stage 3 (1950s and 1960s) and 5 (1970s). Benelux was more advanced at the start and showed less evolution, the biggest category moving only from stage 4 in the 1950s and 1960s to stage 6 in the 1970s. The Federal Republic of Germany showed a major concentration of FURs in stage 3 in the 1950s, the time of major postwar reconstruction of its cities; by the 1960s this had become stage 4, by the 1970s stage 5; and by the late 1970s, the biggest single group of FURs was already in stage 6. Finally, the United Kingdom showed an identical pattern, with steady movement of the biggest cluster from stage 3 in the 1950s, to stage 4 in the 1960s, stage 5 in the 1970s, and stage 6 in the second half of the 1970s.

The obvious conclusion from this comparison would be that the United Kingdom and the Federal Republic, as examples of mature industrial-urban nations, had led the movement through the stages to eventual urban decline; Benelux was slightly behind them, Italy and France were some way behind. This progression would lead naturally to the question, Were these differences related in some way to differences in the size composition of FURs in the different Western European countries? Evidence on this point (table 4.2) is somewhat inconclusive. Major differences in size distribution of FURs are evident between one country and another: Germany in particular has a large number of urban regions with one million and more people, though this may to some degree be a statistical artifact resulting from the large administrative blocks from which German FURs have been built since the Gemeinde reform of the early 1970s. Also evident is that there is no clear correlation between size and movement between size categories—as Hall and Hay (1980, chap. 5) earlier found in their statistical exercise. Such a relationship appears to hold for the United Kingdom, where by the 1970s the largest FURs had mostly moved into the sixth stage; but in the Federal Republic, France, and Italy they tended still to be in stage 5 or even, in the case of France, in stage 4. All that appears is a general tendency, in each of the national divisions, for larger FURs to be farther advanced along the path to decentralization than smaller ones. This finding at least is in accordance with the model.

In the United Kingdom, most of the largest urban regions were in decline, many of them seriously, by the 1970s. But in Germany the picture was mixed, with some declining and some exhibiting vigorous growth. In Benelux, France, and Italy the largest FURs were growing, and in the latter two cases most were showing very strong growth. The conclusion is inescapable that only the United Kingdom had reached the position of negative returns to urban scale—the conclusion of Hall and Hay (1980) for the period to 1975. The more recent work of Cheshire et al. seems to confirm that it was still the case for the whole decade of the 1970s.

The question naturally arises as to why only British cities found themselves in this position. Was it because Britain, as the oldest-industrialized country of

Europe, had reached the end of the process before any other country? Or was it because the declining British cities had some special characteristics? Table 4.3, derived from data of Cheshire, Hay, and Carbonaro, identifies for the 1970s all those FURs in the different national divisions that had reached stage 6 of the model. In the United Kingdom and the Federal Republic of Germany, it is immediately evident that a very high proportion of all such FURs occurred in large agglomerated urban regions such as the British conurbations or German Ruhrgebiet: just over half the British cases fell into this category, and 5 of the 14 German cases were Ruhr cities. Further, both the French and both the Benelux cases were in the coalfield zone straddling the two countries, also a heavily agglomerated zone. Against this, the 15 Italian cases were nearly all freestanding cities. Another way of looking at these cases is that—again with the exception of Italy—most are old industrial cities, many based on coalfields, which suffered exceptionally serious contraction in their traditional coal-mining and heavy metal-manufacturing industry base during the 1970s. Cheshire and colleagues' detailed analysis of employment changes for 23 selected case study cities, two-thirds (14) of them chosen from the lists in table 4.3, shows this fact to have been the case. Associated with these declines in basic industry came a contraction in the consumer-services sector that depends on it. In contrast, the producer service sector (banking, finance, business services, and rentals) generally showed compensating gains in the core cities; the significant exceptions were Manchester, Liverpool, and West Berlin (Cheshire et al., 1984, sec. 7).

Cheshire and colleagues indeed conclude that decline of urban regions is closely related statistically to decline of the wider regions around them; the urban problem reflects the regional problem. Indeed, as they recognize, the two can be regarded as substantially identical. The central problem is a sudden contraction in the manufacturing base of the traditional industrial regions of the coalfields and the 19th-century port cities. Thus urban decline problems are concentrated through Europe in a narrow band of 19th- and early 20th-century industrial cities from Turin and Genoa in northwestern Italy, through eastern and northern France, the Saar and Ruhr valleys, and southern Belgium to the British midlands and northwest and northeast England, and at last north to Glasgow and west to Belfast (Cheshire et al., 1984:145).

Conclusions

We can thus conclude, provisionally, that during the 1970s the European urban system exhibited certain features similar to its American counterpart, but also some that were not:

(1) *Suburbanization* was fairly general throughout the Functional Urban Regions of the EEC. In France, 64 of 81 FURs were exhibiting various stages of decentralization; in Italy,

TABLE 4.3
Functional Urban Regions in Stage Six
(Decentralization During Loss)—
Selected EEC Countries, 1970-1980

France	*Italy*	*Benelux*	*Germany*	*United Kingdom*	
Valenciennes	Genoa	Charleroi	Berlin	Liverpool	Belfast
Douai	Udine	Liège	Essen	Manchester	Glasgow
	Bologna		Duisburg	Newcastle	London
	Ferrara		Schweinfurt	Nottingham	Newport
	Venice		Wuppertal	Birmingham	Hull
	Vercelli		Saarbrücken	Sheffield	Dundee
	Siena		Trier	Coventry	Oldham
	Trieste		Kassel	Brighton	
	Mantua		Bochum	Edinburgh	
	La Spezia		Braunschweig	Rhondda	
	Sabona		Koblenz	Greenock	
	Alessandria		Wolfsburg	Blackpool	
	Pavia		Heidenheim	St. Albans	
	Pisa		Dortmund	St. Helens	
	Carrara				

SOURCE: Cheshire et al. (1983).

64 of 84; in Benelux, 33 of 34; in the Federal Republic of Germany, 46 of 53; in the United Kingdom, 61 of 78 (during 1975-80).

(2) *Deurbanization* was more localized. In the sense of having reached stage 6 of the model, in France only 7 of 81 FURs were in this state; in Italy, 23 of 84; in Benelux, 4 of 34; in Germany, 27 of 53; in the United Kingdom, 30 of 78. Deurbanization can be said therefore to have affected mainly the Federal Republic, the United Kingdom, and (to a more limited degree) Italy.

(3) *Negative returns to urban scale* are only evident in the United Kingdom (U.K.) and, to a certain extent, the Federal Republic. In the U.K. all but one of the 8 biggest FURs (one million and over) were in decline by the 1970s; in the Federal Republic, 6 of 21 were. But in Benelux, Italy, and France none of the biggest FURs was in decline. On the contrary, in Italy, and above all in France, several were exhibiting extremely vigorous growth.

(4) *Older-to-newer region* movement does appear to be affecting parts of Europe, in particular the central highly industrialized zone identified by Cheshire et al. As in the United States, this central industrial zone—consisting mainly of 19th-century industrial cities based on either coalfields or port activities—appears to be losing out to more peripheral regions. The process however mainly affects the United Kingdom and the Federal Republic of Germany. The major traditional industrial centers in France, for instance, do not so far appear to be seriously affected.

Interpretation

These conclusions are not markedly different from those Hall and Hay (1980) drew on the basis of their earlier study. The important question concerns their interpretation. Is it true that, as Hall and Hay suggest, the European nations are arrayed on some kind of continuum, with the United Kingdom at one end and France (together with southern and eastern Europe) at another? If so, we can expect that eventually, at different dates, all of these countries will eventually arrive at the stage of deurbanization and negative returns to urban scale. The fairly regular progression in table 4.1, for all countries—whereby, within each matrix, the largest category moves progressively downward and to the right—lends support to this view. Yet continuing high rates of growth for the large FURs in most countries suggest some doubts. It might be that for a recent half-decade, 1975-80, we should see some change; but the overall shifts (table 4.1) do not give much preliminary support to this thesis.

The alternative view is that—as Cheshire and colleagues suggest—the so-called urban problem is really the older regional problem in new guise. Several simultaneous processes are taking place. One is decentralization, leading to growth of the ring at the expense of the core; this is now almost universal and logically arises from the desire of businesses and families for more space. In itself, provided there is sufficient space in the ring, this need not lead to decline of the entire Functional Urban Region; but in densely urbanized conurbations, where the commuter field of a large central city is constricted by the existence of neighboring fields, decline will be the eventual result. The other process is the structural decline of basic industries including manufacturing and associated goods-handling services (transport, docks), which particularly affects certain older, industrialized cities and regions. The analytical problem, as in the United States, is that the two processes affect the same kind of city: the large, older-industrialized central cities of conurbations (Glasgow, Liverpool, Manchester; Charleroi, Liège; Dortmund, Essen). Therefore it is difficult to determine whether the phenomenon of urban decline is primarily one of negative returns to scale, or one of regional decline.

In the United Kingdom, of 21 FURs in stage 6 in the 1970s, seven were a million and over in size; another three were between 500 thousand and a million. Most were central FURs (London, Birmingham, Liverpool, Manchester, Sheffield, Newcastle, Glasgow). The group also included some smaller FURs within clusters of FURs where spatial constraints may have proved important (Oldham, St. Albans, St. Helens). But there were also freestanding FURs where no such constraint appeared to operate and where the problem was probably structural decline (Belfast, Newport, Rhondda, Dundee). Similarly, in the Federal Republic of Germany, the 14 FURs in stage 6 included six with a million or more people plus four with between half and a million. A significant number of the largest were in the Ruhrgebiet (Essen, Duisburg, Wuppertal, Bochum, Dortmund) though others were freestanding and had no apparent spatial constraint (Schweinfurt, Trier, Kassel, Braunschweig, Koblenz, Wolfsburg, Heidenheim). In Italy the position was completely different: the 15 stage 6 cases were predominantly small or medium-sized, and few were in densely agglomerated areas.

The tentative conclusion seems to be that regional-structural effects were predominant but exacerbated by spatial constraints in the conurbations. This finding seems to be reinforced by the conclusion of Cheshire and colleagues that there is a close statistical association between employment decline in a FUR and in the level 2 EEC region to which it belongs (Cheshire et al., 1984:111).

Toward Further Research

It is not possible to go much beyond this point at present, because of the very limited amount of analysis of employment changes existing for Europe. The very partial sample analysis by Cheshire et al. (1984:17) is extremely suggestive for further work—it concludes that the main factor in urban decline is contraction of the manufacturing base, insufficiently compensated by expansion of the tertiary sector. A body of important work, most of it British, on causes of job loss (Fothergill and Gudgin, 1982; Massey, 1985; Massey and Megan, 1979, 1982) concentrates mainly on the traditional manufacturing sector.

There is far more limited work on job growth, either in the few still-expanding manufacturing subsectors (Hall and Markusen, 1985) or in the tertiary sector. There is a strong suggestion that new, high-technology manufacturing industries tend to develop in areas quite different from those where older manufacturing industry is in decline. New industries seem to seek a particular kind of skilled labor—highly-educated, nonunionized, and geographically mobile—that may be attracted to areas of high amenity. They may also be drawn to nearby university or government research facilities. Though occasionally such high-technology concentrations may develop in older, industrial regions, provided they satisfy these requirements (for instance, the Greater Boston area in the United States), often they locate in previously-unindustrialized areas such as Silicon Valley or the Dallas-Fort Worth area. Similarly, in Europe this kind of industry seems to be drawn to such nontraditional areas as the so-called M4 Corridor west of London, the French Riviera, or Bavaria-Wurttemberg.

The geography of tertiary growth is even less well-established. For the United States, Noyelle and Stanback (1984) have produced a number of interesting analyses and a preliminary classification of tertiary locations. It is possible and desirable to subdivide the tertiary sector into a number of subsectors, with rather different locational patterns:

(1) *Distributive services* include the complex of transportation, communications, and public utilities (TCU); and wholesaling. In general this group has tended to remain roughly static in employment while recording sharp increases in output. Like manufacturing it is prone to substitution of capital for labor (containerization, direct-dial telephone systems).
(2) *The corporate complex* includes central administrative offices; the FIRE (finance, insurance, real estate) group and corporate services (all other producer services except FIRE). Together these constitute the producer services (or business services) which minister to manufacturing and other service sectors. They have grown extremely rapidly in employment and output because of the increasing demand of other sectors for services in marketing, finance, and administration, which are often then contracted to specialized organizations.
(3) *Retailing* and *mainly consumer services* are sectors which have again shown a tendency for capital-labor substitution in recent decades.
(4) In contrast, *nonprofit services* and *government* have shown very large increases because of expansion of educational, health, and welfare services. These sectors may have experienced some recent growth curtailment because of cuts in public spending.

Noyelle and Stanback have analyzed for the United States the geographical pattern of changes in these categories of service employment. Critical here is the role of the three fastest increasing categories: the complex of corporate activities, government, and nonprofit services. One principal finding is the continuing importance of the highest level (4 national, 19 regional) service sectors for FIRE, other corporate services and education; the largest strengthened their position in central administrative offices and health. This finding suggests that both producer services and nonprofit services cohere very strongly to these high-order centers. So-called specialized service centers also did well in corporate services, health and education, and government. These centers include a category called functional nodal, which principally consists of older manufacturing centers in which major corporations carry out headquarter and other nonproduction activities. But they shared with the pure production centers a sharp drop in employment growth after 1969.

Overall, the Noyelle-Stanback conclusions are striking: the fastest growing metropolitan areas during 1959-69 and 1969-76 were those based on residence-resort-retirement functions, followed by industrial-military complexes, then by government-educational areas. The so-called diversified service centers—basically the major service-providing cities—had a medium growth rate, but after 1969 positions reversed: the larger ones (national and regional centers) fell behind the smaller subregional ones. Finally, in both subperiods the two worst-performing kinds of area were pure manufacturing centers and functional nodal centers (manufacturing centers with corporate headquarters).

Noyelle and Stanback's conclusions, though interesting, apply to an open capitalist economy. There are no parallel studies so far for Europe. We could speculate that there, national frontiers will still exert a major influence on the hierarchy of centers and thus on the pattern of service employment growth; clearly, by definition, Europe has many more national-level service centers than the four Noyelle and Stanback identified for the United States. But certain features—the poor performance of the pure production centers and specialized nodal centers—are very likely, from the evidence so far, to be similar in Europe.

Probably, too, behavior of urban hierarchies is different from one nation-state to another. Very partial evidence suggests that Liverpool and Manchester, two major United Kingdom regional service sectors, have suffered a serious loss in producer-service employment (Cheshire et al., 1984:108; cf. Roger Tym and Partners, 1981). If this feature is replicated in other British regional centers, it may contrast with a growth of such services in certain centers at the next-lowest level of the British urban hierarchy: what Smith (1968) called Grade 3A as distinct from the Grade 2 (regional) centers. It appears doubtful that this feature is replicated elsewhere in Europe; the Cheshire-Hay-Carbonaro case studies suggest it is not true of such FURs as Dortmund, Essen, Liège, Nancy, or Saarbrücken, which are comparable regional centers.

The major research priority now, therefore, is not to replicate the American work precisely, but to test its validity in the European context. We need first to produce a classification of European FURs, based partly on their main economic support, partly on their position in the service-performing hierarchy. Noyelle and Stanback's taxonomy may need modification for this purpose, though it would provide a starting point. Then we need to study the changing economic fortunes of these categories in terms of major industrial sectors, but with a special categorization of services comparable to that of Noyelle and Stanback's, which in turn is derived from Singelmann (1979). Ideally we should try to do this investigation for the subperiods 1960-70 and 1970-80. If it proves impossible for the entire list of FURs, then a sample should be studied. (Noyelle and Stanback were only able to study 140 SMSAs—just under half the total array). Larger FURs in the Cheshire-Hay-Carbonaro study, having a population of 330 or more in 1980 and numbering just under 200, might be suitable.

Such a study, despite considerable technical difficulties of comparison, could produce several useful results. First, it could help establish whether the geographical pattern of service provision varies from one country to another, depending on historically inherited differences in the shape of the urban hierarchy. Second, it could throw light on shifts in this pattern—in particular, seeking to understand whether the pattern of provision is becoming to any extent internationally standardized. It would be of particular interest to know, for instance, whether increasing concentration of capital is leading to the very highest-order national centers (London, Paris, Rome) winning out over second-order national centers of smaller nations, as well as over similarly sized major provincial cities in larger nations (cf. Daniels, 1983; Goddard and Smith, 1975). And in parallel, it would be interesting to know whether such processes of concentration are similar from one country to another—in a highly-centralized country like France as against a decentralized economy like the Federal Republic's, for instance (cf. Strickland and Aiken, 1984).

Implicit in this emphasis on the service base of urban economies is the notion that in advanced industrial countries the tertiary sector is now the real engine that drives economic advance. This idea is contrary to traditional ones, but it seems the only reasonable hypothesis corresponding to the known facts: in national economies where some two-thirds of all employment is now in the tertiary sector, this sector is constantly growing at the expense of the manufacturing sector. At the same time, in some European countries, overall employment is actually contracting. There is doubtless more work to be done on this process of manufacturing contraction and on the few exceptions to it. But the interesting and less-understood phenomenon is growth of the service economy and the extent to which it will compensate for employment contraction in traditional manufacturing industries.

Bibliography

Berg, Leo van den, Roy Drewett, Leo H. Klaassen, Angelo Rossi, and Cornelius H.T. Vijverberg. 1982. *Urban Europe: A Study of Growth and Decline*. Vol. 1. Oxford, England: Pergamon for the European Coordination Centre for Research and Documentation in Social Sciences.

Berry, Brian J. L. 1970. "The Geography of the United States in the Year 2000." *Transactions of the Institute of British Geographers* no. 51:21-53.

Cheshire, Paul, Dennis Hay, and Gianni Carbonaro. 1983. *A Survey of Urban Areas in the EEC: Problems of Decline and Growth 1971-75-81.* Regional Policy and Urban Decline: The Community's Role in Tackling Urban Decline, Interim Report no. 3.i. Reading, England: Joint Centre for Land Development Studies, Faculty of Urban and Regional Studies, University of Reading.

Cheshire, Paul, Dennis Hay, Gianni Carbonaro, and Nick Bevan. 1984. *Regional Policy and Urban Decline: The Community's Role in Tackling Urban Decline.* Draft Final Report. Reading, England: Joint Centre for Land Development Studies, Faculty of Urban and Regional Studies, University of Reading.

Daniels, Peter W. 1983. "Business Service Offices in British Provincial Cities: Location and Control." *Environment and Planning* A 15:1101-20.

———. 1984. "Business Service Offices in Provincial Cities: Sources of Input and Destinations of Output." *Tijdschrift voor Economische en Sociale Geografie* 75:123-39.

Duijn, Jacob J. van. 1983. *The Long Wave in Economic Life.* London: Allen & Unwin.

Fothergill, Stephen, and Graham Gudgin. 1982. *Unequal Growth: Urban and Regional Employment Change in the UK.* London: Heinemann.

Freeman, Christopher, ed. 1984. *Long Waves in the World Economy.* London: Frances Pinter.

Goddard, John B., and I. J. Smith. 1975. "Changes in Corporate Control in the British Urban System, 1972-1977." *Environment and Planning* A 10:1073-84.

Gottmann, Jean. 1961. *Megalopolis: The Urbanized Northeastern Seaboard of the United States.* New York: Twentieth Century Fund.

Hall, Peter G. 1984. *The World Cities.* 3d ed. London: Weidenfeld & Nicolson.

Hall, Peter G., and Dennis Hay. 1980. *Growth Centres in the European Urban System.* Berkeley: University of California Press.

Hall, Peter G., Harry Gracey, Roy Drewett, and Ray Thomas. 1973. *The Containment of Urban England.* 2 vols. London: Allen & Unwin; Newbury Park, CA: Sage.

Hall, Peter G., and Ann Markusen, eds. 1985. *Silicon Landscapes.* Boston: Allen & Unwin.

Hauser, Philip M. 1981. "The Census of 1980." *Scientific American* 245(5):53-61.

Kasarda, John D. 1980. "The Implications of Contemporary Redistribution Trends for National Urban Policy." *Social Science Quarterly* 61:373-400.

Kawashima, Takeyoshi, and Piotr Korcelli, eds. 1982. *Human Settlement Systems: Spatial Patterns and Trends.* IIASA Collaborative Proceedings Series, CP-82-S1. Laxenburg, Austria: International Institute for Applied Systems Analysis.

Kondratieff, Nikolai D. 1935. "The Long Waves in Economic Life." Translated by W. F. Stolper. *Review of Economics and Statistics* 17:105-15.

Massey, Doreen B. 1985. *Spatial Divisions of Labour: Social Structures and the Geography of Production.* London: Macmillan.

Massey, Doreen B., and Richard Meegan. 1979. *The Geography of Industrial Reorganization.* Progress in Planning, vol. 10, pt. 3. Oxford, England: Pergamon.

———. 1982. *The Anatomy of Job Loss: The How, Where, and Why of Job Decline.* London: Methuen.

Noyelle, Thierry J., and Thomas M. Stanback, Jr. 1984. *The Economic Transformation of American Cities.* Totoway, NJ: Rowman & Allanheld.

Roger Tym and Partners. 1981. *Capital of the North: The Business Service Sector in Inner Manchester/Salford.* Report to the Manchester/Salford Inner City Partnership. London: Roger Tym and Partners.

Rothwell, Roy, and Walter Zegveld. 1981. *Industrial Innovation and Public Policy: Preparing for the 1980s and 1990s.* Contributions in Economics and Economic History no. 42. Westport, CT: Greenwood.

Singelmann, Joachim. 1979. *From Agriculture to Services: The Transformation from Industrial Employment.* Newbury Park, CA: Sage.

Smith, R.D.P. 1968. "The Changing Urban Hierarchy." *Regional Studies* 2:1-19.

Sternlieb, George, and James W. Hughes, eds. 1975. *Post-Industrial America: Metropolitan Decline and Inter-Regional Job Shifts.* New Brunswick, NJ: Center for Urban Policy Research, Rutgers-The State University of New Jersey.

Strickland, Donald, and Michael Aiken. 1984. "Corporate Influence and the German Urban System: Headquarters Location of German Industrial Corporations 1950-1982." *Economic Geography* 60:38-54.

Vernon, Raymond. 1966. "International Investment and International Trade in the Product Cycle." *Quarterly Journal of Economics* 80:190-207.

Vining, Daniel R. 1982. "Recent Dispersal from the World's Industrial Core Regions." Department of Regional Science, University of Pennsylvania, Philadelphia.

Vining, Daniel R., and T. Kontuly. 1977. "Increasing Returns to City Size in the Face of an Impending Decline in the Size of Large Cities: Which Is the Bogus Fact?" *Environment and Planning* A 9:59-62.

Vining, Daniel R., and Anne Strauss. 1977. "A Demonstration that the Current Deconcentration of Population in the United States Is a Clean Break with the Past." *Environment and Planning* A 9:741-58.

5

Large Cities in Eastern Europe

Jürgen Friedrichs

Analyzing the urban change and problems of cities in socialist countries is a challenge. Data problems are considerable. Moreover, a Western approach may be misleading by ignoring specific political-economic contexts of these countries, for instance, stronger planning of land uses, priorities in spatial allocation of investment, and even rent control. These nations like to be referred to as "socialist" with respect to the claimed absence of unemployment, crime, low rents, and low transportation costs. Yet, these cities are not "socialist cities" in a spatial sense, as the introductory discussion in French and Hamilton (1979) suggests. Except for a short period in the early 1920s (see Gradow, 1971; Kopp, 1979) there are no specific socialist types of land use, distribution of new housing, internal organization of residential blocks, or location of companies. Even the principal goal of socialist city planning—to locate new residential areas close to working areas—has been pursued in Western planning too, and failed, since most residents did not have jobs in the adjacent working areas.

As a conceptual tool, I use the term *persisting problems* to designate a complex of structural deficiencies. Two such persisting problems are obvious to any scholar of urban development in socialist countries: continuous housing shortage and effects of planning vis-à-vis claimed efficiency.

AUTHOR'S NOTE: This chapter was prepared with the assistance of Raissa Biller who graduated in economic geography from Lomonosov University in Moscow. The study was partly funded by the Joint British American Tobacco-University of Hamburg Foundation. I gratefully acknowledge comments on an earlier version of the chapter by Jens Dangschat and Klaus Kiehl.

Such persisting problems may arise from two sources: (1) the dynamics of long-range change at the national or urban level, for instance, declining birth rates, migration, or transformation of the employment structure; (2) unanticipated consequences of planning, because of misunderstanding the interdependence of elements of the urban system. The analysis which follows elaborates the latter reasoning by focusing on the sequence of four interrelated problems: migration, housing, decentralization, and infrastructure.

Overview

For practical reasons, the study has been restricted to four Eastern European cities: Moscow, Budapest, Warsaw, and Prague. Instead of using different cities to document problems, the selective focus on four cities allows for a more thorough and comparative analysis. See table 5.1 for data on all nine capital cities of Eastern Europe.[1]

Choosing capital cities is done under the presumption that—at least in this case—changes and problems of urban development can best be observed in the largest cities. This argument cannot be elaborated here in full.

For comparative purposes, analysis is not done by city, but by dimensions: population and housing, employment and location of jobs, and infrastructure. As has been argued in prior publications, these entities represent the basic dimensions of social ecology: population, economy, and technology (Friedrichs, 1978, 1983, 1985). Throughout the study, tables are compiled as similar data were available to aid in comparative interpretation. For this reason, more detailed data on specific problems have been omitted if they were only available for one or two of the cities, for instance on segregation.[2]

Basic characteristics of the four cities appear in table 5.2. Despite the variation over all characteristics, each city is the largest in its country and a primate city by concentration of functions and size difference from the next-largest city. In 1980, the second-largest cities were Leningrad (57.8 percent of Moscow's population), Miskolc (10.2 percent of Budapest's), Lodz (52.4 percent of Warsaw's), and Bratislava (32.2 percent of Prague's population). Moscow is not only the capital of the USSR, but may be viewed as the capital of the other socialist countries since it is, among other things, headquarters of the Eastern Economic Alliance.

Population

All four study cities have grown in population during the past three decades. In absolute figures Moscow grew most dramatically, expanding by about 100 thousand inhabitants annually. In percentage, Warsaw grew the most (see table 5.3). Population increase in all cities except Moscow was caused by incorporations (annexation) in the 1970-80 period. (Even Moscow

TABLE 5.1
Capital Cities—Eastern Europe, 1980

Country	*Capital*	*Population (in thousands)*	*Area (km²)*
USSR	Moscow	8099	879
Hungary	Budapest	2059	525
Romania	Bucharest	1754	1369
Poland	Warsaw	1596	485
Yugoslavia	Belgrade	1455[a]	184
Czechoslovakia	Prague	1182	496
Bulgaria	Sofia	1057[b]	1282
German Democratic Republic	Berlin	1158[b]	403
Albania	Tirana	198[c]	–

SOURCE: Länderberichte des Statistischen Bundesamtes der Bundesrepublik Deutschland.
a. 1981.
b. Sofia agglomeration.
c. 1978.

expanded beyond the beltway by adding two new districts, not included in table 5.3.)

Population growth has for decades constituted a major problem for Moscow, more recently for Warsaw, and to a lesser extent for Budapest. The figures by far exceed plans, so plans must be adjusted continuously. For Moscow, Vydro (1976) projected a population of eight million in 1980, assuming an annual growth of 100 thousand persons. In 1983, Soviet demographers estimated a population of 9.6 million by the year 2000 for Moscow (Shuper, 1983:208). More recent projections for Warsaw arrive at an estimate of 1.65 million for 1990 (Warsaw Town Planning Office, 1980:8). Evidently, these estimates seem realistic given trends of the last few years; furthermore, they indicate a certain degree of resignation in coping with the problem of immigration.

Estimates for Prague assume a growing population as well; the Statistical Office projected 1.139 million inhabitants for 1990, and 1.166 million for the year 2000 (SLDB, 1970:114; 1982:160). However, age structure of the city's population and present migration suggest lower figures.

See table 5.4 for more detailed data on population growth. Except for Budapest, birth and death rates are rising. In all cities, birth rates fall below and death rates rise above the respective national rates. There is only a small natural increase or negative natural balance. The latter process corresponds to the pattern of change in large cities in Western Europe. In a stage model of demographic urban development, this situation can be interpreted as the late stage of urban development in highly industrialized countries, as in West Germany, France, or Great Britain.

Under such conditions, population increase mainly results from migration gains—considerable in Moscow and Warsaw, very low in Budapest and Prague. Moscow gained about 90 thousand residents annually in the period under study. The ratio of gains from natural increase to migration has

TABLE 5.2
Basic Characteristics—Budapest, Moscow, Prague, Warsaw, 1980

Characteristic	*Budapest*	*Moscow*	*Prague*	*Warsaw*
Area (km^2)	525	879	496	485
Population (thousands)	2,059	8,099	1,182	1,596
Density (per km^2)	3,923	9,214	2,383	3,289
Number employed (thousands)	1,029	5,230	661	815
Administrative districts	22	31[a]	10	7[b]
Region				
Population (thousands)	2,483	14,629	2,333	2,319
Area (km^2)	1,673	46,879	11,498	3,788

SOURCE: Compiled from various government and other publications; detailed source information available from author.
a. Excluding Zelenograd, 27 km northwest of Moscow; in 1984 a new district, Solntsevskiy, was added, adjacent to district 14.
b. 1982.

changed from 1:3 in 1960 to 1:9 in 1982. Several efforts were undertaken to limit the population influx. The major instrument was to allow for residence (*propiska*) only if the migrant could show a working permit in a Moscow-based company. Irrespective of the "grey" influx, companies had a large share in this migration by attracting badly needed qualified labor from other republics of the USSR. About 45 percent of all migrants are attracted by companies (Goltts and Lappo, 1981:20). This in turn can be partly explained by the age structure of the Moscow population, the city having absorbed practically the entire labor force potential of the male and female population (cf. Lappo et al., 1980:89).

Warsaw basically faces the same problems as Moscow. The ratio of population growth through natural increase versus migration was 4:6 in 1960, but 1:9 in 1981 (WWUSW, 1982:39). Again, programs have been implemented to limit migration to Warsaw: first, a formal restriction of migration to Warsaw began as early as 1954; second, a program of deglomeration, steered future development to regions outside the Warsaw Conurbation; and third, in 1982 new housing construction was stopped—only new estates under construction are to be completed. By the end of 1983, the restriction was abandoned because it was not successful and people did not want to leave Warsaw because they were afraid of not being permitted to migrate back.

But, as with Moscow, these programs had little success. Based on the common experiences of Moscow and Warsaw, it may be concluded that any such restrictions will fail to the extent that disparities between the capital city and other cities in the country are large with respect to job opportunities and quality of life. Such disparities also indicate regional inequalities within the two countries.

Budapest and Prague have grown very little in population. The much lower net migration gains can be interpreted by the hypothesis above. However, in Budapest permission to apply for a dwelling has since 1950 been

TABLE 5.3
Population (in thousands) and Population Change—
Budapest,, Moscow, Prague, Warsaw, 1960-1982

Characteristic	*Budapest*	*Moscow*	*Prague*	*Warsaw*
Population				
1960	1,805	6,242	1,133	1,139
1970	1,945	7,077	1,141	1,315
1980	2,059	8,099	1,182	1,596
1982	2,064	8,302	1,186	1,629
Percentage Change				
1960-70	7.8	13.4	0.7	15.4
1970-80	5.9	14.4	3.6	21.4
1980-82	0.2	2.5	0.3	2.1
Plan				
Date	1971	1971	1979	1971
Reference year	1985	1985/90	2000	1985
Estimated population[a]	2,150	7,500	1,350	1,550

SOURCE: Compiled from various government and other publications; detailed source information available from author.
a. 1971 projections did not account for incorporations.

TABLE 5.4
Population Characteristics—Budapest, Moscow, Prague, Warsaw,
1970-1980

Characteristic	*Budapest*	*Moscow*	*Prague*	*Warsaw*
Birth rate				
1970	12.0	11.8	11.4	10.1
1980	11.5	13.6	12.7	13.8
Death rate				
1970	12.3	9.5	14.0	8.7
1980	15.1	11.7	14.9	10.6
National population balance				
1970	−0.3	2.3	−2.6	1.4
1980	−3.6	1.9	−2.1	3.2
Migration balance				
1970	10,048	16,500	4,913	14,860
1980	9,146	15,400	1,836	14,979
Net population change and percentage due to net migration				
1975	8,278	104,000	8,341	25,700
	34.0	84.6	68.6	81.3
1980	2,496	88,000	−703	19,500
	397.5	82.5	261.2	74.1
1982	3,101	99,000	3,248	17,300
	353.0	87.4	213.1	80.0
Age structure, 1980 (percentages)				
0-15	18.0	18.5	19.9	17.5
16-59	61.8	56.8	59.7	67.1
60+	20.2	24.7	20.4	15.4

SOURCE: Compiled from various government and other publications; detailed source information available from author.

restricted to persons having worked in Budapest for at least five years. The more equal distribution of jobs and quality of life is not only a goal of the Communist Party of the CSSR (as it is with reference to Marx in all socialist countries) but also has been reinforced by the two national parties of Czechoslovakia (CSR) and Slovakia (SSR). As a result, efforts have been more successful in recent years, and job opportunities have been more evenly spaced over the nation's large cities. Prague even has a problem of attracting labor since salaries vary little across the country (8.3 percent above the national average in 1980). A housing shortage and high pollution have further adverse consequences.

Expansion of the Central Administrative Area (in terms of capitalistic societies, the Central Business District or CBD) and the population influx have lead to an ecological process well known from analyses of urban development: a dispersion of population toward the peripheral districts and even into the suburban areas. This situation is well documented by showing decreases of population in the inner districts and increases in the outer districts in all four cities. Some districts have in just one decade experienced a population increase of more than 60 percent.

In all four cities, inner districts lost population between 1970 and 1980, whereas outer districts gained:

	Thousands			
	Budapest	*Moscow*	*Prague*	*Warsaw*
Inner districts	−2.3	−522.0	−61.9	−24.8
Outer districts	+60.6	+1,009.0	+103.3	+305.3

Housing

The major consequence of population increase is a high demand for housing. As many authors have noted, one of the most persisting problems of cities in Eastern Europe is the housing shortage. The main reasons for the excess demand for housing are one or more of the following processes:

(1) *Quality of existing housing stock.* Many buildings from the 19th century or turn of the century need to be replaced. Little or too late restoration has further increased deterioration. Construction budgets have predominantly been directed toward new construction—the historic core of Polish cities being an exception.

(2) *High concentration of jobs and cultural facilities.* Concentration of employment opportunities and cultural advantages has resulted in extremely positive migration balances, which in turn increased housing demands beyond construction capacities and planning goals. All efforts to limit the influx of population have so far failed.

(3) *Increased numbers of households.* Household increases are attributed to new young households, as in Western countries, and mainly to overcrowding of existing dwellings shared by several families.

Processes 1 and 3 can be traced for all four cities, the second for all but Prague. In all four cities—most dramatically in Moscow and Warsaw—population increase aggravates the existing housing demand induced by processes 1 and 3 above. The combination of destruction of older buildings, high inmigration, and household partition will continue for the next decade, even if new construction could provide sufficient dwellings for migrants. The same holds true, except for the migration factor, in Budapest and Prague.

Most new construction was—and still is—done in the areas at the periphery of the central city. As in the large cities of Western Europe, high demand for housing after the Second World War could only be met by building estates on large sites of previously undeveloped land. Many of these new estates are immense, the largest being Chertanovo in southwest Moscow, planned for 300 thousand inhabitants, and Warsaw's Ursynow-Natolin, originally planned for 150 thousand. Most of the buildings have 9 to 13 stories.

Not unlike their counterparts in Western Europe, these new large estates, which are really cities inside the city, have problems endemic to their planning: low accessibility, insufficient infrastructure, and homogeneous age structure. Generally, infrastructure follows with a time lag after residential blocks and streets have been completed. Construction of service facilities (including telephone) is delayed while population increases (Fedotovskaya, 1981:63); one consequence is that residents must do the main part of their shopping in the city center. Further, most construction is with prefabricated components. Transport damages, corrosion, and bad construction create subsequent repair and maintenance problems. In spite of these disadvantages, this type still dominates over traditional brick construction, since costs and construction time are lower, and—as in Warsaw and Prague—there is pressure to continue operating the factories for prefabricated elements to ensure employment of the workers. Data on housing stock and type of new construction appear in tables 5.5 and 5.6.

Moscow has—according to official statistical sources—no private construction. Moreover, communal and cooperative housing are no different in terms of average dwelling size. In contrast, Bater (1980:167) reports cooperative apartment blocks have more living space and are more homogeneous by residents' incomes and occupations than communal blocks. Moscow still has privately owned dwellings, but private dwelling space continuously drops. For instance, private housing in the outer district Voroshilovskiy was projected to drop from 2.1 percent to 1.7 during 1976-80 (TSitsin, 1978:177). Budapest and Prague have a considerable amount of private construction and differences in average size of dwelling by type of management are remarkable.

In Warsaw, from 1976 to 1980, a total of 64,915 dwellings with a total space of 3,605,500 square meters was constructed. Of these 8.9 percent were built by the city, 88.4 percent by cooperatives, the remaining 2.7 percent privately. Average dwelling size was 54.5 square meters for the combined communal and cooperative construction, as opposed to 92.7 square meters for the dwellings (calculations by Dangschat, 1985a, based on data in WWUSW, 1977:290; 1978:275; 1979:297; 1981:179).

TABLE 5.5
Housing Stock—Budapest, Moscow, Prague, Warsaw, 1960-1980

Characteristic	Budapest	Moscow	Prague	Warsaw[a]
Dwellings (thousands)				
1960	536	880	352	307
1970	637	1,867	383	408
1980	727	2,542	448	514
Average size (m^2)				
1960	54.0	67.5	31.9	–
1970	55.0	53.1	51.3	42.0
1980	59.0	52.6	57.4	44.6
Average number persons/dwelling				
1960	3.4	7.0	3.4	3.7
1970	3.1	3.9	2.8	3.2
1980	2.8	3.2	2.6	2.9
M^2/person				
1960	–	6.9	10.9	–
1970	–	9.7	12.1	13.8
1980	19.0	10.0	14.0	15.2

SOURCE: Compiled from various government and other publications; detailed source information available from author.
a. 1980 data is actually from 1978.

Despite the housing shortage, construction has dropped markedly in Moscow and Warsaw, aggravating existing housing problems. For Warsaw in 1950, 1960, and 1970, Ciechocińska (1975:208, 213) and Dangschat (1985a, for 1978) have calculated the following deficits:

	Thousands			
	1950	1960	1970	1978
Households	302.3	396.1	466.8	594.2
Dwellings	199.0	307.5	408.0	514.0
Deficit	103.3	88.6	58.9	80.7

Similarly, Budapest has a considerable but decreasing deficit, in contrast to Warsaw:

	Thousands		
	1960	1970	1980
Households	659	748	807
Dwellings	535	640	727
Deficit	154	108	96

SOURCE: Calculations from Kiehl (1985) based on data in Preisich (1973:175; HKSH, 1982:4).

TABLE 5.6
New Dwellings by Type of Management—
Budapest, Moscow, Prague, Warsaw, 1982

Characteristic	*Budapest*	*Moscow*	*Prague*	*Warsaw*
Total				
Number	16,848	71,700	8,855	8,502
M^2		3,976,800	426,000	501,000
Average size		55.5	48.1	58.9
Communal				
Number	4,025	65,200	3,036	7,944
M^2		3,616,000	139,000	441,000
Average size		55.5	45.8	55.5
Cooperative				
Number	4,336	6,500	4,941	a
M^2		360,800	221,000	
Average size		55.5	44.7	
Private				
Number	8,487	0	878	558
M^2			66,000	60,000
Average size			75.2	107.5

SOURCE: Compiled from various government and other publications; detailed source information available from author.
a. Included in communal category.

Budapest is estimated to need 300 thousand new dwellings in the next 20 years, one-third to replace old ones below standard. However, if new construction continues at the present level, households seeking a new dwelling will have to wait for six years (Kiehl, 1985:712).

In Moscow, housing construction dropped by 46.7 percent between 1970 and 1980. To calculate the deficit for Moscow is difficult. As an approximation, the number of persons in families was calculated and subtracted from the total population; the remainder equals one-person households. To this figure the number of families was added, thus providing the number of households. From these calculations the following deficits emerge:

	Thousands		
	1961	*1970*	*1979*
Households	2,214.0[a]	2,698.0[a]	3,323.0[b]
Dwellings	880.0	1,781.0	2,479.0[c]
Deficit	1.333.9	917.1	844.1

SOURCE: Calculations from data in (a) SUGM (1972:103); (b) VPN (1982:78); (c) SUGM (1981:155); Saushkin and Glushkova (1983: 81); SUGM (1980:8); and Vydro (1976:28).

In Prague, the level remained fairly constant. When interpreting the Prague data, note that the city has—compared to Warsaw—lately begun to restore or renovate old housing stock in the districts of Praha 1 and 2. Given the city's tight budget, the government is troubled by the task of implementing renovation programs for the historic city core while continuing to construct new housing at the periphery (cf. Sykora, 1980). From the projected 190 thousand dwellings to be built or renovated in 1980 to 2000, half will be renovation. There is a considerable deficit, when calculated as in the preceding cases:

	Thousands		
	1961	*1970*	*1980*
Households	390.8[a]	472.2[b]	513.8[d]
Dwellings	351.8[a]	383.2[c]	448.0[e]
Deficit	39.0	89.0	65.7

SOURCE: From (a) SLDB (1970:60, 65, 70); (b) SLDB (1982:82); (c) FSU (1971:91); (d) SLDB (1980:82); (3) FSU (1981:99).

The situation becomes even more complicated when the diversity of applicants is taken into account. Local planning authorities have to attribute priorities to specific groups or interests. For Prague, this can be documented in more detail (personal information):

Plan 1981-85:	*48,600*	*New Apartments*
Planned distribution	15,000	communal—given to national executive companies (*narodni vybor*)
	14,800	cooperative—destined to stabilize demand from companies for their employees
	2,200	for the 29 cooperatives
	5,800	for cooperatives with members doing part of the constructions themselves
	4,500	expansion of "family homes" (personal property)
	6,300	given to ministries and distributed among government officials

Thus only a small portion of new construction in Prague is available to meet current demand and reduce the deficit.

Employment and Decentralization

All four cities have followed the well-known pattern of structural change in the economy from a high percentage in secondary to a high percentage in tertiary-sector employment (see table 5.7). Increases in the primary sector in Warsaw and Prague are through incorporation.

TABLE 5.7
Employment by Economic Sector–
Budapest, Moscow, Prague Warsaw, 1970-1980
(in thousands and percentages)

Characteristic and Year	*Budapest* N	*Budapest* %	*Moscow* N	*Moscow* %	*Prague* N	*Prague* %	*Warsaw* N	*Warsaw* %
1970								
Primary	21	2.4	0	0	7	1.2	3	0.5
Secondary	607	70.2	1667	40.7	211	36.3	308	43.6
Tertiary	237	27.4	2430	59.3	364	62.5	395	55.9
1980								
Primary	37	5.0	0	0	8	1.3	4	0.5
Secondary	465	62.3	1773	36.8	224	35.4	347	42.6
Tertiary	244	32.7	3045	63.2	402[a]	63.3	464	57.0
Percentage Change								
Primary								
1960-70		50.4		–		50.9		–58.8
1970-80		75.6		–		18.4		15.0
Secondary								
1960-70		–13.2		6.8		5.0		24.1
1970-80		–23.4		–10.5		6.5		12.7
Tertiary								
1960-70		19.0		12.9		13.6		39.7
1970-80		3.3		51.7		9.9		17.6

SOURCE: Compiled from various government and other publications; detailed source information available from author.
a. "Other" and "Nonclassified" have been added to tertiary sector.

Using a modified Fourastié model adapted for urban analysis, the threshold for a postindustrial city is 64 percent of all gainfully employed in the tertiary sector (Dangschat et al., 1985; Friedrichs and Kiehl, 1985). Judged by this criterion, both Moscow and Prague are by now in this stage of urban development. In contrast, Budapest and Warsaw still have a considerable share of their employed in the secondary sector because of a reindustrialization phase both cities experienced after World War II, especially Warsaw. Further, Budapest still has a very high concentration of industry (see below).

For a more detailed analysis of shifts in tertiary employment, see table 5.8. Despite some difficulties in the statistical data which do not permit identical categories nor a breakdown of the "other" category, some tentative comparative conclusions may be drawn. Tertiary employment is not to a large extent employment in governmental agencies—it accounts for less than 10 percent, except for Prague. The fastest growing employment is in "science/research" and in the mixed category of "health, etc." Moscow, in particular, has experienced high growth rates in total tertiary employment and in all categories. The high percentage of employees in the "hotel/food" category in Prague does in part reflect the historic and present function of the city as a center of tourism.

A major policy to restrict inmigration and problems related to population growth has been to decentralize industries from inner city areas to the

TABLE 5.8
Tertiary Sector Employment by Economic Category—Budapest, Moscow, Prague, Warsaw, 1970-1980 (in percentages)

	Budapest			*Moscow*			*Prague*			*Warsaw*		
Economic Category	*1970*	*1980*	*Percentage Change*	*1970*	*1980*	*Percentage Change*	*1970*	*1980*	*Percentage Change*	*1970*	*1980*	*Percentage Change*
Transport, communication	27.6	19.8	21.5	16.0	15.2	18.8	13.6	12.1	−1.4	14.1	12.8	6.7
Hotel, food	–	–	–	15.5	14.6	17.6	23.7	25.0	16.4	–	–	–
Commerce	19.8	24.9	37.7	–	–	–	–	–	–	21.8	20.6	11.2
Communal organizations	–	–	–	8.6	7.7	13.0	9.7	5.6	−36.1	11.8	11.8	17.3
Science, health, research, education	–	–	–	29.7	31.0	31.0	8.2	8.9	19.8	9.1	11.4	48.2
Culture, sports	23.4	28.0	29.0	18.6	19.5	31.5	22.5	23.2	13.9	28.3	27.2	13.0
Administration, law	–	–	–	7.6	8.1	33.3	13.5	14.7	20.5	7.4	5.7	−9.5
Other, nonclassified	29.1	27.3	3.2	4.0	3.9	22.5	8.8	10.5	32.5	7.5	10.5	64.5
Total	100.0	100.0	9.3	100.0	100.0	25.3	100.0	100.0	10.6	100.0	100.0	17.6

SOURCE: Compiled from various government and other publications; detailed source information available from author.

periphery and suburbs (Lappo et al., 1980). In addition, new enterprises were to locate in the suburbs (the *oblast*). This policy is illustrated with several examples.

Moscow

Among the wide range of industries located in Moscow, two are of particular importance: machine construction and light industry, especially textiles. Textile production in Moscow and the Moscow region amounts to 50 percent of the total production in the USSR. Moscow's share is 15 percent.

Industry is spatially dispersed and includes a large amount of small enterprises. Since 1973, "production associations" have been formed, each comprising several larger and smaller companies to allow for better central planning. Many companies use old machinery, but modernization of equipment has been restricted to large factories. This situation may be illustrated by a decision of the Communist Party in 1981 to decentralize 30 light industry factories during the current Five-Year-Plan period. However, according to Kibalčhich (1983:35) hundreds of factories require relocation and—as a concomitant consequence—modern machinery and buildings. During the five years from 1976 to 1980 only 5 percent of the machinery of enterprises in Moscow was replaced; 40 percent of all blue collar labor is manual labor, thus requiring too high an amount of labor (*Pravda*, 21 Jan., 1981:2) and a ratio of one skilled to six unskilled workers (Bitunov, 1979:42). These data indicate a general problem of all four countries: the comparatively low rate of adoption and investment in new production technologies.

Already in 1971 the comprehensive plan stated a change for Moscow industry: limitation of the number of establishments, restriction to industries with low pollution, and relocation of almost a thousand establishments to the 66 industrial zones in Moscow. Gosplan's more recent publication on "Methods of Formulating Five Year Plans of the Economic and Social Development of the USSR" explicitly concludes for 1981 to 1985 and 1986 to 1990 there should be neither expansion of existing nor location of new industries—except for consumption goods—in Moscow and the Moscow Region (U.S.S.R., 1980:455). Any expansion of industry, large factories, or scientific institutions shall be done by setting up new branches located outside of Moscow and Leningrad (U.S.S.R., 1980:739).

Interestingly, these rules also state that production may be expanded, but neither spatially nor by increasing the number of employees (U.S.S.R., 1980:455). The latter argument can be derived from the unfavorable age structure of Moscow (see table 5.4), resulting in a high demand for external labor. A case in point is the textile industry: two-thirds of the employed were reported not to come from Moscow (Gorlov et al., 1983:27-32).

Deconcentration and fostering suburban growth seems as difficult as the New Towns program intended to relieve London. Goltts and Lappo (1981:269) describe the example of Zelenograd. Planned to be a "sputnik" of Moscow, industry was to be located there to give jobs to the local residential population and thereby reduce commuting to Moscow. However, the authors find that only one-third of the industry came from Moscow and 20 percent of the population had migrated to the town; yet the population size is double the planned figure. Further, Zelenograd attracts labor from Moscow—according to the 1970 Census, 14 thousand daily commuters.

Budapest

As in Moscow, a decentralization policy for Budapest was proposed as early as 1960. At that time Budapest had an extremely high share of 43 percent of Hungary's industrial employment. It was decided to have no further new industry in Budapest, nor in the 43 surrounding communities which make up the suburban area. But from 1968 to 1971, only 18-20 thousand jobs were decentralized from Budapest out of a planned figure of 70 thousand. One reason for this failure probably was the economic reform in 1968, which for the first time allowed workers in Hungary to choose their place of work, inducing a wave of migration (20 percent of all employed in Hungary changed jobs in 1968/69).

The Third Comprehensive Plan for Budapest relaxed the decentralization policy insofar as an "intensive" and an "extensive" development was proposed. The former refers to higher quality production and productivity (goals, included as well in the Five-Year Plan 1971-1975). The latter refers to strengthening only those industries in Budapest depending on qualifications of the local labor force. The plan even proposes 4.7 hectares of land for industrial expansion; further, an expansion of the CBD was projected 12 years ago (Preisich, 1973).

Because of lack of data on the suburbs, decentralization can only tentatively be analyzed by changes in the composition of the labor force in Budapest. It should be kept in mind that any decentralization policy for Budapest faces the problem of a very high concentration of jobs in the capital, foremost in the metal, chemical, textile, and food industries. For instance, 90 percent of all exports in the metal industry come from Budapest. In addition, Budapest has 33 percent of all employed in the tertiary sector, including 81 percent of all research institutes.

In the decade from 1960 to 1970, the share of industrial employees remained almost constant in Budapest, but grew in the 43 suburbs from 33 to 55 percent. In the following decade, loss of employment in industry in Budapest amounted to 146,000 workers (29 percent), compared to an overall decrease of employees of 81,406 (–7.3 percent). A further indicator is the percentage of industrial employment Budapest held of the nation's total industrial employment: it dropped to 34 percent in 1970, down to 27 percent in 1980. These data might be interpreted as a successful decentralization, but changes are mainly because of modernization in industry. Budapest has lost 180 thousand jobs in manufacturing in the 1970-80 decade by rationalization of production. This interpretation is supported by the increase in Budapest's share of the nation's total value of industrial production from 41 to 44 percent in the same period.

Warsaw

In 1965, the Central Committee of the Polish Communist Party decided to limit growth of the labor force in Warsaw. It was decided to develop a plan for "Conurbation Warsaw" (Warsaw plus a suburban area of 4,496 square kilometers), and to decrease disparities in the standard of living between Warsaw and the suburban area (Warsaw, Praesidium, 1971:20-21). The Comprehensive Plan up to the year 1985 was enacted in 1969. Measures to be taken were expansion of industries and location of new industries in the 30-60-kilometer zone around Warsaw, especially in the subregions in the northwest, west, and east (see Ciechocińska, 1973, 1978; Dziewonski, 1976). In the northwest, the town of Plock is supposed to grow from 43 to 120 thousand residents; in the east, Siedlce from 33 to 70 thousand residents (Warsaw, Praesidium, 1971; Warsaw, Town Planning Office, 1980).

The most recent Comprehensive Plan for Warsaw Province/Agglomeration for the Year 1990 (Warsaw, Town Planning Office, 1980), approved in 1978, reaffirmed these goals, and proposes to concentrate regional growth along two major *bands* (axes): southwest to northeast and northwest to southeast. Warsaw thus applies one of the two basic concepts to control suburban growth, the concept of axes, like Hamburg in West Germany. The other concept is to steer growth to satellite towns or new towns, as in Moscow and London.

To evaluate these programs, changes in employment figures and amount of commuters are used as indicators. Data in table 5.9 show a growth in employment for Warsaw and the suburban zone. Note, however, that during the decade both areas gained in employment until 1977; thereafter the figures dropped. Thus, from 1977 to 1980, Warsaw lost more than the suburbs. More important, Warsaw's share of the labor force in the conurbation has grown slightly: from 81.5 percent in 1970 to 82.7 in 1977 to 82.5 in 1980.

The overall pattern conceals major changes in employment by economic sectors (table 5.10). Warsaw lost employment in the secondary sector; these losses make up half the total loss in employment in the 1977-81 period. Warsaw gained in health and social services and science. In contrast to the plan, employment in the suburban area has grown in the primary (+4,023) and tertiary sectors (+17,455), but lost in the secondary (–8,450), the sector to be relocated from Warsaw. Data on commuting, presented in the following section, also suggest that planning objectives have not been reached.

In a careful evaluation of data and Polish literature, Dangschat (1985a) concludes that the decentralization program has failed when measured against initially stated goals. Even if it may be too early to resume this program, its success is doubtful since it is based on several interrelated, simultaneous advances. First, population has to be redistributed closer to

TABLE 5.9
Employment—Warsaw, Warsaw Suburbs, and Warsaw Conurbation, 1970-1980 (in thousands)

	Warsaw	*Suburbs*	*Conurbation*
Year			
1970	706	160	866
1977	830	174	1004
1980	815	173	988
Change			
1970-80	+109	+13	+122
Percentage	15.4	8.1	14.1
1977-80	−15	11	−16
Percentage	−1.8	−1.6	−1.6

SOURCE: Calculated from government publications; detailed source information available from author.

work places in the suburbs, but in 1980 Warsaw had a 78 percent share of the conurbation's population. Second, the housing supply has to be improved in the suburbs and the rest of the country, since the supply in Warsaw has been better than in other parts of the country (Leszczycki and Lijewski, 1977:49). Third, migration to Warsaw is supposed to stop in the suburban zone, whereas now suburban residence is often just an intermediate stage of a final migration to Warsaw (Poleszak and Rakowski, 1980).

Prague

Prague has no decentralization because of the comparatively high deconcentration of industries in the CSSR. For instance, even in the 1930s two-thirds of all industrial enterprises were located in cities with less than two thousand inhabitants. After the Second World War, larger industrial enterprises were relocated to smaller cities and only a few larger ones, but not to Prague (Musil, 1977:233, 244; cf. Musil, 1980). Further, there was less investment directed to new industries in the larger cities; instead, investment was geared to modernizing industries in larger cities. As a result, Prague has a decreasing share of national industrial production (SLDB, 1980:13). Planning for Prague (Master Plan 1975; Comprehensive Plan in the process of adoption) toward increasing the city area led to an incorporation in 1975 to facilitate new housing construction at the periphery. This, in turn, is part of a program to separate industrial and residential land uses functionally in currently still mixed areas. The goal at present is instead to attract young people to Prague (CSR.CSU, 1981:85, 183).

To assess these processes, data for two indicators are available: population and labor force distributions in the central city and the rest of the Prague metropolitan area. Between 1970 and 1980, the central city's share of the metropolitan population increased only slightly from 47.5 percent to 50.7 and the percentage of gainfully employed in the Prague Region has remained

TABLE 5.10
Employment by Sector—Warsaw, Warsaw Suburbs, Warsaw Conurbation, 1970-1980 (in percentages)

Sector/Year	*Warsaw*	*Suburbs*	*Conurbation*
Primary			
1970	0.5	5.5	1.4
1980	0.5	7.4	1.7
Secondary			
1970	43.6	56.0	45.9
1980	42.6	46.9	43.3
Tertiary			
1970	55.9	38.6	52.7
1980	57.0	45.7	55.0
Percentage Change 1970-80			
Primary	14.9	45.7	
Secondary	12.7	–9.4	
Tertiary	17.6	28.3	

SOURCE: Compiled from Warsaw, Wojewodzki Urzad Statystyczny (1976: 102; and 1981: 43).

constant with 57.0 percent and 56.2 percent, respectively. These data are conclusive, since Prague gained little in population during the decade, whereas the suburban area lost slightly (calculations from data in FSU, 1971:86, 136; SR Praha, 1981:18, 21).

Infrastructure

The problem of housing shortage is attacked basically by two policies: new construction at peripheral areas of the central city and diverting development to cities in the suburban zone ("deglomeration"). For Moscow, the latter program has led to considerable efforts in expansion of suburban cities (*sputniki*). Not only population, but also industry, administration, and research institutes have been located in these 72 cities. The same holds true for the Warsaw decentralization policy, but there decentralization is directed toward the zone adjacent to the Warsaw Conurbation.

Even the green belt surrounding Moscow has been intruded upon by nonresidential uses. As much as 70 percent of new construction for housing is expected to be in the green belt area (Goltts and Lappo, 1981:26). Initially, the belt was to be preserved according to the 1971 Plan and new developments to be restricted to areas at greater distances from Moscow. (This plan very much resembles planning for London, in particular the 1944 Abercrombie plan to conserve the green belt in favor of new town development.) However, by 1970, administration and research institutions were located in the green belt area. Already from 1939-59, several towns were "formed" in this area (Lappo et al., 1980:112).

Major problems resulting from dispersal and planned suburbanization are high numbers of commuters and subsequent transport problems. By the end of the 1960s, one of eight employed in the capital's enterprises and institutions lived in the suburbs (Lappo et al., 1980:147). In 1970, approximately 500 thousand persons commuted to Moscow. By the mid-70s the figure had increased to 600 thousand (Danilova et al., 1981:56; cf. Hall, 1977:168). Average commuting time within Moscow was 45 minutes (Byalkovskaya, 1981:24), from the suburbs to Moscow 92 minutes (Saushkin and Glushkova, 1983:230). The true figures may even be higher, estimated from time-budget data, according to which blue- and white-collar employed Moscow residents spend 10-12 hours per day for work and travel to it (Dmiterko, 1984:130).

A more detailed assessment of Moscow's transport situation can be drawn from an analysis by Barbash and Gutnov (1979). The authors conducted two factor analyses, first with 56, then 11 variables from the domains of public transport, land use, and housing. Spatial units were 423 "transportation areas." (Unfortunately, the reference year for the data is not stated; but from the text may be estimated as the early 70s.) In the final solution eight factors emerged, the three main ones defined as distribution of jobs, distribution of dwelling space, and ease of long-distance accessibility by public transport. When plotting the factor scores, a clearly reverse pattern emerged, exhibiting an inner-city versus outer-city dichotomy of land use. Furthermore, accessibility of the peripheral areas was very low (metro and bus), indicating the above-mentioned problems of commuting time.

Transport is mainly public and has constantly increased in volume; see table 5.11. More specifically, the problem is the low rate of private car ownership in all four cities, especially Moscow (see table 5.12). Official statistics do not supply information on the number of private cars per resident. According to Hamilton (1976:20), there were 150 thousand private cars in the Moscow metropolitan area. Using this figure as a baseline, annual private car sales for Moscow were added (SUGM, 1981:131), resulting in an estimate of 608 thousand private cars in 1980, or 76 per 1,000 inhabitants. This estimate seems adequate when compared to Shestokas et al. (1984:12, 25) who state that the Baltic Republic and Moscow have the highest share of private car ownership in the USSR—60-80 at the beginning of the 80s. The planned figure for 1990 of 150-200 (Cherepanov, 1981:105) does not seem realistic.

The latter problem of low accessibility of suburban areas is of concern for the Budapest planning authorities: to substitute for bus transport to suburbs by extending railway routes. Again, this goal is a consequence of the decentralization policy.

The decentralization policy has—as in cities in capitalist countries—resulted in a spatial disparity of places of work and residence, especially when analyzed by type of job. In the case of Budapest, in 1980, 44.8 percent of all employed in the secondary sector lived in inner city areas, more or less next to the historic industrial districts. In the process of decentralization new industrial areas were developed in the outer districts. In the outer areas 230 thousand jobs are located, one-third in large industrial enterprises.

Location of old and new industries at the periphery and in the suburbs is

TABLE 5.11
Public Transport—Moscow, 1940-1982

Mode of Transport (percentages)	*1940*	*1965*	*1970*	*1975*	*1980*	*1982*
Metro	13.0	29.6	32.2	33.4	38.4	39.8
Bus	7.4	27.9	30.0	30.2	29.6	30.3
Trolleybus	7.5	17.8	15.5	15.5	13.4	12.2
Tramway	65.0	15.6	12.5	10.8	8.7	7.9
Suburban train	7.2	9.0	9.8	10.1	9.9	9.8
Number of taxis (thousands)	10.1	10.1	14.2	16.2	17.1	17.4
Number of passengers (millions)	2.844	4.489	5.055	5.889	6.033	6.075

SOURCE: 1940 data from Hamilton (1976:26); other years calculated by author from Statisticheskoye Upravleniye Goroda Moskvy data (1976a:61; 1983:77).

paralleled by an increase in tertiary jobs in the inner city areas. In 1980, 740 thousand jobs were located in these areas; 720 thousand employed (out of a total of 1.029 million) lived in this area; of these, 470 thousand commuted to other city areas. In contrast, 490 thousand commuted to inner city areas. The CBD had 120 thousand jobs, of which 90 percent were in the tertiary sector. The residential population of the CBD was 50 thousand; of these, 28 thousand were in the labor force, but only 6 thousand had their jobs in the CBD, whereas about 100 thousand employed persons commuted daily to the CBD.

As a result of the decentralization policy in Warsaw, the number of commuters and commuting distances have increased considerably. Kaminski (1974:153) reports an annual increase of 17 percent; Ciechocińska (1978:604) gives a lower figure: she calculates the total increase from 1964-68 as 18 percent. The majority of long-distance commuters come from the southwestern (Skierniewice) and the eastern (Siedlce) parts of the *woiwodships* adjacent to the Warsaw Conurbation.

As in Moscow, residents have to rely mainly on public transport—suburban trains, buses, and tramways. Because of a high concentration of traffic lanes in the city center, up to ten tram lines have to use the same lane, resulting in severe traffic jams at peak hours. Not until 1983 was construction of a metro system started.

Unequal distribution of transport facilities was demonstrated in a study by the Warsaw City Planning Agency. The agency combined several variables on transport density per area and per resident and number of commuters into one transport index using 1978 census figures. Dangschat (1985a:933) plotted distribution of the values for the 82 planning districts. Results show low index values for outlying residential areas, mostly new housing estates; among them the largest, Ursynow-Natolin, with 100 thousand inhabitants, is served directly by only two bus lines.

Transport problems in Prague are not as dramatic as in Moscow and Warsaw. In 1980, only 6,798 employed commuted daily to the suburbs, but

TABLE 5.12
Car Ownership, Commuters, and Average Commuting Time—
Budapest, Moscow, Prague, Warsaw, 1980

Cars, Commuters, Communing Time	*Budapest*	*Moscow*	*Prague*	*Warsaw*
Private cars	274,599	600,000	193,000	250,599
Number/1000	133	60-80	163	157
Commuters				
From suburbs	205,062	600,000[a]	65,439	180,000[c]
To suburbs	26,789	150,000[a]	6,798	18,000
Average commuting time (minutes)	90	92[b]	–	–

SOURCE: Compiled from various government and other publications; detailed source information available from author.
a. Data refer to the mid-1970s.
b. 1970 data.
c. 1976 data.

65,493 commuted into Prague. As in the other cities (and, in fact, most large cities in Western Europe), problems are aggravated by the still lagging expansion of the metro system to peripheral new housing estates. Despite tremendous efforts to expand the metro system, major constraints are costs of approximately 1.2 billion Kcs per kilometer—the city's total budget being 12 billion Kcs. One program to alleviate rush hour problems has been to stagger working hours.

In contrast to Warsaw, renovation and conservation of the historical section of Prague began in 1970. The almost completely conserved areas of Praha 1 and 2 confront the city with several problems: to renovate the buildings and renew the entire infrastructure (tramway, sewage, electricity) at the same time. Further, renewal of water supply and sewage pipes is an urgent problem in itself, since approximately 30 percent of the water is lost by the bad condition of the 1869 pipe system.

Similar communal tasks are listed in the Third Comprehensive Plan for Budapest, especially expansion and renovation of the sewage system; further, in 1970 only 84 percent of all 636 thousand dwellings were connected to the public water system (Preisich, 1973).

A final problem of infrastructure for Prague is that 60 percent of the private dwellings are still heated by coal, which is cheap and its resources expected to last for the next 40 years. However, this fuel creates high pollution and may be judged an adverse factor in living in Prague.

Planning

Planning is split into territorial and economic planning. With respect to the four capital cities under study, the basic dilemma in socialist countries is coordinating sectoral (economic) and territorial planning. For instance, in

Hungary, sectoral planning is officially subordinated to territorial planning (Palotas, 1965:203).

In the Khrushchev period (1956-65), planning by territorial units (*oblasti*) was introduced, ensuring a larger impact and control over land uses, but at the same time increasing competition among the oblasti. After 1965, the USSR returned to the principle of planning by sectors, which led to competition among them (departments) and between sectors and Mossoviet, Moscow's city government, for land and labor force.

Until 1976, each industrial department planned and *constructed* separately. They thus had the power of territorial changes and land uses, including housing. The Mossoviet was unable to coordinate this development, because the departments had power and financial means. Because of their interest in putting a premium on location (Bater, 1980:78), their planning has often run against Moscow's territorial interests (see Kibalčhich, 1983).

The situation changed with the tenth Five-Year-Plan, 1976-80. Each department now had to specify by a separate line in the sectoral plan its planning as far as Moscow was concerned. Gosplan, the supreme national planning institution, had to approve all plans. But even this reconciliation of sectoral and territorial planning altered little the diversity of planning. More recently, with the 11th Five-Year-Plan, 1981-85, a specific chapter in each department's territorial plan must be devoted to Moscow and Leningrad. Coordination and control of the plans of all sectors (departments) is now done by Mossoviet, which in turn, after consultations with the departments, transfers the plan to Gosplan. Effects of this change in the procedure remain to be seen. Presumably, despite coordination of sectoral and territorial plans the change will be insignificant, because of the persisting power of the departments and their advantages by initially formulating the plan.

When discussing the problems of sectoral and territorial planning, another difficulty became apparent: the complicated and hence time-consuming regulations of consultation among the numerous organizations involved in each plan. Further, once a plan has been established, it becomes very difficult to adjust it to any changes in the actual situation.

For instance, the scheme for the Moscow Oblast Plan 1971-90 was approved in 1973. The plan was based on norms of construction and services, such as size of shops or demand for transportation. Norms like these probably change every ten years, but once included in current plans do not allow for adjustment to changing demands. Such "standards" do exist in many countries, especially for construction and necessary infrastructure of new housing estates (see for West Germany: Borchard, 1972; Spengelin et al., 1972). Standards are based on assumptions of demand and current service ratios, for example hospital beds per 1,000 residents or square feet of sales space in food shops. However, deviations in age structure and changes in income over time require changes in the initial plan and of the standards themselves in general for a better infrastructure than planned. Undoubtedly, the same problems have been documented for new housing estates in cities in capitalist countries in

Europe, but market forces and a higher rate of car ownership have compensated for the initial deficits in a shorter period than in socialist cities.

Conclusions

From the preceding analysis several similarities in what have been designated "persisting problems" emerge: restricting inmigration, providing more and better housing, decentralizing more efficiently, solving transport and commuting problems. When tentatively ordered in a temporal sequence, the following conceptual model seems appropriate:

migration → housing shortage → decentralization → transport problems

There are several similarities in the processes of urban change:

- population losses in the inner city areas and gains in the peripheral areas (*dispersion*)
- increases in the suburban share of the total population in the metropolitan region (*suburbanization*)
- changes in the location of industries, especially manufacturing: decentralization to the outer city areas and suburbs, location of new industries in the suburbs or even the zone adjacent to the agglomeration
- increasing distance of journey to work
- similar decisions of industrial (sectoral departments) on location of enterprises, analogous as well to location decisions in capitalist countries, despite explicit reference to location theory and land values
- a trend toward revising the pattern of distribution of blue-collar workers and their jobs: increase in tertiary jobs in the inner city and secondary jobs in outer city areas with an opposite pattern of distribution of residential population by social stratum (as observed for the U.S., see Kasarda, 1980, 1983).

The ecological processes listed above (Berry and Kasarda, 1977) are similar to those in cities in Western Europe (Dickinson, 1963; Friedrichs, 1978, 1985; Hall, 1977; Klaasen et al., 1981), but it may be misleading to infer similar causes. Because of the lack of precise theories for comparative urban development in socialist and capitalist countries, the following propositions are approaches toward more refined propositions.

A first approach to understanding the causes is to analyze the *differences* among the four cities with respect to the extent the problems occur. Prague has no inmigration problem and has plans to attract young qualified labor. The main reason for this difference with the other three cities presumably is a certain degree of competition between the CSR and SSR and the national policy not to give priority to Prague in national economic investment. Both facts result in a more equal spatial distribution of jobs and living conditions in the large cities of the CSSR, and even better opportunities to build a private house in smaller cities. For the same reasons, a decentralization policy is neither formulated nor necessary.

The housing shortage is particularly dramatic in Moscow and Warsaw. Thus, city size does not seem to be the major variable accounting for the extent of this shortage. In Moscow, the shortage dates to the prerevolution period; it was later aggravated by industrialization of the area and heavy inmigration.

This cumulative deficit can hardly be reduced. First, inmigration continues, as disparities in standard of living and jobs prevail. Second, inmigration of qualified labor will continue—like the factual failure of decentralization—since the national economy is given priority over spatial planning. Thus national economic growth is achieved more easily by expanding existing centers of production and administration—the major center being the capital city.

In Warsaw, destruction of large parts of the city in World War II reduced the housing stock considerably and inmigration increased the deficit. Finally, the critical state of the national economy restricts investment and further aggravates housing problems. Hence, the main explanation may be found in the national economy.

Solutions to cope with the housing shortage vary among the four cities. Moscow has a considerably lower share of cooperative construction and no private construction of dwellings. This policy may prevent the obvious differences in dwelling size found in Budapest, Prague, and even Warsaw. Construction by the state as well is favorable to renters through low rents and rent control. On the other hand, the high rate of savings in all four countries has led to greater purchasing power (see Alton et al., 1983). This purchasing power could be used to stimulate cooperative or private housing, alleviating the problem of housing deficit, done in all cities except Moscow. Unfortunately, the negative consequence of the latter policy is a dual housing market, as in Budapest and Prague with the "grey" market offering dwellings at substantially higher rents than the officially controlled ones.

A second approach is to seek explanations by referring to the *planning process*. One inference may be drawn from the conceptual model presented above. The programs implemented to solve a problem are insufficient and create as a consequence—be it anticipated or not—new problems (in urban planning in both socialist and capitalist countries). Housing shortage may serve as an example; it is alleviated by construction of huge new complexes at the periphery, but both infrastructural facilities and transportation are inadequate, imposing high costs of time and money on inhabitants. The same holds true for the decentralization policy, which creates additional problems of growth of the satellite towns substantially exceeding the targeted population figure.

The latter reasoning implies a more general and more radical question: Under which conditions can spatial planning be successful? The diverse agents in a city are competing for land, space, and access. Urban forces may thus be described by a market model, irrespective of type of national economy. Any spatial planning (not including regulations, like zoning laws) will only be successful when in accord with the outcomes of the process of competition and bargaining. We have insufficient theoretical and empirical knowledge to predict the results of the interrelated processes. Consequently, any social technology, as implicitly used by planners, is based on the existing body of knowledge, which is fragmentary, even if cumulative evidence is available for certain problems. Under these conditions, planning must fill the gaps using a mixture of normative statements and mostly implicit, unvalidated assumptions. This process in turn increases the probability of false projections. Obviously, these deficiencies are not peculiar to planning in socialist countries; they can be documented for capitalist countries, too—for instance for urban renewal programs. What accounts for the difference are the additive effects of long-term, overly comprehensive, bureaucratic, and time-consuming planning in socialist countries.

A final approach refers to the state of the national economy and national priorities in directing investment to sectors and regions. Bater (1980:56) has phrased it the "dilemma of regional equity and growth vs. national efficiency and growth." The aim of international competitiveness, which increasingly dominates policies in socialist countries as in capitalist ones, almost inevitably forces governments to direct investment into already powerful urban centers. The underlying assumption is that as the nations under study compete for world markets and power, one of the major spatial effects will be a more pronounced urban hierarchy, strengthening the largest cities and the capital city in particular.

In addition, the relatively low investment in new technologies may have the advantage of low unemployment rates, but at the same time productivity has dropped, compared to other highly industrialized countries, thus requiring more (manual) labor, often not sufficiently available. Furthermore, the low priority given to consumer goods production has as one consequence the problems of mass transport and high individual commuting costs.

In summary, national investment policies, disparities in living conditions among the large cities, and planning procedures seem to be the most adequate explanations for the persisting problems in the four Eastern European capital cities studied.

Notes

1. The editors wish to note that Professor Friedrichs has had access to extensive data sources in Soviet block countries, many of them unavailable to scholars in the West. To conform to other chapters, numbers in tables in this chapter have been rounded to thousands. Specific data sources are not given because of their complexity and nonavailability for verification. The interested reader is encouraged to correspond with Professor Friedrichs concerning more detailed statistical information and precise sources.

2. Many data are drawn from a project, of which I am senior investigator, on urban development in socialist and capitalist countries, especially the monographs on Moscow (Osterwold, 1978), Budapest (Kiehl, 1985), and Warsaw (Dangschat, 1985b; Dangschat and Wendl, 1978).

Bibliography

Alton, Thad Paul, Krysztof Badach, Elizabeth M. Bass, and Gregor Lazarcik. 1983. *Money Income of the Population and Standard of Living in Eastern Europe 1970-1982*. Occasional Paper no. 78. New York: L. W. International Financial Research, Inc.

Barbash, N. B., and A. E. Gutnov. 1979. "Gradostroitel nyye aspekty territorial noy organizatsii Moskvy." *Izvestiya Akademii nauk SSSR, Seriya geograficheskaya*, 1979, no. 2:53-67 (in Russian).

Bater, James H. 1980. *The Soviet City: Ideal and Reality*. Explorations in Urban Analysis, vol. 2. Newbury Park, CA: Sage.

Berry, Brian J.L., and John D. Kasarda. 1977. *Contemporary Urban Ecology*. New York: Macmillan.

Bitunov, Vladimir Vladimirovich. 1979. "Moskva—narodnokhozyaystvennyi kompleks." *Moskva: Znaniye, Seriya ekonomika i organizatsiya proizvodstva*, 1979, no. 12 (in Russian).

Borchard, Kurt. 1972. *Orientierungswerte für die städtebauliche Planung*. 2d ed. Munich: Deutsch Akademie für Städtebau und Landesplanung (in German).

Byalkovskaya, V. 1981. "Eknomischeskiye voprosy razvitiaya narodnogo khozyzystva Moskvy." *Voprosy ekonomiki*, 1981, no. 6:23-32 (in Russian).

Cherepanov, V. A. 1981. *Transport v planirovke gorodov*. Moscow: Stroyizdat (in Russian).

Ciechocińska, Maria. 1973. *Deglomeracja Warszawy 1965-1970; Wybrane problemy zatrudnienia pracowników zakladów deglomerowanych*. Polksa Akademia Nauk Komitet Przestrzennego Zagospodarowania Bulletin no. 80. Warsaw: Pánstwowe Wydwnictwo Naukowe (in Polish).

———. 1975. *Problemy ludnosciowe aglomeracji warszawskij*. Warsaw: Panstwowe Wydawnictwo Naukowe (in Polish).

———. 1978. Wplyw wielkich zakladow przemyslowych na rozwoj aglomeracji Warszawskiej w latach 1945-1975." In *Wielkie zaklady przemyslowe Warzawy w rozwoju historycznum*, edited by Jozef Kazimiersiki et al. Warsaw: Pánstwowe Wydawnictwo Naukowe (in Polish).

CSR.CSU. *See* Czech Socialist Republic. Český Statistický Úřad.

Czech Socialist Republic. Cesky Statisticky Urad. 1980. *Rozbory. Praha a ostatni relka mesta CSSR v letech 1970 az 1980*. Prague (in Czech).

Dangschat, Jens. 1985a. "Warschau." In *Stadtentwicklungen in West- und Osteuropa. See* Friedrichs, 1984.

———. 1985b. *Soziale und räumliche Ungleichheit in Warschau*. Hamburg: Christians (in German).

Dangschat, Jens, and Norman Wendl. 1978. "Warschau." In *Stadtentwicklungen in kapitalistischen und sozialistischen Ländern. See* Friedrichs, 1978.

Dangschat, Jens, Jürgen Friedrichs, Klaus Kiehl, and Klaus Schubert. 1985. "Phasen der Landes- und Stadtenwicklung." In *Stadtentwicklungen in West- und Osteuropa. See* Friedrichs, 1985.

Danilova, I. A., T. D. Ivanova, and V. M. Moiseyenko. 1981. "Aktual ňyye problemy migratsionnoy situatsii v g. Moskve." In *Rasseleniye i dinamika naseleniya Moskvy i Moskovskoy oblasti*. Moscow: Moskovskiy filial Geograficheskogo obschchestva SSSR (in Russian).

Dickinson, Robert Eric. 1963. *The West European City*. 3d ed. International Library of Sociology and Social Reconstruction. London: Routledge & Kegan Paul.

Dmiterko, D. Ya. 1984. *Sotsialňoye planirovanye kak aspekt sotsialnoy politiki*. Moscow: Moskovskiy Universitet (in Russian).

Dziewonski, Kazimierz. 1976. "Changes in the Processes of Industrialization and Urbanization." *Geographia Polonica* 33(2):39-58.

Federalni statistický úřad, Český statistický úřad, Slovenský štatistický úřad. 1971. *Statistická ročenka Ceskoslovenske Socialisticke Republiky 1971*. Prague: SNTL-Nakladatelstvi technicke literatury Nakladatel stvo ALFA, n. p., Bratislava (in Czech).

———. 1981. *Statistická ročenka Ceskoslovenske Socialisticke Republiky* 1981. Prague: [Statistical Office] (in Czech).

Fedotovskaya, T. A. 1981. "Nekotoryye problemy sotsial ňo-ekonomicheskogo razvitiya Moskvy i sotsial ňaya adaptatsiya migrantov." In *Rasseleniye i dinamika naseleniya Moskvy i Moskovskoy oblasti. See* Danilova et al., 1981.

French, Richard Anthony, and F. E. Ian Hamilton. 1979. "Is There a Socialist City?" In *The Socialist City. Spatial Structure and Urban Policy,* edited by Richard Anthony French and F. E. Ian Hamilton. New York: Wiley.

Friedrichs, Jürgen. 1983. *Stadtanalyse. Soziale und räumliche Organisation der Gesellschaft.* 3d ed. Opladen, West Germany: Westdeutscher Verlag (in German).

———, ed. 1978. *Stadtentwicklungen in kapitalistischen und sozialistischen Ländern.* Reinbek bei Hamburg: Rowohlt.

———, ed. 1985. *Stadtentwicklungen in West- und Osteuropa.* Berlin-New York: de Gruyter (in German).

Friedrichs, Jürgen, and Klaus Kiehl. 1985. "Ökonomische Phasen der Stadtentwicklung" *Kölner Zeitschrift für Soziologie und Sozialpsychologie* 37:96-115 (in German).

FSU. See Federalni statistický úřad.

Goltts, Grigorii Abramovich, and G. M. Lappo. 1981. "O proyektnoy razrabotke problem rasseleniya v Moskovskom regione i roli ekonomiko-geograficheskogo podkhoda." In *Rasseleniye i dinamika naseleniya Moskvy i Moskovskoy oblasti.* See Danilova et al., 1981.

Gorlov̀, W. N., Evgeniĭ Naumovich Pertsik, S. E. Khanian, and A. T. Kruschchev. 1983. "Economiko-geograficheskiye problemy razvitiya Moskovskogo regiona. Vestnik MGU." *Geografiya,* 1983, no. 2:27-32 (in Russian).

Gradow, G. A. 1971. *Stadt und Lebensweise.* East Berlin: VEB Verlag für Bauswesen (in German).

Hall, Peter G. 1977. *The World Cities.* 2d ed. London: Weidenfelt & Nicholson.

Hamilton, F. E. Ian. 1976. *The Moscow City Region.* Problem Regions of Europe. London: Oxford University Press.

HKSH. *See* Hungary. Kozponti Statisztikai Hivatal.

Hungary. Központi Statisztikai Hivatal. 1982. *Budapest Statisztikai Zsebkönyve 1981.* Budapest (in Hungarian).

Kaminski, M. (1974. "Auslastung des Netzes der polnischen Staatsbahnen im städtischen verkehr Warschaus." In *Internationale Transport Annalen.* Berlin: Ost (in German).

Kasarda, John D. 1980. "The Implication of Contemporary Redistribution Trends for National Urban Policy." *Social Science Quarterly* 61:373-400.

———. 1983. "Caught in the Web of Change." *Society* 21:41-47.

Kibalčhich, Oleg Alekseevich. 1983. "Voprosy strukturnogo isucheniya Moskovskogo stolichnogo regiona." In *Moskovskiy stolichnyi region: Vzaimodeystviye strukturnykj elementov.* Moscow: Akademiya nauk SSSR, Moskovskiy filial Geograficheskogo obschchesva SSSR (in Russian).

Kiehl, Klaus. 1985. "Budapest." In *Stadtenwicklungen in West- and Oesteuropa. See* Friedrichs, 1985.

Klassen, Leonardas Hendrick, Willem T. M. Molle, and Jean H. P. Paelinck, eds. 1981. *The Dynamics of Urban Development.* New York: St. Martins Press.

Kopp, Anatole. 1979. *Architecture et mode de vie.* Grenoble: Presses Universitaires (in French).

Lappo, G. M., A. Chekishev, and A. Bekker. 1980. *Moscow, Capital of the Soviet Union.* Moscow: Raduga.

Leszczycki, Stanislaw M., and Teofil Lijewski. 1977. *Polen. Land. Volk. Wirtschaft.* Hirt's Stichwortbücher. Vienna: F. Hirt (in German).

"Moskva za kol'tsevoy." 1984. *Stroitel̀stvo i arkhitektura Moskvy,* 1984, no. 8:28-29.

Musil, Jiri. 1977. *Urbanizace v socilistických zemích.* Sociologická kniznice. Prague: Svobada (in Czech).

———. 1980. *Urbanization in Socialist Countries.* Armonk, NY: Sharpe.

Osterwold, M. 1978. "Moskau." In *Stadtenwicklungen in kapitalischen und socialistischen Ländern. See* Friedrichs, 1978.

Palotas, Z. 1965. "Der Raumordnungsplan Ungarns." *Informationen des Instituts für Raumforschung* 15(6).

Poleszak, E., and W. Rakowski. 1980. "Niektóre prawidlowości wystepujace w zakresie przemieszczen ludności w strefie podmiejskiej Warszawy." *IGS Biuletyn* 23(2):64-77 (in Polish).

Preisich, Gabor. 1973. *Budapest Jövője*. Budapest: Múszaki Könyvkiadó (in Hungarian).

Saushkin, IUlian Glebovich, and Vera Georgievna Glushkova. 1983. *Moskva sredi gorodov mira: Ekonomikogeograficheskow issledovanie*. Moscow: "Mysl'" (in Russian).

Scitani lidu, domu a bytu: Hlavni mesto Praha, 1970. 1970. Prague: Mestska sprava ceskeho statistickeho uradu v hl. m. Praze (in Czech).

Scitani lidu, domu a bytu: Hlavni mesto Praha, 1980. 1980. Prague: Mestska sprava ceskeho statistickeho uradu v hl. m. Praze (in Czech).

Scitani lidu, domu a bytu: Hlavni mesto Praha, 1982. 1982. Prague: Mestska sprava ceskeho statistickeho uradu v hl. m. Praze (in Czech).

Sheshtokas, V. V., Vitautas Piiaus Adomavichius, and P. V. Yushkyavichius. 1984. *Garazhi stoyaniki*. Moscow: Stroyizdat (in Russian).

Shuper, V. A. 1983. "Ispolžovaniye teorii tsentralňykj mest pri razrabotke prognozov rosta gorodov v Moskovskom stolichnom regione. *Izvestiya Vsesoyuznogo geograficheskogo obschchestva SSSR*, 1983, no. 3:203-9 (in Russian).

SLDB. *See* Scitani lidu, domu a bytu.

Spengelin, Friedrich, et al. 1972. *Funktionelle Erfordernisse zentraler Einrichtungen als Bestimmungsgröbe von Siedlungs- und Stadteinheiten in Abhängigkeit von der Gröbenordnung und Zuordnung*. Bonn-Bad Godesberg: Schriftenerihe des BMBau 03.003 (in German).

Statisticheskoye upravleniye goroda Moskvy. 1972. *Moskva v tsifrakh (1966-1970 gg.)*. Moscow: Izdatelstvo "Statistika." (in Russian).

———. 1976-83. *Moskva v tsifrakh*. Moscow (in Russian).

Statistická ročenka 1981 Praha. 1981. Prague: Mestska sprava ceskeho statistickeho uradu v Praze (in Czech).

SUGM. *See* Statisticheskoye upravleniye goroda Moskvy.

Sykora, M. 1980. "K nekteryn zasadam koncepce bytove politiky v 7. petiletce." *Architektura ČSR*, 1980, no. 7:319-93 (in Czech).

TSitsin, Petr Georgievich. 1978. *Organizatsiia upravleniia ekonomikoi gorodskogo raiona*. Moscow: Ekonomika (in Russian).

U.S.S.R. 1980. *Metodicheskie ukazaniia k razrabotke gosudarstvennykh planov ekonomicheskogo i sotsialnogo razvitiya SSSR*. Moscow: Gosplan SSSR (in Russian).

VPN. See Vsesoyuznaya peripis naseleniya.

"Vsesoyuznaya perepis ňaseleniya." 1982. *Vestnik Statistiki*, 1982, no. 10:72-80.

Vydro, Morits IAkovlevich. 1976. *Naselenie Moskvy*. Statistika dia vsekh. Moscow: Statistika (in Russian).

Warsaw. Praesidium of the People's Council of the Capital City, Warsaw Town Planning Office, and Chief Architect of Warsaw, eds. 1970. *Warsaw 1970-1985*. Warsaw.

Warsaw. Town Planning Office. Chief Architect of Warsaw, ed. 1980. *Planning for Warsaw Agglomeration*. Warsaw.

Warsaw. Wojewódzki Urząd Statystyczny w m. st. Warszawie, ed. 1976-82. *Rocznik Statystyczny Województwo Stolecznego Warszawskigo*. Warsaw (in Polish).

WWUSW. See Warsaw. Wojewódzki Urzad . . . .

6

Great Cities of Eastern Asia

Yue-Man Yeung

Eastern Asia, broadly defined as that part of the western Pacific rim between Korea and Indonesia, is one of the fastest growth regions in the world in the postwar period, economically, demographically, and in terms of urbanization. As a reflection of the remarkable economic upsurge experienced in many of the countries under review, Japan, Hong Kong, and Taiwan witnessed a 13-, 11-, and 14-fold increase, respectively, in per capita gross national product in the two decades prior to 1980. Demographically, approximately 463 million people have been added to the region from 1960 to 1980, although in relation to the world population, Eastern Asia's share stood consistently at 32 percent in 1960 and 1980 (see United Nations, 1982, table 4). Over the same period, the region's urban population was augmented by some 195 million. While in 1970, the entire developing world had 72 cities with more than one million inhabitants, by 1980 Eastern Asia alone had 60 "million cities." Many were gargantuan urban agglomerations of many millions (figure 6.1).

Rapidly urbanizing Eastern Asia has engendered, particularly in the great cities, a myriad of problems which planners and policymakers are still having to come to terms with. The term *great city* follows conventional use as first employed by Ginsburg (1955) in his landmark statement on the phenomenon of functional dominance of primate cities in Southeast Asia. This chapter

---

AUTHOR'S NOTE: I am grateful to my many friends in the cities studied for counsel and providing data and reference material. I especially appreciate the timely research assistance of Yvette Luke Ying at the Centre for Contemporary Asian Studies. Margo Monteith, International Development Research Centre, Ottowa, helped graciously with repeated bibliographic searches.

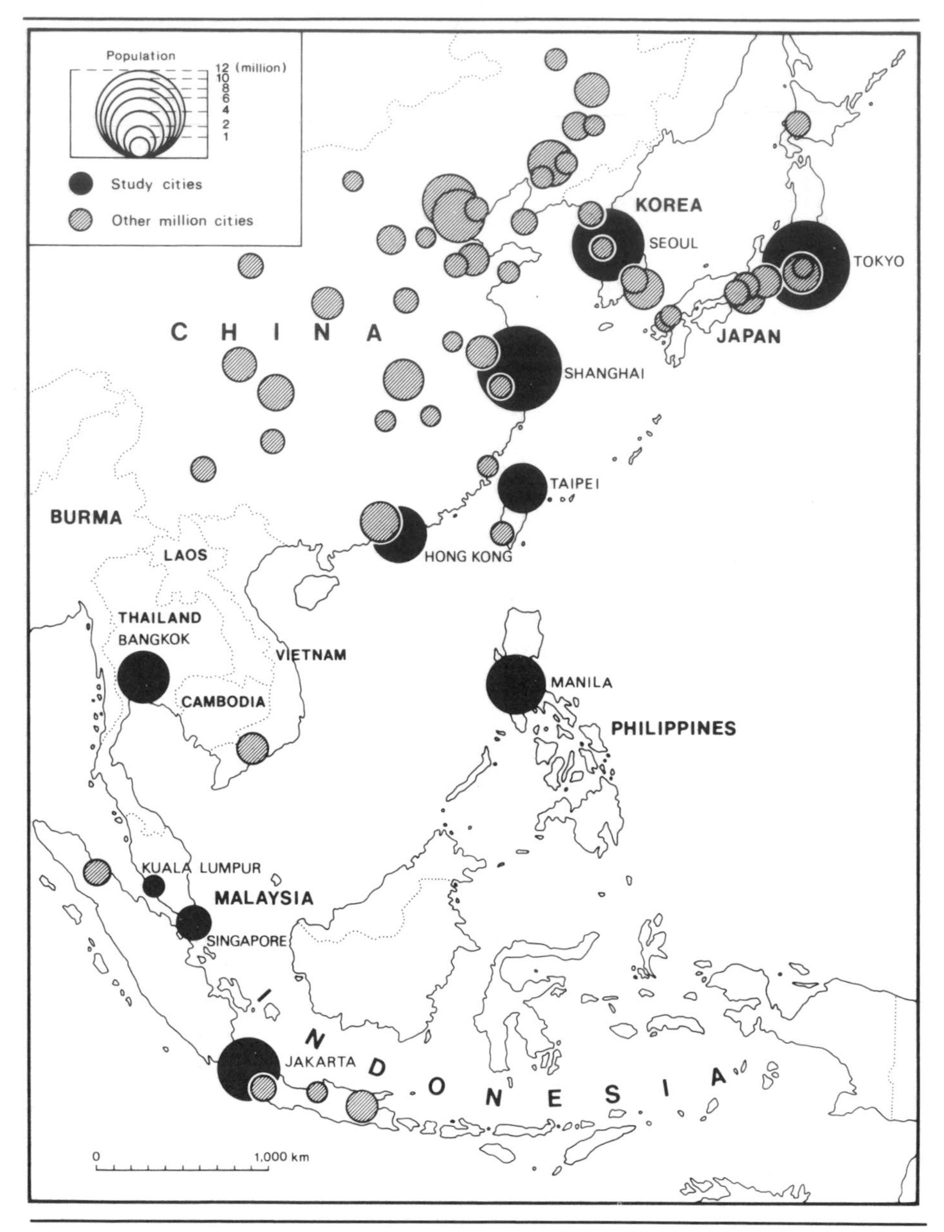

Figure 6.1: The Million Cities of Eastern Asia, c. 1980

focuses on one of the critical dimensions of development in Eastern Asia in the period 1960-80 by examining trends of urban growth, salient problems faced by the cities, and attempted solutions and policies. I then discuss policy options open to the cities before concluding with tentative constructions of urban scenarios to the end of this century.

To capture the heterogeneity and diversity of Eastern Asia, ten large cities—Shanghai, Hong Kong, Tokyo, Seoul, Taipei, Jakarta, Kuala Lumpur, Manila, Singapore, and Bangkok—were selected for analysis. Each represents the largest urban agglomeration of the countries under study and can, therefore, best personify the urban problems confronting the region. The diversity of the region is underlined by the inclusion of socialist and huge China, laissez-faire and city-state Hong Kong, several culturally Confucius-based countries, and the heartland of the Malay world. The study of urban problems and policies must heed this diverse background in one of the most culturally rich, historically ancient, and economically vibrant regions of the world.

Urban Growth Patterns

Measured by levels of urbanization, Eastern Asia can be roughly divided into two subregions. East Asia, represented by countries north of Hong Kong, is characterized, with the obvious exception of China, by a higher-than-world-average level of urbanization. This pattern was already clear in 1960, merely accentuated in 1980 (table 6.1). The doubling of South Korea's level of urbanization during this period is most noteworthy, mirroring the circumstances that led to its extraordinary urban and economic growth since the end of the Korean War. By contrast, Southeast Asia, save for the city-state of Singapore, may be depicted as one of the least urbanized regions, consistently below world average, even developing country average, in both 1960 and 1980. As I have maintained elsewhere (1976), an urban transformation has not taken place in this part of Asia.

Levels of urbanization, however, leave unspecified the magnitude of change of the urban population. Substantial urban growth can occur if the urban population grows in tandem with the total population, leaving the level of urbanization relatively unchanged. Singapore aside, this situation did happen in all countries in Southeast Asia, which saw a doubling of the urban population in the two decades since 1960 (table 6.2). The urban population has been growing slightly faster than the total population, but an urban transition as defined by Davis (1965) has not been in progress in Southeast Asia. In East Asia, on the other hand, the most dramatic has been the tripling of South Korea's urban population in the 1960-80 period, from 6.9 million to 21.1 million, an increase of 204.1 percent. Equally important, one should note that twice as many people in China lived in urban places in 1980 compared with 1960. Despite a modest increase in the level of urbanization, 128.5 million inhabitants were added to the Chinese cities.

At still another level of data aggregation, table 6.3 portrays population profiles of the ten cities included in this study. Because of highly varied definitions of metropolitan areas, both population and density figures must be read with extreme care. Overall, the figures should be seen as indicative of the magnitude of change and comparative reading of density figures must be

TABLE 6.1
Levels of Urbanization—Eastern Asia and the World, 1960-2000 (in percentages)

Place	*1960*	*1970*	*1980*	*1990*	*2000*
China	18.7	21.7	25.7	31.5	39.1
Hong Kong	89.1	89.7	90.3	91.4	92.6
Japan	62.5	71.4	78.3	83.0	85.9
North Korea	40.2	50.1	59.7	67.4	72.9
South Korea	27.7	40.7	54.8	65.2	71.4
Taiwan	58.4	62.4	66.8	–	–
Indonesia	14.6	17.1	20.2	25.2	32.3
Malaysia	25.2	27.0	29.4	34.2	41.6
Philippines	30.3	32.9	36.2	41.6	49.0
Singapore	77.6	75.3	74.1	75.0	78.5
Thailand	12.5	13.2	14.4	17.5	23.2
World	33.9	37.4	41.1	45.8	51.2
More developed regions	60.3	66.4	70.6	75.6	77.8
Less developed regions	21.4	25.2	29.4	36.8	40.4

SOURCES: United Nations (1982: Table 1); for Taiwan, Liu (1983: 3).

duly qualified. Even then, one may note that while Kuala Lumpur and Jakarta have more than doubled their populations, Seoul, Taipei, and Bangkok tripled theirs. Of the ten cities, Seoul grew the fastest, with an increase of 126 percent in the decade of the sixties, whereas Kuala Lumpur vastly expanded in the succeeding decade with a 108 percent growth. With respect to population density, all cities except Kuala Lumpur increased. Expansion of metropolitan area accounted for the decrease. In other cities population density increase sometimes is quite marked. As cautioned earlier, at first glance density figures could be misleading. For example, inclusion of sizeable rural areas in the metropolitan area of Shanghai resulted in incredibly lower density figures compared with other cities. In reality, Shanghai and Hong Kong are two of the most densely populated cities in the world. In 1983, Hong Kong and Shanghai recorded in their urban areas an overall density of 28 thousand and 27,762 persons per square kilometer, respectively, and the highest density in Shamshuipo of Hong Kong reached 165 thousand persons per square kilometer, while the Nanshi district of Shanghai similarly recorded 101,500 persons per square kilometer.

Rapid population increases in several study cities were accompanied by extensive boundary expansion with net gains of people and area. By annexing three areas in 1963 and two in 1973, the Seoul administrative area reached its present size of 627 square kilometers. Similarly, Taipei quadrupled its city area to 272 square kilometers in 1968, with corresponding expansion of its metropolitan area, both of which have since remained unchanged. The most recent areal redefinition occurred in Kuala Lumpur in 1974 when its area was more than doubled to 243 square kilometers as it became the Federal Territory. Just prior to the study period, in 1958 Shanghai also gained about ten times its area to reach its present size by acquiring jurisdiction over ten adjacent

TABLE 6.2
Total Urban and Largest City Populations, Growth Rates, and Share of Urban Population in Largest City—Eastern Asia, 1960-2000

	Country					Largest City							
	Urban Population (millions)			Growth Rate[a] (percentages)		Population (millions)			Growth Rate[a] (percentages)		Share of Urban Population[a] (percentages)		
Country/City	1960	1980	2000	1960-80	1980-2000	1960	1980	2000	1960-80	1980-2000	1960	1980	2000
China/Shanghai	127.5	256.0	491.9	100.7	92.2	7.7	15.0	25.9	94.8	72.7	6.0	5.9	5.3
Hong Kong	2.7	4.6	6.5	68.4	40.0	—	—	—	—	—	—	—	—
Japan/Tokyo	58.8	91.3	111.1	55.3	21.7	10.7	20.0	23.8	87.6	19.0	18.2	22.0	21.4
North Korea/ Pyongyang	4.2	10.7	19.9	152.4	86.0	0.6	1.3	2.2	102.0	74.6	15.0	12.0	11.3
South Korea/ Seoul	6.9	21.1	36.2	204.1	72.0	2.4	8.5	13.7	254.4	61.2	34.1	40.3	37.9
Taiwan/Taipei	6.3	11.9	—	88.2	—	1.1	2.2	—	102.1	—	17.5	18.8	—
Indonesia/ Jakarta	14.3	29.9	64.1	109.9	114.3	2.9	7.0	14.3	141.4	104.3	20.3	23.4	22.3
Malaysia/ Kuala Lumpur	2.1	4.1	8.8	100.5	114.1	0.4	1.1	2.6	198.9	130.7	18.0	26.9	28.9
Philippines/ Manila	8.5	17.8	37.8	109.3	112.0	2.3	5.7	10.5	147.6	84.2	26.9	31.8	27.8
Singapore	1.3	1.8	2.3	39.6	31.5	—	—	—	—	—	—	—	—
Thailand/ Bangkok-Thonguri	3.4	6.8	15.9	98.5	135.2	2.2	4.9	9.9	126.4	102.0	65.1	72.0	62.3

SOURCE: For Taiwan, Liu (1982:4; 1983:3); United Nations (1980: Table 48; 1982: Tables 2 and 8).
a. Computed from source tables, not from rounded figures presented here.

TABLE 6.3
Population Profiles of Selected Great Cities—
Eastern Asia, 1960-1981

City	Year	Population (millions)	Growth Rates Decadal (%)	Growth Rates Annual (%)	Area (km²)	Population Density (per km²)
Bangkok	1960	1,577	+37	+3.2	–	–
	1970	2,157	+118	+8.1	–	–
	1980	4,697			1,565	3,001
Hong Kong	1961	2,668	+30	+2.6	184	14,499
	1971	3,457	+23	+2.1	197	17,547
	1981	4,241			198	21,418
Jakarta	1960	2,907	+56	+4.6	–	–
	1970	4,546	+43	+3.6	604	7,527
	1980	6,503			656	9,914
Kuala Lumpur	1967	450	+0.3	+0.03	93	4,839
	1970	452	+108	+7.6	93	4,857
	1980	938			243	3,860
Manila	1960	2,462	+61	+4.9	636	3,879
	1970	3,967	+49	+4.1	636	6,237
	1980	5,926			636	9,317
Seoul	1960	2,445	+126	+8.5	268	9,125
	1970	5,536	+51	+4.2	613	9,032
	1980	8,367			627	13,344
Shanghai	1962	10,225	+1	+0.2	5,895	1,735
	1970	10,368	+5	+0.5	–	–
	1980	10,889			6,059	1,797
Singapore	1960	1,646	+26	+2.3	582	2,828
	1970	2,074	+16	+1.5	586	3,539
	1980	2,414			618	3,906
Taipei	1960	1,546	+79	+6.0	884	1,749
	1970	2,769	+58	+4.7	1,070	2,588
	1980	4,381			1,070	4,095
Tokyo	1960	9,684	+18	+1.7	2,023	4,787
	1970	11,408	+2	+0.2	2,141	5,328
	1980	11,618			2,156	5,389

SOURCE: Variously derived official statistics.
NOTE: Wherever possible, metropolitan areas were used in constructing profiles.

counties. This expansion was the last of a series of administrative boundary changes affecting that urban complex since 1949.

Hand in hand with these areal expansions have been important changes in the internal structure of some cities. In almost every case, there has been very large extension of built-up areas. Tokyo may be considered an extreme case in which the daily commuting distance reached as far as 40 kilometers in 1970 (Hall, 1977:229). This distance was extended by a further 10 kilometers in

1984. In part as a result of the successful development of commuting transport links, the daytime and nighttime populations in the central parts of Tokyo have become highly imbalanced. In 1980 the daytime population in the three central wards of Chiyoda, Chuo, and Minato exceeded the nighttime population by 6.8 times (TMG, 1984b:26). In Seoul, imbalanced development of another type has occurred. Because of its relative proximity to the Demilitarized Zone and North Korea, the area north of the Han River in Seoul has developed more slowly in recent years. Consequently, accelerated development south of the Han resulted in a doubling of the population in the six years between 1973 and 1979, as opposed to a stable or declining population north of the river (Kwon, 1981a:320). Over the past 20 years, some study cities have developed mass rapid transit systems, leading to improved areal and functional integration within their administrative territories. In this respect, the opening of a subway system in Hong Kong 1 October 1979, along with the vital cross-harbor tunnel open since 3 August 1972, has greatly assisted in integrating what previously had been a fragmented and complex urban structure (Lo, 1972). In Seoul, too, the opening of a subway system in 1974 has progressively improved travel times and the general traffic situation.

Finally, the study cities have grown by natural increase as well as rural-urban migration. Nowhere is the latter phenomenon more poignant than in Seoul, where urbanward migration accounted for 82 percent of the total population increase in the period 1966-70. Since then, the contribution of migration has been much reduced, thanks to a host of population diversion and decentralization strategies (Kwon, 1981b:80). In Taipei inmigration accounted for 51 percent of the total population increase between 1968 and 1973 (Tsay, 1982:23). However, the migration factor has become less significant as a component of population increase since 1970, so much so that, by 1977, net migration contributed only 8 percent of Taipei's population increase (Hsu and Pannell, 1982:29). Elsewhere, there were also signs that the cities under review grew more by natural increase than inmigration in recent years. Tokyo's is probably the only case where since 1967, population outflow has exceeded inflow. In 1980, Tokyo was the only prefecture recording a population decrease (TMG, 1984a:68).

Whatever the sources of growth dynamism, the cities studied are and will remain, the most important urban centers in their respective countries. Their pivotal roles nationally and internationally in spearheading development are clearly evidenced in their disproportionate share of the total urban population in their countries. Greater Bangkok and Seoul, for instance, constituted 72 and 40 percent, respectively, of their total urban population in 1980 (table 6.2). Dominance of the study cities over the economic, social, political, and other spheres in their countries is brought home vividly in the random indicators compiled in Table 6.4. While their importance in national life is indisputable, they have, as well, a greater-than-usual share of urban problems. Concentrated development and population have magnified many of these problems to crisis proportions, which the next section addresses.

TABLE 6.4
Selected Indicators of Functional Dominance in Great Cities—Eastern Asia, 1972-1984

City		*Measures of Centrality*
Shanghai	Accounted for	12.5% of national industrial output
		11.0% of national heavy industrial output
		14.0% of national light industrial output
		5.8% of state industrial staff and workers
		10.3% of new state industrial investment
	Produced	30% of China's television sets and wristwatches
		50% of China's cameras
	Conducted	23% of China's foreign trade
	Had	48 universities and higher educational institutions with 70,000 students and 250 research centers with about 240,000 scientific and technical personnel
Tokyo	Accounted for	60% of Japan's top business leaders
		60% of total invested capital
		33.3% of deposited banking accounts
		33.3% of university graduates
		21% of tertiary industry income
		27% of retail and wholesale sales
		15% of sale of manufactured products
		15% of working population in secondary industry
		33.3% of department store sales
		25% of entertainment admissions
		20% of Japan's universities
		50% of university students
		60% of heads of industrial and commercial enterprises
Seoul	Housed	78% of headquarters of business firms
		90% of large business enterprises
	Accounted for	28% of the nation's value added by manufacturing
		25% of the nation's total manufacturing employment
		32.3% of all manufacturing establishments
		65% of all loans and deposits
		55% of all colleges and universities
		50% of all medical doctors and specialists
		50% of the national wealth
		27.9% of South Korea's GNP (1977)
Bangkok	Accounted for	26.8% of Thailand's GNP
		77.7% of Thailand's banking
		48.8% of transport and communication installations
		33.3% of the country's manufacturing industry
		33.3% of the country's construction industry
		33.3% of foreign trade
		34% of all motor vehicles
		56.5% of electricity consumed
		74% of telephones installed
		8 of Thailand's 12 universities

SOURCES: For Shanghai, *Business China* (1982: 44); Henderson (1984: 43); for Tokyo, Honjō (1972: 12; 1975: 346); Tokyo Metropolitan Government (1984a: 31); for Seoul, Chu (1980: 438); for Korea, South (1982: 28); Kwon (1980: 35; 1981b: 74-75); for Bangkok, Kanjanaharitai (1981: 203-4).

A Mosaic of Urban Ills

The litany of problems in the great cities of Eastern Asia is highly similar to that in other parts of the Third World: chronic housing shortages, traffic snarls, deteriorating basic services, increasing social inequality, omnipresent squatter settlements and slums, widespread poverty, limited employment opportunities, and so on. As each of the study cities differs in the extent and details it suffers from these afflictions, what follows will be synopses of each urban situation.

Shanghai

The largest socialist city in the world, Shanghai epitomizes the contradictions inherent in the development of Chinese cities since 1949. It suffers from acute housing shortages, poor infrastructure and public services, and environmental pollution (Wu, 1984). About half the population lives in lane housing—usually of two or three stories, lacking basic facilities, and located in alleys or small lanes (*China Daily*, 1984), and another 16 percent live in slums built of temporary materials. Average per capita housing space was 4.3 square meters in the early 1980s, only a minor improvement on living conditions in 1949. Shanghai is probably the most crowded city in China, with only 24.6 square meters per person in the central city area (Lee et al., 1983). The housing crisis is exacerbated by uncontrolled and indiscriminate industrial development. Between 1962 and 1980, about 1.6 million square meters or 15 percent of the total residential floor area was lost to factories (Tao, 1981). This unfortunate situation resulted from the long-accepted policies of "production first, livelihood second" and "construction where there is room."

Similarly, the ideological bias against investment in "nonproductive sectors" since 1949 has led to the sad neglect of infrastructure development. Only 16 percent of the population in the urban districts is served by sewers, another 16 percent by septic tanks, the rest by night soil collection. Extreme traffic congestion resulted in the average traffic speed being halved from 30 kilometers per hour in 1949 to 15 in 1979 (UNDP, 1979).

Serious cases of water, air, and noise pollution have also been reported. In the central city, daily industrial waste and domestic sewage amount to 500 thousand tons, the majority of which is discharged into the Huangpu River untreated (Yan and Tang, 1984:71-72). Measured by sulphur dioxide, nitrogen dioxide, and suspended particulates, the standard acceptable air quality standards have all been exceeded. Poor air quality is not confined to the urban area as acid rain is known to affect large parts of the surrounding rural area (Yan and Tang, 1984).

Finally, some 200 thousand youngsters leave middle schools every year, in addition to returned youths from the rural areas. Because of the limited absorptive capacity of the economic structure, many have to wait for years before jobs are assigned.

Hong Kong

Situated at the doorstep of China as a borrowed place on borrowed time, Hong Kong is a prime example of laissez-faire capitalism. Its remarkable postwar economic growth has been matched to a certain degree by significant progress in urban management and infrastructure development. Public housing, for example, accommodated over 2.4 million inhabitants or 45 percent of the total population in 1984. Each year close to 180 thousand people are allocated new public housing. These massive efforts still lag behind demand, for even in 1984, there was still an estimated total of 500 thousand living in squatter colonies, as opposed to 275 thousand in 1973 when the Ten-Year Housing Program was launched (Hong Kong Housing Authority, 1984). In the peak 1978-79 property boom, the annual rate of price increase for various sizes of private housing was in the region of 70 percent (Kwok, 1983:335).

Planners and administrators find themselves in a quandry, for carefully drawn up plans are often overtaken by events beyond their control. In particular, successive waves of legal and illegal immigrants from China have rendered many development plans short of meeting targets. Nevertheless, over the years substantive improvements have been made in basic service provision which are no longer critical issues in most areas. A recent four-district study showed that it was largely for improving the quality of life that local leaders organized themselves (Lau et al., 1983).

Extremely high population densities in the urban area separated by the harbor imply thorny traffic problems. Between 1966 and 1981, the number of cars registered soared from 93,000 to 310,916. In the period 1976-81 private cars increased by 13 percent per year but road mileage increased by only 2 percent (Fong, 1984; C.-K. Leung, 1984). Although the population is heavily dependent on public transport, extensive car ownership has created problems for transportation planners. In land-scarce Hong Kong, the price of land until recently rose precipitously.

Finally, in recent years Hong Kong has been plagued by a rising incidence of crime, drug addiction, and other social problems symptomatic of a progressively sophisticated urban society.

Tokyo

By certain definitions, Tokyo is arguably the largest city in the world. It occupies 1 percent of Japan's land and constitutes about 10 percent of its population. Described as a city of paradoxes, Tokyo is a symbol of Japan's postwar economic miracle, with by far the highest per capita gross national product in Asia and every evidence of high-technology urban sophistication. Yet its inhabitants do not generally enjoy the same level of amenities available in lesser Asian cities.

Tokyo is probably the most innovative with urban solutions, but many could not be implemented for lack of funds (Blair, 1974:34; Hall, 1977). Housing, commuting, traffic, and pollution rank as the major problems besetting the city. Ten years ago, some 28 percent of the families lived in substandard, overcrowded houses and 34 percent complained about housing problems (Honjo, 1975). The existence of over one million wooden rental apartments reflects the gravity of the housing situation (TMG, 1984a:26). Overcrowing and high land prices have been pushing home-seekers farther from the city, causing rapid suburbanization and large-scale commuting.

Fantastic advances have been made in mass transport but planners cannot solve traffic jams caused by three million automobiles (1980). Every day over 2.2 million people commute from neighboring prefectures to offices and schools in Tokyo. Concomitant to this pattern of development is the "emptying out" of the central city area at night.

In 1981 only 76 percent of the population was served by a sewerage system. Pollution of every type remained serious, although air pollution caused by sulphur dioxide concentrations gradually decreased after peaking in the mid-1960s (TMG, 1984b:155).

After the disastrous fire of 1923, planners in Tokyo have been very conscious of how vulnerable the city is to natural disasters. Yet not a great deal has been accomplished in this direction. The biggest challenge to the city is to lure more people away from Tokyo so that the feared paralysis of basic services will not materialize.

Seoul

Richard Meier once described Seoul as the product of the most intense and compressed process of urbanization in the world, with continuous inmigration flooding the city for 16 years. Some 22 percent of South Korea's population lived in Seoul in 1982, compared with 5 percent in 1955 (Kwon, 1981b:73). Unabated urban growth has led to intolerable population and traffic congestion, serious industrial pollution, environmental deterioration, housing shortages, and inadequate public services (Kim, 1980; Kwon, 1981a). As a measure of crowding, there were 2.51 persons per room in Seoul in 1975, hardly an improvement on the 1970 figure of 2.67 (Kwon, 1981a:308). *Panjachon* or squatter settlements predominate the hills surrounding Seoul. One reason for the difficulty in housing provision is the astronomical increase in land prices that skyrocketed by almost 28 times from 1963 to 1979 (Kwon, 1980:38).

Being a monocentric city with a high concentration of activities within the Central Business District (CBD), Seoul has the problem of unbalanced provision of public services in different parts of the city. This imbalance is further aggravated by the uneven development, since 1970, across the northern and southern part of the Han River. Considerable residential relocation to the south of the river has taken place, unaccompanied by the provision of community and basic facilities (Kwon, 1980, 1981a).

One inevitable consequence of growth has been separation of home and work place. In the area within two kilometers of the CBD, daytime population is more than four times the resident population. This situation has brought further strains on the transport system which has to move 500 thousand people daily in and out of Seoul (Kwon, 1981a:318). The opening of the subway system since 1974 has helped mobility but a threefold increase in automobiles since then has counteracted any improvement made in urban traffic. Between 1975 and 1983 the number of automobiles soared from 83,661 to 253,647 (SMG, 1983:88).

Taipei

In Taipei one thorny problem has been the speculative land price rise. Between 1952 and 1976, the average land price increased 4.3 percent in Taiwan, as contrasted with a 265-fold spiral in Taipei. During the same period, the average commodity price increased by 3 times in Taiwan versus 76 times in Taipei (Tang, 1982). The increase in land price, far surpassing that in commodities, has been the result of primarily speculative investment and improper planning control. Far worse, the government has no mechanism to capture part of the land price increase, thus providing an impetus to land speculation. Another problem is restriction on land use conversion. For example, the government permits conversion only of low-grade agricultural land to other uses, but restricts similar conversion of high-grade arable land. Even then, in the period 1956-67, some 628 hectares of vacant or arable land, constituting 9.5 percent of Taipei's land area, was converted to residential, commercial, or industrial uses (Hsu and Pannell, 1982:32). Consequently, average land price already accounts for over half of housing value (Chen, 1984:166). This situation has led to a very tight housing situation saddled with problems of insufficient supply, concentrated ownership, and a deteriorating living environment. In 1975, the total housing stock in Taipei was owned by 5.6 percent of its inhabitants, with 11,187 each owning at least three dwelling units accounting for at least 48,999 dwelling units (Tang, 1982). Public housing is negligible, totalling merely 31,794 units in 1981.

The hodgepodge of urban development has also given rise to a confused transport situation in which road construction has lagged far behind the growth of motorized vehicles. Between 1960 and 1980 the number of automobiles exploded from 5,506 to 179,106, while that of motorcycles soared from 6,704 to 350,921. The proliferation of small vehicles since the mid-1950s has been a real problem. Widespread vehicle ownership has contributed directly to air pollution, particularly that of suspended particulates and sulphur dioxide. The reported incidence of water and noise pollution has also been high. In a recent survey in Taipei, 82.4 percent of the respondents regarded noise pollution as very serious (Tang, 1982).

Jakarta

Often described as a vast conglomeration of *kampungs* (villages), Jakarta is a city with monumental obstacles to overcome. It sorely lacks infrastructure, housing, jobs, and a sound tax base. Within the urbanized area of the city, as much as 70 percent of the residential space is occupied by *kampungs*, providing shelter for some 80 percent of the total population (Erni and Bianpoen, 1980). Most of the housing units within this residential environment are temporary or semipermanent structures. Various sources have estimated that around 60 thousand new dwelling units would be required annually to meet demands, whereas the Jakarta Office of Public Works approved in a year only 5 percent of this number of new housing units (Sethuraman, 1976:33-34).

The sad state of urban services is revealed by 80 percent of all dwelling units having no electricity and 70 percent having to depend on private or public wells, as opposed to 30 percent using piped water—part of which may be purchased from vendors. A recent survey showed that residents cited supply of drinking water, access to telephones, family planning services, and recreational facilities as highly unsatisfactory. Less dissatisfying were waste disposal, sewers, and street lighting, while shopping facilities, public transport, and religious services fared better (Krausse, 1982:61).

Economically, Jakarta is ill-prepared to provide employment for its hordes of rural immigrants. One study found that nearly 43 percent of the economically active population was employed in the informal sector where low requirements of capitalization, skills, and education permit the employment of many (Moir, 1978). The continual influx of these migrants has not been welcomed by planners and administrators. Some measures they adopt, such as becak-free zones where muscle-powered transport is forbidden and banning hawkers in certain busy areas, are directly inimical to the livelihood and existence of low-income groups.

Finally, despite its status as the National Capital Special Region, Jakarta has a narrow tax base to carry out most of the needed urban functions. Although the central government's contribution to Jakarta's revenue increased markedly by the early 1970s, in the final analysis those who make the laws and design development policies will have to agree to tax themselves at a higher rate if local fiscal power is to increase (Specter 1984b:27).

Kuala Lumpur

Called the "superliner" city, Kuala Lumpur has witnessed progressive and intensive expansion westward toward Port Klang, some 35 kilometers away, over the past three decades. The Klang Valley is now strewn with "dormitory suburbs" with Kuala Lumpur acting as the heart of the conurbation. Rapid and less-than-careful development has increased the risks of erosion, flooding, and water pollution, the last two phenomena especially affecting Kuala Lumpur (Aiken and Leigh, 1975).

The city has since the Japanese Occupation period had to deal with a squatter problem. In 1979, 240 thousand inhabitants—22 percent of the population—lived in squatter settlements (Leong, 1981:273). About 40 percent of these squatters were Malays and 77 percent of them were inmigrants. This situation has given a different complexion to Kuala Lumpur, which is beginning to tip in favor of *bumiputras* (Malays) in population composition, government spending, and urban redevelopment. Since its implementation in 1970, the effects of the New Economic Policy are most clearly seen in the Federal Territory with increasing Malay participation in every facet of life. Some local politicians even went so far as to assert that government development plans were transforming Kuala Lumpur into a city in which the Chinese majority was becoming a minority (Specter, 1984a:24). Low-cost housing is being provided by the government but during the Third Malaysia Plan (1976-80), only 58 percent of the plan target was met. Although the government still takes full responsibility for providing low-cost rental housing, it has encouraged the private sector to assume a greater role in providing low-cost dwellings for sale at a maximum price of M $25,000 (U.S. $10,800).

Traffic congestion is most serious in Kuala Lumpur's older areas with the doubling of automobile ownership between 1973 and 1980. Despite only 30 percent of the population's being dependent on public transport, the government is committed to the twin transit system of aerobus and light rail transit to solve its traffic problems (Specter, 1984a).

Manila

Metropolitan Manila, created in 1975 to cover an area of 636 square kilometers consisting of 4 cities and 13 municipalities, has been a bold management innovation in the Philippines in attempting to come to grips with problems this primate city faces. The official diagnosis of major problems besetting the city is as apt today as a decade ago when this list was compiled:

- Scattered indiscriminate subdivision and urban development
- Serious traffic and transport bottlenecks and congestion
- Inadequate and uncoordinated transport services
- Coexistence of very high densities in core areas with haphazard low-density sprawl in the suburbs
- Deficient water supply, sewage, and drainage systems
- Extensive annual flood damage (1972 was the most disastrous)
- Land and housing shortages reflected by formidable slums and squatter areas
- Lack of community facilities, open spaces, and recreational areas
- Environmental pollution
- High land values and inordinate land speculation (NEDA, 1975).

Identifying problems is one thing; solving them is quite another. Basic service needs of the population outstrip the capability of government agencies to supply them. In sewer facilities, for instance, 46 percent of Manila is served by an antiquated system installed in the early 1900s. Much of the sewage is being discharged untreated into waterways that eventually drain into Manila Bay. Flooding and drainage also still pose serious problems despite government programs in flood control. A recent study of low-income communities in Metro Manila found only 15.2 percent of surveyed households had water directly piped into their houses and 17 percent depended on water peddlers for their supply (Aquino, 1983). The National Housing Authority designed different programs to improve housing for the poor but 28 percent of the population still lived in squatter settlements (Viloria, 1981:287).

Singapore

Widely known as the garden city of Asia, Singapore had by the early 1980s, solved many of its urban problems. In 1983, 75 percent of the total population lived in public housing, unemployment was negligible, basic services were available to the vast majority of the population, and the economy was healthy. As the population becomes more affluent, problems of a social nature take a higher priority. These include ways to provide for an aging population, better recreational facilities, and community participation.

There is, by and large, general acceptance of the high-rise, high-density environment, despite critics' questioning the pace of development (Soon, 1969). The gradual disappearance of the old Chinatown west of the Singapore River, for example, has been lamented by many Singaporeans as well as romanticized foreign visitors. In place of the former shophouses and low-rise, tile-roofed structures, the new environment is one of tall, modern, glass-facaded office buildings, hotels, and apartments.

The only major problem the city faces is that of ever-increasing traffic caused by the rapid rise in car ownership. Between 1961 and 1975 automobile ownership more than doubled. Since 1975, when a congestion price scheme was put into practice, the situation has much improved. Overall, Singapore is perhaps the one city in which the usual problems afflicting the other cities are within the most manageable dimensions.

Bangkok

Finally, although Bangkok is endearingly named the city of angels, its citizens have to fight daily battles with horrendous traffic jams, a polluted environment, and poor housing. Traffic jams in Bangkok are arguably among the worst in Asia. During peak hours cars run at 12 kilometers per hour and buses at 9. It has been estimated that costs of traffic congestion were 30 million baht per day in 1978 (Medhi Krongkaew and Trongudai, 1983:39-40;

also Kanjanaharitai, 1981). It was also estimated that if congestion could be reduced to half, savings on gasoline alone could amount to at least 750 million baht a year. A 1975 survey showed that some of the main roads had air pollution levels more than twice the standard danger limits (Muqbil, 1982).

Water and noise pollution are also serious, while waste disposal is no less easy for a city of almost five million. Existing facilities can handle half of the waste.

About 25 percent of the population lives in squatter settlements, the biggest of which is in Klong Toey. Moreover, in 1982 the National Housing Authority identified 410 slum areas involving 551 thousand inhabitants—12 percent of the total population. Planning controls in Thailand are very weak, so that the Bangkok City Planning Department can only advise rather than implement. It takes an individual act of parliament to obtain land for a specific public purpose (Kanjanaharitai, 1981:218). Consequently, development in Bangkok is practically free from any planning intervention, but is guided by the alignment of public infrastructure, land values, and private initiative. The resultant pattern of land uses is highly mixed, lacking definable areas of special activities. Related to this style of planning is the increasing monopolization of the land market by real estate developers and professional land speculators (Tanphiphat, 1982).

Most serious of all, Bangkok is a sinking city of its own making. The excessive use of deep artesian wells, drawing about 1.15 million cubic meters per day, has caused soil subsidence by as much as 0.6 meters in the southeastern parts of the city and 0.4 meters in the central areas. The introduction of multistory buildings of heavy construction also contributes to this land subsidence threat. Because Bangkok lies virtually at sea level, the smallest rate of sinking could result in catastrophe (Donner, 1978:771; Kanjanaharitai, 1981).

Urban Strategies and Innovations

A generalization that may be drawn from the foregoing paragraphs is that the large cities of Eastern Asia have grown too quickly and too big, not out of any deliberate urbanization policy but as the consequence of population movements and dynamics over which the cities did not have full control. In this section I review briefly the important urban strategies that have been implemented, with an emphasis on those considered innovative in the region. Rather than follow a city-by-city format, urban policies will be discussed comparatively across cities.

Decentralization Policies

The metropolitan governments under study have for years been vexed at their rapid growth and have devised variants of decentralization policies to slow, divert, and control future growth (Simmons, 1979). First, we note two rather radical policies adopted in China and Jakarta, with quite different results.

In China a policy called *hsia fang*, literally the sending down of secondary school graduates to rural areas, provides the first example of government-sponsored, large-scale population transfer from urban to rural areas. Throughout China this movement involved 10 to 15 million individuals in the period 1969-73 (Prybyla, 1975). The movement could only be successful given the Chinese government's strong administrative controls (travel permits and ration cards), massive media propaganda, and political exhortation. Shanghai was an active participant in this movement, a decision that no doubt accounted for stabilization of its population in the region of ten million since 1959.

Even more desperate has been Jakarta's declaration, in August 1970, of a "closed city" policy to new jobless settlers. Immigrants are required to show evidence of employment and housing accommodation before they are issued residence permits. They are further required to deposit with the city government for six months the equivalent of the return fare to their point of origin. The drastic measure was intended primarily for its psychological impact rather than physically to prevent people from coming to Jakarta (Critchfield, 1971). Evidence to date indicates that the policy has produced mixed results. On the one hand, Suharso et al. (1975) have suggested that there has been no noticeable diminution of the rate of growth from migration. An average of 648 migrants continued to enter Jakarta every day. On the other hand, Papanek (1975) has found that since a high "price" is attached to the residency card, the lowest income groups tend to ignore the regulation altogether, whereas others who need it contribute toward bribery and corruption.

Relocation

In terms of actually decentralizing metropolitan growth, at least four strategies may be cited. First is the sequence of structural, land-use, and industrial relocation plans adopted by Seoul since 1964 to decentralize growth in the metropolis. Each plan was found wanting in fully meeting the targets of controlling growth, to be succeeded by more encompassing designs, culminating in the Capital Regional Plan adopted in 1981 for ten years. The plan involved 30 local and provincial governments in the Capital Region affected by the regional land-use plan, but the absence of a single metropolitan government may present a real obstacle in implementation (W. Kim, 1983).

In 1973, in desperation to slow the tide of inmigration, new industries in Seoul were subject to a "new citizen tax." Property and acquisition taxes payable by new firms were several times higher than those for established industries. In addition, differential school fees and new college entrance examinations for candidates outside the Seoul area were introduced (Hwang, 1979:8). Furthermore, as most migrants have been attracted by jobs provided by industries, an active industrial relocation has been pursued with vigor since 1977. Relocatable industries were identified by size of firms, type of production, interindustry linkages, labor availability, and other factors. It was reasoned that spatial rearrangement of manufacturing is a precursor to population dispersal (Kwon, 1981b). Consequent upon these decentralization strategies, Seoul's population growth slowed to 4 percent per year by 1975, as compared with 7.3 percent in the period 1960-66. Likewise, Seoul's share in the number of industrial firms and employees, university enrollment, and gross regional product has declined appreciably (Hwang, 1979:12-13).

New Towns

Second, several study cities have chosen to develop new towns, satellite towns, and small/medium towns to divert metropolitan growth to these areas. Seoul's first attempt at decentralization through a satellite town was the establishment of Sungnam, 25 kilometers to the south, in the late 1960s to relocate squatters from the city. However, the new town did not have sufficient industries to provide jobs, with the result that many squatters moved back to or commuted to Seoul. Realizing Sungnam as a financial drain, Seoul quickly imposed the new town on the neighboring province of Kyonggi in 1973 (Kwon, 1981a:324). The satellite town approach, nevertheless, was not forsaken, for in the basic guidelines for the capital region in 1971, the development of ten satellite towns was proposed in areas within a radius of 30 kilometers from Seoul. More recently, in 1978, five medium-sized cities with populations between 200 thousand and one million were designated as priority investment centers to provide alternative migration destinations to Seoul (S.-U. Kim, 1980:66).

In Shanghai, too, a policy of satellite towns, designed as balanced and self-contained communities, has been in practice since 1958 in efforts to disperse population and industries. As conceived by Shanghai's planners, the optimal distance between the central city and satellite settlements lies between 20 and 70 kilometers, far enough from being absorbed by Shanghai and near enough to retain needed mutual coordination. These towns range in population between 50 and 200 thousand (Fung, 1981:289-90). Similarly in Tokyo, satellite towns were planned at distances 17 to 45 miles from central Tokyo to absorb decentralized population and employment, but these did not develop for the most part into self-contained communities without large-scale commuting (Hall, 1977:232-25).

On a smaller scale, the city-states of Hong Kong and Singapore have also decentralized population and employment from congested core areas through development of new towns since the early 1960s. In both cases public housing has been a salient component in this development and responsible for improving living conditions in these communities. By 1981, 18.7 percent of Hong Kong's population lived in the new towns (W. T. Leung, 1983:211). While these new towns allow room for new physical and structural designs, estate management has become more challenging and complex (Fung, 1983; Lim et al., 1983).

New Capitals

Third, there have been proposals to build brand new capital cities, much along the model of Brasilia, Canberra, and Islamabad, to relieve pressure on existing ones. During the reign of President Park Chung Hee in South Korea, for example, a plan was broached to establish a new capital city to reduce Seoul's sociopolitical attraction by removing from it central administrative and executive functions. After the abrupt turnover of political leadership in 1979, the idea of a new capital city no longer received political support (Kim, 1981:68). The latest proposal for twin cities came surprisingly from Malaysia, where the prime minister recently revealed plans for constructing a new city at Janda Baik, at present a quiet retreat 30 kilometers from Kuala Lumpur. Initially the new city will be under the administration of Kuala Lumpur but will grow fast to become a city of 100 thousand by 1990 and 500 thousand by the year 2000 (Specter, 1984a:24-25).

Green Belts

Finally, if construction of new urban communities is viewed as a positive strategy to channel new growth, designation of green belts may be regarded as a measure to restrict uncontrolled metropolitan sprawl. Several cities under review have adopted this strategy to limit undesirable growth. In Tokyo a seven-mile wide green belt was delineated in the 1956 development plan right beyond the defined built-up area of ten miles in all directions from the Tokyo Central Station. This belt was abandoned a decade later as it was unable to hold development, and was replaced in 1965 by a new suburban development area extending beyond 30 miles from the city center (Hall, 1977:236). In Seoul the green belt was formally instituted in 1972 through revision of the City Planning Act to contain its physical growth. By 1982 the total area amounted to 369.5 square kilometers, out of which 66.5 percent was development-restricted (Park, 1981:35; SMG, 1983:50). Likewise in Bangkok, green belts covering agricultural land 12 kilometers wide along the eastern and western flanks of Bangkok's built-up areas were specified in 1981 prohibiting any building more than 100 meters from either side of existing roads (Tanphiphat, 1982:36).

Somewhat different in purpose but not in function has been the preservation of a vegetable production belt around the central city and satellite communities in Shanghai (Fung, 1981:292-94). Food production for the city has thus been successfully safeguarded. In 1959 Shanghai produced three-quarters of the fruits and 97 percent of the vegetables its inhabitants consumed (White, 1981:260).

Policies for Urban Well-Being

Having reviewed macroantimetropolitan growth policies, I now shift attention to some of the more important policies designed for improving urban well-being.

Shelter

With respect to shelter—one of the intractable problems evidenced by the data presented earlier—it has been maintained elsewhere that the 1970s was the most important decade insofar as policy innovations and physical construction in Asia are concerned (Yeung, 1983). Following the lead provided by Singapore, Hong Kong (1972), Thailand (1973), Indonesia (1974), and the Philippines (1974) established unified housing bodies, thus minimizing many of the difficulties of overlapping responsibilities and competition for scarce financial resources arising from multiagency delivery of low-cost housing.

While Singapore and Hong Kong have accomplished impressive success in public housing, most other countries have adopted a combination of slum upgrading, sites and services, and core housing approaches. True, most of these approaches were not innovated in the region, but they have been adjusted and adapted to suit local circumstances, such as the Kampung Improvement Program (KIP) in Indonesia, and the Zonal Improvement Program (ZIP) and Bagong Lipunan Sites and Services Program (BLISS) in the Philippines. Many of these programs receive outside funding, notably from the World Bank, Asian Development Bank, and United Nations organizations.

Nonetheless, it is probably the large-scale public housing developments in Singapore and Hong Kong which are unparalleled in any city of the world and which have the most to offer in terms of experience and innovations for other similarly situated metropolitan areas. To a limited degree, Seoul and Taipei have also adopted this high-rise, high-density mass housing to provide shelter for the low-income population.

Transport

Over the past two decades, many study cities have taken steps to strengthen transport systems and improve traffic. Most basic to the former has been construction of expressways, elevated roadways, ring roads, and other design measures to achieve mode separation, higher speed, and direct connection. External factors such as Tokyo's hosting the Asian and Olympic Games and the same games hosted by Seoul in 1986 and 1988 have been instrumental in speeding construction of transport and other infrastructural facilities.

With urban traffic becoming ever-congested, the region has witnessed in the past ten years several cities adopting mass rapid transit systems of different kinds. Most notable were subway systems opened in Seoul in 1974 and in

Hong Kong in 1979. Both systems are still being expanded. Singapore, Kuala Lumpur, and Manila are also committed to rail-based mass transit systems which are under construction. Shanghai is the only mega-city in the world without a mass transit system; the situation may change soon as the city is assessing the feasibility of constructing one.

As new systems are being developed, roles of intermediate transport systems will have to be defined. For example, the electric, elevated metrorail system being developed in Metro Manila will compete directly with the present system dominated by jeepneys. Both systems run along Taft Avenue, one of Manila's busiest thoroughfares to the central station (*SCMP*, 1984). In Jakarta and Bangkok traditional low-cost travel modes will likely continue to play an important role, given no immediate plans for any mass rapid transit system (Ocampo, 1982). In light of China's new economic policy allowing for individual initiative, intermediate transport modes appeared to have become quite active in Shanghai. Along with advances in physical transport provision have been noteworthy innovations in improving traffic. Bus lanes, air-conditioned buses, and one-way traffic systems are some of the recent experiments carried out in Singapore, Hong Kong, Bangkok, and Kuala Lumpur.

To curb automobile ownership, Singapore and Hong Kong are two cities that have adopted the most comprehensive car restriction policies through substantial increases in registration fee, road tax, fuel prices, and parking charges. These measures were only partly effective, leading to the adoption, since 1975, in Singapore of the first area licensing scheme in the world. Cars are restricted in entry to the central city area of Singapore in the morning peak hours between 7:30 and 10:30 a.m. unless they meet requirements of car pooling or payment of a daily or monthly fee. From all accounts results have been highly successful (Watson and Holland, 1976). Toward the same end, Hong Kong tested an electronic road-pricing system which charged motorists selectively at times and places where congestion needed reducing. The scheme would require some 200 toll sites, fitting vehicles with electronic number plates, and installing enforcement cameras (Fong, 1984). Completed in June 1985, the pilot test was technically successful, but the scheme was socially unacceptable and has been shelved temporarily.

Other Strategies

This overview of urban strategies and innovations is intended to be indicative rather than comprehensive. Equally important, Eastern Asian cities have made praiseworthy contributions in other spheres, such as the enlightened package of policies for assisting the informal sector in Kuala Lumpur, successful and systematic urban renewal in Singapore, metropolitan growth management in Manila, collective enterprises as a means to cope with youth unemployment in Shanghai, and pollution abatement measures implemented in Seoul, Shanghai, and Tokyo. Unfortunately, space does not allow further discussion of these subjects.

Policy Issues and Options

Underlying most of the strategies and measures reviewed above are some critical policy issues and options. It will be instructive to recount some of these insofar as they pertain to most of the cities studied. Discussion of these issues should bear in mind the varied socioeconomic contexts in which the cities are set.

City Size

First is the question of city size. Pervading many decentralization policies reviewed earlier was accepting that the city had grown too large. At its peak, Seoul grew at 7 to 9 percent annually and Tokyo's population rose by an average of 329 thousand per year from 1955 to 1960 (Hall, 1977:225; Hwang, 1980:34). Bangkok also grew to 40 times the size of Thailand's next largest city, Chiengmai. There was no prescription of optimal size of the city in each case but a good sense of pragmatic governance prevailed to the extent that the cities could not go on growing at those rates without severely compromising the quality of life or level of basic services. Indeed, the urban environment in some of the cities studied has visibly deteriorated over time, a development that added urgency to the need to control and divert future urban growth. For this reason, the anti-big city philosophy was enthusiastically embraced by several of the cities represented, particularly Shanghai, Tokyo, Seoul, Bangkok, and Jakarta, which have already become, or are on the way to becoming very large cities in Eastern Asia. Although there is theoretically no size limit to growth, more rapid growth than a city is prepared for invariably results in urban diseconomies. It is therefore common practice for cities to plan development according to a population target over a certain period, but such a target is more often than not exceeded by events not foreseen in the plan.

Cities and Development

The next issue is the city's role in national and international development. Here the cities represented differ the most as they are informed by varied ideologies. The choice of making Shanghai a productive instead of a consumer city led, inter alia, to uncoordinated growth of industries at the expense of infrastructure construction, emphasis on self-sufficiency in fruits and vegetables in the city region, and assumption of a critical position in spearheading the industrialization of China (Fung, 1981:283). Similarly, accelerating Malay urbanization has taken place in Kuala Lumpur since implementation of the New Economic Policy in 1970 designed to eradicate poverty and restructure the economy. A different urban ecology, with an accent on Malay participation, is emerging and the cultural role of Kuala Lumpur as a center of Malay urbanism is being promoted (Specter, 1984a; Yeung, 1982). These cities are two examples of how very different policy options have led to varied functional roles and spatial patterning of activities within cities.

Other cities, notably Hong Kong and Singapore, play very pivotal roles in regional and international economies. As an aspiring global city, Singapore offers many types of professional services to surrounding countries, as well as being used as a base for some types of regional activities like oil exploration. With the adoption of an increasingly open economic policy in China, Hong Kong's contribution to China's modernization programs, in particular development of special economic zones, will increase. As the third largest financial center and third largest container port in the world, Hong Kong's role in international development is clearly underlined. In fact, Hong Kong, Singapore, Bangkok, and Kuala Lumpur, with their large overseas Chinese communities, are becoming a genre of world cities where circuits of capital are not geographically restricted, giving rise to a changing economic order around the Pacific rim. Overseas Chinese investment in real estate in and across cities in the region is a momentous phenomenon (Goldberg, 1984).

Viewed from another perspective, development of many of the study cities through transnational corporations has produced other effects. It is argued that unequal terms of exchange within the international economy have led to further increases in urban primacy, distortion in previous patterns of urban hierarchy, and increasing centralization of activities within cities (Walton, 1982). It would appear that Bangkok and Manila, and to a degree, Jakarta, Taipei, and Seoul, are prime candidates that fit this description of development.

Food and Fuel

As Eastern Asian cities continue to grow and as their surrounding rural area is being encroached upon, the issues of food and fuel will be of increasing concern. On one hand, there is the question of changing food demand structures with increasing affluence. On the other hand, the extent of domestic production versus foreign imports of food has to be rationalized and decided upon. Generally speaking, foodgrains for the cities under study are brought in, either from within the country or abroad, but cities are to different degrees self-sufficient in vegetables, fruits, eggs, poultry, and the like. Singapore and Hong Kong, for example, have concentrated their efforts on certain types of food and have become efficient producers for their needs. In other cases, the city has not formulated clearly defined development plans, resulting in loss of arable land and hence food production. Urban agriculture in Asian cities is a little-studied subject that given proper attention is likely to yield tangible benefits.

Meeting fuel requirements for Eastern Asian cities entails careful analysis of the sources of supply, fuel alternatives, and relative costs before any energy policy can be designed.[1] Such a policy is not in place in any of the cities reviewed despite lingering memory of the energy shocks of the 1970s.

Finances

In terms of the administration of cities, one of the dilemmas confronting planners and decision makers is lack of funds. Even Tokyo could not carry out many of its public projects for this reason. During the boom years prior to the early 1970s, economic buoyancy was sufficient to carry many of the ambitious projects through. Strains quickly developed and, by 1976, Tokyo was said to be in a financial crisis (Hall, 1977:237). More glaring in inadequate budgetary provision is the case of Jakarta. In 1959 Jakarta had a municipal budget less than 10 percent of Singapore's, for a population twice as large (Hanna, 1961:5-6). Bangkok's fiscal powers are not much stronger as property, land, and cars are grossly undertaxed. Only four taxes[2] come under the Bangkok Metropolitan Administration. In 1981 the average revenue of the city government was 741 baht per person, one-third the national average (Medhi Krongkaew and Tongudai, 1983).

Lack of financial resources often bedevils well-intentioned plans. The experience of the National Housing Authority (NHA) in Thailand, formed in 1973, is illuminating. The Five-Year Plan drawn up in 1976 called for construction of 120 thousand dwelling units at a rate of 24 thousand units a year. The high cost of this conventional approach quickly led to its being downgraded, complemented by a much heavier emphasis on slum upgrading as something that would benefit most people and at costs the government and the poor could afford. In the Fifth Plan Period (1982-86), NHA no longer provides for rental housing; only sites and services projects and slum upgrading will be pursued (Kanjanharitai, 1981:214-17; Tanphiphat, 1982: 45-46).

In contrast to the above-cited cities, greater fiscal autonomy is realized in Hong Kong and Singapore by virtue of their power to tax and acquire land. Not surprisingly, their ability to carry through public works projects is correspondingly higher.

Services

The inability of the cities to provide needed basic services raises the next policy dilemma: Who is responsible for service provision—the government or the people? When a city government is financially hamstrung in providing essential services, people often organize themselves to improve the situation. In several cities included in this study, participatory urban services premised on people's initiatives and resources have been successfully carried out.[3] In Seoul the *Saemaul Udong* (new village movement) initiated in rural areas was extended to urban slum communities. In low-income communities in Manila and Jakarta similar mechanisms of self-help delivery of urban services exist to alleviate the hardship of life in these communities. Leadership is a critical variable which means success or failure of these people-based efforts (Yeung and McGee, in press).

The Poor

Finally, the remarkable economic growth experienced in Eastern Asian cities has not been equally shared. A large proportion of the population remains mired in poverty and will likely remain so in the foreseeable future. Many policy options taken in the cities do not have the interests of urban poor taken into account. Hyung-Kook Kim (1981:70) thus criticized Seoul's population dispersal plans as being pursued in their own right without the welfare of the urban poor or inmigrants taken into account. Uneven distribution of basic services in high-income versus low-income areas is also a common feature in Eastern Asian cities. At a more general level, development plans within a city are not always compatible with interests of the urban poor. For example, a new shopping center may be constructed at the expense of relocating existing vendors, a new expressway can be constructed only if certain slum housing is demolished, and modernization of urban transport may mean phasing out traditional modes of transport. Each policy alternative affects the livelihood of the urban poor, and good policy decisions must take into consideration conflicting interests of different income groups and strike a balance among them. Most cities, however, do have some policies geared toward improving the lives of the urban poor. Poverty-redressing policies in Manila (Viloria, 1983), Kuala Lumpur, and Seoul are noted for their concerted efforts to combat poverty[4] and are mostly organized by sectors rather than being location-specific.

Fathoming Urban Futures

A decade ago Hicks (1974:3) introduced her study of large world cities with this assessment:

> All over the world the great cities are in trouble. The problem of how to deal with the large urban concentrations of the modern world has not yet been solved. It is a problem which besets not only the advanced countries, but afflicts all areas with dense populations and consequently large cities. The troubles seem to be particularly severe (if one can particularise) in Japan, India and the U.S.A.; very likely also in China . . . .

This diagnosis of the problem is still accurate for the large cities of Eastern Asia not only today but also likely to the end of this century.

According to population projections of the United Nations, the developing countries will loom exceptionally large in the share of very large cities by 2000 A.D. Of the 25 largest urban agglomerations in the world, 18 will be in Asia alone. Eight Eastern Asian cities will be in this class, including Shanghai (25.9 million), Tokyo/Yokohama (23.8), Jakarta (14.3), Seoul (13.7), Manila (10.5), and Bangkok/Thonburi (9.9) covered in this study (United Nations, 1982:Table 8). However, there are some comforting signals from the United Nations statistics. They show a slowdown of urbanization in developing countries as a whole, giving rise to speculation that an urban turnaround in these countries is possibly afoot. For East Asia and South Asia, 34.2 and 36.8 percent, respectively, of the total population will reside in urban places by the year 2000, still below the less developed regions' average and world average of 40.4 and 48.2 percent, respectively. Between 1980 and 2000, cities in East Asia and South Asia will grow at 1.0 and 1.9 percent, respectively. There are also signs that migration patterns have turned toward smaller urban places rather than large metropolises (United Nations, 1983).

Along with the robust economic forecast for Eastern Asia, there is little doubt that the cities under review will continue to play vital roles in their national and international economies. Hofheinz and Calder's speculation (1982:251) well sums up the thoughts of many regarding sustained growth of the region:

> But it is one thing to doubt whether Eastasia can continue to grow at the same phenomenal rate, and quite another to believe that its fortunes will be reversed. Short of a world war or some other cataclysmic event that interrupts the flow of commerce and raw materials, it is hard to conceive of a dramatic decline in Eastasian growth and performance, and the possibility exists of a considerable and sustained upward thrust. Given the deep-seated ills of Western societies, Eastasia may gain against the West even if, in comparison with past performance, it only stands still.

With such an optimistic outlook, officials of most large Eastern cities have envisioned their futures to the end of this century. Tokyo (TMG, 1984a), Seoul (W. Kim, 1980), and Kuala Lumpur (Dewan Bandaraya, 1982), to cite only three examples, have approached the subject by preparing an official perspective view of the future, an academic discussion, and a formal development plan, respectively. Certain common elements can be distilled from these urban futures.

First, there is an increasing realization that problems of the great Eastern Asian city cannot be solved within the context of the metropolis. The metropolitan area will become a large, integrated regional complex, with the central city being its core. Beyond the regional complex, a national urbanization policy should be in place in which other spatial components should be articulated with the metropolitan region. The futility of the approach of metropolitan growth management divorced from national urban development goals is widely accepted. At least on paper and with a vision, the great Eastern Asian city should strive to create what W. A. Robson and D. E. Regan idealized—an ordered, coherent, decentralized metropolitan region.

Second, within the metropolitan-regional framework rural-urban migration will probably be moderated, given the likelihood of an urban turnaround alluded to earlier. Assuming continued rapid economic development, rapid technological, social, and functional change will impinge on every aspect of metropolitan life. With concomitant growing affluence and persistent poverty, administrators and planners will forever be torn by the trade-off between efficiency and equity goals. Notwithstanding political rhetoric to the contrary, the interests of the urban poor are by and large not incorporated into existing plans.

Third, despite the application of comprehensive private car restraint programs in some cities, the general tendency is toward a continuing and greater emphasis on the automobile. Even bicycle-dominated Shanghai is changing in this direction. At the same time, within the next 15 years, mass rapid transit systems of some sort will probably be installed in Shanghai, Taipei, Kuala Lumpur, Singapore, and Manila. While the former will aggravate the already serious condition of traffic congestion, the latter will improve mobility within these cities.

At the industrial city level, a few remarks of special interest on future development may be of interest. In view of China's recent more open and relaxed economic policy and the signing of the joint Sino-British accord, Shanghai and Hong Kong are destined to play more vital roles in China's development. Shanghai has been designated one of the 14 open cities along the coast of China, signalling probably the beginning of a period of rapid growth for the coastal areas of the country, and then beyond. With its excellent harbor, strategic location, and infrastructural facilities, Shanghai can be catalytic in ushering in a new period of economic growth to that part of China. Similarly, Hong Kong is expected to bring its development experience to bear on assisting China's modernization programs through the special economic zones and open coastal cities to speed progress. Barring any unforeseen political twists and turns in China, Hong Kong's contribution to China's economic development goals can only be enhanced toward 1997 and beyond.

Concerning future needs for social institutions and physical structures in the great Eastern Asian cities, Tokyo may well be the most pressed. By the year 2000, Tokyo will have 12 percent of its population aged 65 and above, as compared with 7 percent in 1984. Tokyo is already making plans to cope with its aging population with adjustments in welfare service, employment structure, medical care, and so on in light of changing needs (TMG, 1984a: 11-14). By the end of this century, tertiary industries will employ over 70 percent of the labor force, vertical land-use zoning will operate in the inner city where land values are high, and disaster prevention living areas will have been designed.

Both Seoul and Taipei will develop multiple-core city structures to decentralize development. Urban renewal will have begun in Taipei, with anticipation of improved infrastructural facilities. However, water supply may be a difficult problem to deal with. One perspective of Seoul by 2000 A.D. is to predict the ascendancy of transactional metropolitan development. The central city will be dominated by abstract, information-oriented functions that operate in offices and in skyscrapers that form the skyline (Corey, 1980:66).

Much along the orientation of the metropolitan region, Jakarta has developed a master plan which will link it to three other neighboring cities, Bekasi, Bogor, and Tangerang. Called Jabotabek, it will have a projected population of 25 million as it enters the 21st century. During Repelita IV (1984-88) almost Rps 900 million (U.S. $904.5 million) is earmarked for expenditure on Jabotabek. The central government has committed to provide 75 percent of the funds, a decision which indirectly means increased taxes to finance the plan (Specter, 1984a:26-27).

Kuala Lumpur of 2000 A.D. will look substantially different from what it is today. It will be more modern, Malay-dominated, and densely developed. Government policies since 1970 have consistently promoted Malay urbanization focused on the Federal Territory, which is gradually witnessing a population shift in favor of Malays. Increasing *bumiputra* (Malay) participation is most notable in the development and redevelopment of the central planning area where, by virtue of the established policy, there must be 30 percent participation by Malays in ownership and operation of most economic activities. By administrative fiat and contrived change, an ethnically and functionally different city is being shaped.

The urban scenarios that have been lightly touched on are for the most part purposeful vistas with which to perceive the future in Eastern Asia. Their realization, like any development plan, depends on the interplay of a complex of political, economic, human, and fiscal factors. Given the special administrative status of many of the cities studied and their growing importance in national and international economies, the outlook of their ability to fulfil their respective urban futures is moderately sanguine.

Notes

1. Two parallel research groups are interested in urban food and fuel issues in Asia. One group was to be organized by the Resource Systems Institute at the East-West Center which jointly organized with Nihon University a May 1984 Tokyo meeting to mount a multicountry research project. Another, coordinated out of Urban Resource Systems in San Francisco, tried to organize a similar research network focused on urban agriculture in Asia. A meeting with International Development Research Centre funding was held in Singapore in mid-1983. Both networks are yet to get off the ground.

2. These taxes are house and land, land development, signboard, and animal slaughter. They constitute a small faction of the total revenue of the Bangkok Metropolitan Administration which is heavily supported by the central government.

3. The International Development Research Centre supported a five-country project in Eastern Asia on this subject involving South Korea, Hong Kong, the Philippines, Malaysia, and Indonesia. The project has been completed and publication of the findings is in progress.

4. The International Development Research Centre has supported research on this subject in the three cities mentioned. See Viloria (1983) for a summary of the Philippines results.

Bibliography

Aiken, S. Robert, and Colin H. Leigh. 1975. "Malaysia's Emerging Conurbations." *Annals of the Association of American Geographers* 65:546-63.

Aquino, Rosemary M. 1983. *The Delivery of Basic Services in Three Selected Philippine Urban Centers: Implications for a Participatory Management Model.* Manila: Integrated Research Center, De La Salle University.

Blair, Thomas L. 1974. *The International Global Crisis.* New York: Hill & Wang.

Business China 8(31 Mar. 1982):44-45. "Shanghai—Up Close."
Chen, S. H. 1984. "Problems of Urban Development." In *Social Problems in Taiwan,* edited by K. S. Young and C. C. Yeh. 2d ed. Taipei: Che-Liu (in Chinese).
China Daily (26 July 1984):3. "Shanghai Housing Shortage Studies."
Chu, Chong-Won. 1980. "Issues on Housing and Urban Development in Seoul." In *The Year 2000. See* Kim, 1980.
Corey, Kenneth E. 1980. "Transactional Forces and the Metropolis: Towards a Planning Strategy for Seoul in the Year 2000." In *The Year 2000. See* Kim, 1980.
Critchfield, Richard. 1971. "The Flight of the Cities: Dakarta—The First to 'Close'." *Columbia Journal of World Business* 6(4):89-93.
Davis, Kingsley. 1965. "The Urbanization of the Human Population." *Scientific American* 213(3):41-53.
Dewan Bandaraya. 1982. *Kuala Lumpur Draft Structure Plan.* Kuala Lumpur.
Donner, Wolf. 1978. *The Five Faces of Thailand: An Economic Geography.* New York: St. Martin's Press.
Erni, L.L.M., and Bianpoen. 1980. "Case Study: Jakarta Indonesia." In *Politics toward Urban Slums.* Bangkok: United Nations Economic and Social Commission for Asia and the Pacific.
Fong, Peter K. W. 1984. "The Electronic Road Pricing System in Hong Kong." Hong Kong: Centre of Urban Studies and Urban Planning, University of Hong Kong.
Fung, Kai-iu. 1981. "The Spatial Development of Shanghai. In *Shanghai. Revolution and Development in an Asian Metropolis.* Contemporary China Institute Publications. Cambridge: Cambridge University Press.
Fung, Tung. 1983. "Public Housing Management in Hong Kong's New Towns." In *A Place to Live. See* Yeung, 1983.
Ginsberg, Norton S. 1955. "The Great City in Southeast Asia." *American Journal of Sociology* 60:455-62.
Goldberg, Michael A. 1984. "Hedging Your Great Grandchildren's Bets: The Case of Overseas Chinese Investment in Real Estate around the Cities of the Pacific Rim." Working Paper no. 22. Vancouver: Institute of Asian Research, University of British Columbia.
Hall, Peter G. 1977. *The World Cities.* 2d ed. London: Weidenfeld & Nicolson.
Hanna, Willard Anderson. 1961. *Bung Karno's Indonesia: A Collection of 25 Reports Written for the American Universities Field Staff.* New York: American Universities Field Staff.
Henderson, J. V. 1984. "Urbanization: International Experience and Prospects for China." Department of Economics, Yale University, New Haven, CT: Mimeo.
Hicks, Ursula K. 1974. *The Large City: A World Problem.* New York: Halsted.
Hofheinz, Roy, Jr., and Kent E. Calder. 1982. *The Eastasia Edge.* New York: Basic Books.
Hong Kong Housing Authority. 1984. *A Review of Public Housing Allocation Policies: A Consultation Document.* Hong Kong.
Honjō, Masahiko. 1972. "Recovering the Tokyo Bay Coastal Area: The Choice for Balance." Background paper. Nagoya, Japan: United Nations Centre for Regional Development.
———. 1975. "Tokyo: Giant Metropolis of the Orient." In *World Capitals: Toward Guided Urbanization,* edited by H. Wentworth Eldredge. Garden City, NY: Anchor Press/Doubleday.
Hsu, Yi-Rong Ann, and Clifford W. Pannell. 1982. "Urbanisation and Residential Spatial Structure in Taiwan." *Pacific Viewpoint* 23:22-52.
Hwang, Myong-Chan. 1979. "A Search for Development Strategy for the Capital Region of Korea. In *Metropolitan Planning: Issues and Policies,* edited by Yung Hee Rho and Myong-Chan Hwang. Seoul: Korea Research Institute for Human Settlements.
———. 1980. "Planning Strategies for Metropolitan Seoul." In *The Year 2000. See* Kim, 1980.
Kanjanaharitai, Paiboon. 1981. "Bangkok: The City of Angels." In *Urbanization and Regional Development,* edited by Masahiko Honjō. Singapore: Maruzen Asia.
Kim, Hung-Kook. 1981. "Social Factors of Migration from Rural to Urban Areas with Special Reference to Developing Countries: The Case of Korea." *Social Indicators Research* 10:29-74.
Kim, Song-Um. 1980. "An Overview of Recent Urbanization Patterns and Policy Measures for Population Redistribution and Resettlement in the Republic of Korea." In *Migration and Resettlement Rural-Urban Policies.* Vol. 2. Manila: Social Welfare and Development Center for Asia and the Pacific.

Kim, Won. 1983. "Land Use Planning in a Rapidly Growing Metropolis: The Case of Seoul." *Asian Economies* (Jan.):5-21.
———, ed. 1980. *The Year 2000: Urban Growth and Perspectives for Seoul.* Seoul: Korea Planners Association.
Korea. South. 1982. *The Second Comprehensive National Physical Development Plan 1982-1991.* Seoul.
Krausse, G. H. 1982. "Themes in Poverty: Economics, Education, Amenities, and Social Functions in Jakarta's Kampungs." *Southeast Asian Journal of Social Science* 10(2):49-70.
Krongkaew, Medhi, and Pawadee Tongudai. 1983. "The Growth of Bangkok: The Economics of Unbalanced Urbanization and Development." Paper presented at the Annual Conference of the Association of Asian Studies, San Francisco, 25-27 March.
Kwok, Reginald Y. W. 1983. "Land Price Escalation and Public Housing in Hong Kong." In *Land for Housing the Poor,* edited by Shlomo Angel, Raymon W. Archer, Sidhijai Tanphipat, and Emiel A. Weglen. Singapore: Select Books.
Kwon, Won-Young. 1980. "Metropolital Growth and Management. The Case of Seoul." Seoul, Korea: Korea Research Institute for Human Settlements.
———. 1981a. "Seoul: A Dynamic Metropolis." In *Urbanization and Regional Development. See* Kanjanaharitai, 1981.
———. 1981b. "A Study of the Economic Impact of Industrial Relocation: The Case of Seoul." *Urban Studies* 18:73-90.
Lau, Siu-kai, Hsin-chi Kuan, and Kam-fai Ho. 1983. "Leaders, Officials and Localities in Hong Kong." Centre for Hong Kong Studies Occasional Paper no. 1. Hong Kong: Institute of Social Studies, Chinese University of Hong Kong.
Lee, Chunfen, Chungmin Yen, and Jianzhong Tang. 1983. "A Spatial Analysis of Shanghai's Economic Development." Department of Geography, East China Normal University, Guanzhou.
Leong, K. C. 1981. "Kuala Lumpur: Youngest Metropolis of Southeast Asia." In *Urbanization and Regional Development. See* Kanjanaharitai, 1981.
Leung, Chi-Keung. 1983. "Urban Transportation." In *A Geography of Hong Kong,* edited by T. N. Chiu and C. L. So. Hong Kong: Oxford University Press.
Leung, W. T. 1983. "The New Towns Programme." In *A Geography of Hong Kong.* See C.-K. Leung, 1983.
Lim, Kok Leong, Keim Hoong Chin, Koon Fun Chin, Leslie Goh, and Sze Ann Ong. 1983. "Management of Singapore's New Towns." In *A Place to Live.* See Yeung, 1983.
Liu, Paul K. C. 1982. "Labor Mobility and Utilization in Relation to Urbanization in Taiwan." *Industry of Free China* (May):1-12.
———. 1983. "Factors and Policies Contributing to Urbanization and Labor Mobility in Taiwan." *Industry of Free China* (May):1-20.
Lo, C. P. 1972. "A Typology of Hong Kong Census Districts: A Study in Urban Structure." *Journal of Tropical Geography* 34:34-43.
Moir, Hazel V. J. 1978. *Jakarta Informal Sector.* Monograph Series. Jakarta: National Institute of Economic and Social Research (LEKNAS-LIPI).
Mugbil, I. 1982. "The Poisoning of Bangkok." *Bangkok Post* (13 June):24.
NEDA. See Philippines. National Economic Development Authority.
Ocampo, Romeo B. 1982. *Low-Cost Transport in Asia: A Comparative Report on Five Cities.* Ottawa: International Development Research Centre.
Papanek, Gustav F. 1975. "The Poor of Jakarta." *Economic Development and Cultural Change* 24:1-27.
Park, Soo Young. 1981. "Urban Growth and National Policy in Korea." Paper presented at the Pacific Science Association 4th Inter-Congress, Singapore, 1-5 September.
Philippines. National Economic Development Authority. 1973. *Regional Development Projects: Supplement to the Four-Year Development Plan, FY 1974-77.* Manila.
Prybyla, Jan S. 1975. "*Hsia-Fang*: The Economics and Politics of Rustication in China." *Pacific Affairs* 48:153-72.
SCMP. See South China Morning Post.

Seoul Metropolitan Government. 1983. *Seoul: Metropolitan Administration*. Seoul.
Sethuraman, S. V. 1976. *Jakarta: Urban Development and Employment. A WEP Study*. Geneva: International Labour Office.
Simmons, Alan B. 1979. "Slowing Metropolitan City Growth in Asia: Policies, Programs, and Results." *Population and Development Review* 5:87-104.
SMG. See Seoul Metropolitan Government.
Soon, Tay Kweng. 1969. "Housing and Urban Values—Singapore." *Ekistics* 27(158):27-28.
South China Morning Post (13 Sept. 1984). "Manilans Queue Up for Metrorail."
Specter, Michael. 1984a. "The 'Small Town' Big City." *Far Eastern Economic Review* 125(27 Sept.): 23-30.
———. 1984b. "A Sprawling, Thirsty Giant." *Far Eastern Economic Review* 123(29 Mar.): 23-30.
Suharso et al. 1975. *Migration and Education in Jakarta*. Jakarta: National Institute of Economic and Social Research. (LEKNAS-LIPI).
Tang, F. Z. 1982. *Research on Questions and Policies of Urbanization in Taiwan*. Taipei: Development Research Committee of the Executive Assembly (in Chinese).
Tanphiphat, Sidhijai. 1982. "Thailand Country Study: Urban Land Management Policies and Experiences." Paper presented at the International Seminar on Urban Development Policies, Nagoya, Japan, 13-18 October.
Tao, Z. J. 1981. "Examining Urban Construction and Urban Finance from the Perspective of the Urban Structure of Shanghai." In *Zhongguo Caizheng Wenti*, edited by Zongguo Cainzheng Bu. Tianjin: Tinajin Kenxue Jixue Chubanshe (in Chinese).
TMG. See Tokyo Metropolital Government.
Tokyo Metropolitan Government. 1984a. *Long-Term Plan for Tokyo Metropolis: "My Town Tokyo"—Heading into the 21st Century*. TMG Municipal Library no. 18. Tokyo.
———. 1984b. *Plain Talk about Tokyo: The Administration of the Tokyo Metropolitan Government*. 2d ed. TMG Municipal Library no. 15. Tokyo.
Tsay, Ching-lung. 1982. "Migration and Population Growth in Taipei Municipality." *Industry of Free China* (Mar.): 9-25.
United Nations. Department of International Economic and Social Affairs. 1980. *Patterns of Urban and Rural Population Growth*. ST/ESA/SER.A/68. Population Studies no. 68. New York: United Nations.
———. 1982. *Estimates and Projections of Urban, Rural and City Populations, 1950-2025: The 1980 Assessment*. ST/ESA/SER.R/45. New York. United Nations.
———. Population Division. 1983. "Urbanization and City Growth." *Populi* 10(2):39-50.
United Nations Development Programme. 1979. *Project of the Government of the People's Republic of China: Municipality of Shanghai*. Mimeo.
Viloria, Leandro A. 1981. "Manila: Creation of a Metropolitan Government." In *Urbanization and Regional Development*. See Kanjanaharitai, 1981.
———. 1983. *Study of Poverty Redressal Programs in Metro Manila*. Report submitted to International Development Research Center. Mimeo.
Walton, John. 1982. "The International Economy and Peripheral Urbanization." In *Urban Policy under Capitalism*, edited by Norman I. Fainstein and Susan S. Fainstein. Urban Affairs Annual Reviews, vol. 22. Newbury Park, CA: Sage.
Watson, P. L., and E. P. Holland. 1976. "Congestion Pricing—the Example of Singapore." *Finance and Development* 13(1):20-23.
White, Lynn T. 1981. "The Suburban Transformation." In *Shanghai: Revolution and Development in an Asian Metropolis*, edited by Christopher Howe. Cambridge: Cambridge University Press.
Wu, C. T. 1984. "Coping with Urban Growth under Socialism: The Case of Shanghai." Department of Town and Country Planning, University of Sydney. Mimeo.
Yan, Z. M., and J. Z. Tang. 1984. "Urbanization and Urban Eco-Environment in Shanghai." *Journal of East China Normal University* 1:68-73 (in Chinese).
Yeung, Yue-Man. 1976. "Southeast Asian Cities: Patterns of Growth and Transformation." In *Urbanization and Counterurbanization*, edited by Brian J.L. Berry. Urban Affairs Annual Reviews, vol. 11. Newbury Park, CA: Sage.

———. 1982. "Economic Inequality and Social Injustice: Development Issues in Malaysia." *Pacific Affairs* 55(1):94-101.
———, ed. 1983. *A Place to Live: More Effective Low-Cost Housing in Asia.* Ottawa: International Development Research Centre.
Yeung, Yue-Man, and Terence G. McGee, eds. 1986. *Community Participation in Delivering Urban Services in Asia.* Ottawa: International Development Research Centre.

7

Levels of Urbanization in China

Sidney Goldstein

Urbanization in China, like China itself, has a long history. Through a good part of the 1,000 years from 800 to 1800, China contained the largest city in the world (United Nations, 1980:5). Indeed, China is also credited with having, in the eighth century, the first city in the world to exceed one million persons—Changan (modern Xian) (Chandler and Fox, 1974). China briefly held the distinction of having the world's only million-plus city a second time, when, after other cities had declined, Beijing attained this size in the late 18th century, only to be succeeded soon after by London.

Neither urbanization nor big cities are as new to China as they are to many other developing countries. Nor are problems of development and modernization. As Leung and Ginsburg (1980) have documented, problems of development and modernization confronted the governments of China more than 100 years before the Communist Revolution, and efforts to cope with these have involved a large variety of approaches. Nonetheless, the modern-

---

AUTHOR'S NOTE: The research reported here was made possible through an Award for Advanced Study and Research from the Committee on Scholarly Communication with the People's Republic of China and support to the Population Studies and Training Center, Brown University, from the Ford Foundation, the William and Flora Hewlett Foundation, and the American Express Foundation. Special thanks go to the Center for Population Research of the Chinese Academy of Social Sciences for assistance provided during fieldwork in China, and to Zhou Junli and Ma Rong, graduate students from the People's Republic of China in the Department of Sociology, Brown University, for assistance in preparing this report. Thanks, too, to Norton Ginsburg and Chang-Tong Wu, who kindly shared their reactions to an earlier draft of the manuscript. This chapter was originally published in a longer version by the East-West Population Institute, East-West Center, Honolulu, in 1985, as Paper no. 93.

ization process initiated by the People's Republic of China (PRC) in the 35 years since its establishment in 1949 has had the most sweeping impact on the nation as a whole. These efforts to modernize have had to confront, on a massive scale, demographic challenges similar to those faced by other developing countries—rapid population growth, substantial increases in the size of the urban and rural populations, and imbalances in geographic distribution and rates of city growth.

In China, as in many other developing countries, considerable differences of opinion have emerged—and sometimes have even been reflected in practice—about urbanization's role in the development process and the extent to which urban growth should be controlled. Since 1949, urbanization policies in China at times conformed to the view that expansion of the modern sector requires some spatial polarization to obtain the economies of scale needed to ensure successful development. In view of the early insistence that urban-rural differentials must be leveled, that more emphasis must therefore be placed on rural development, and that cities tended to be seats of corruption and bourgeois influence, the role of cities, and especially big ones, has long been open to question (see, e.g., Chiu, 1980). Nonetheless, in the years immediately following establishment of the PRC and during the period of the Great Leap Forward (1958-60), China's urbanization level rose substantially, reaching about 20 percent in 1960. Such development was not wholly inconsistent, however, with Maoist views as expressed in 1956, which, while stressing the importance of developing heavy industry especially in inland places, also recognized the potential of existing coastal cities and the need to develop agriculture and light industry (Mao, 1977:1-6).

Thereafter, the underlying negative views about the effects of urbanization became dominant—ones more commensurate with the belief that polarization was inconsistent with long-run regional and national development goals. It was feared that such development would run the risk of exacerbating the problems big cities face in providing livelihood and adequate infrastructure to their residents and producing tremendous loss of human potential in rural areas and smaller urban locations. This view influenced the readjustment that followed in the early 1960s. Millions of urban residents were resettled in the countryside, resulting in a reduction in the urban population from its peak of 130.7 million in 1960 to 116.5 million in 1963. Overlooking problems of definition (see the next section), the reduction was short lived. In 1966, at the beginning of the Cultural Revolution (1966-76), a new urban population high of 133.1 million had been reached, and each successive year since then has seen a rise in the number of people living in urban places (SSB, 1983a:103). By the mid-1970s, the Cultural Revolution's heritage of stressing industrialization and paying minimal attention to urban housing and infrastructural needs had resulted in severe conditions in cities even while their populations grew markedly.

This situation, coupled with continuing rapid growth of the rural population, led China in 1980 to adopt a policy of planned urbanization that involves strict control of the growth of big cities. The strategy grows, in part, out of the belief, based on China's own experience and observation of the situations in other developing countries, that the too rapid growth of big cities gives rise to many problems related to housing, employment, and infrastructure. Concentration of population and industrial activity in China's big cities has already aggravated shortages of land, water, energy, and transportation facilities. In 1981, 65 percent of the nation's industrial output value was located in the 43 largest cities (Li, 1983).

The critical balance between population and arable land is a major consideration affecting urban policies and accounts in part for the decision in the late 1950s to give big cities control of the adjoining rural counties (Koshizawa, 1978:15). Through such control, each big city is able to ensure its own daily vegetable supplies. Each can also control the extent to which arable land is converted for housing and industrial purposes, thereby avoiding serious incursions into the city's ability to feed its population.

The official Chinese position is, therefore, that the number of big cities and the sizes of their populations need correction at the same time that the needs of China's 800 million rural masses must be met. Given this perspective, it is argued that urbanization must be harmonious with industrial and agricultural development. The need to develop all three concurrently lies behind the basic urban policy of strictly limiting the size of big cities, properly developing medium-sized cities, and encouraging growth of small cities and market and agricultural towns (Ye, 1982; Zhuo, 1981).[1] Li (1983:9) has explained the rationale:

> In our circumstances, if only the big cities have good job opportunities, housing, schools, stores and services, and cultural and recreational facilities, then of course people will want to live there. But if all these things are developed in small and medium sized cities, people will be much more willing to live in these places and the population pressure on big cities will be eased.

In contrast to the big cities, medium- and small-sized cities are seen as having more space for industrial development and housing needs for the resident labor force. Such cities are, therefore, increasingly becoming the locus of new efforts at industrialization. A potential danger here lies in the rapid pace at which some medium-sized cities are growing; the challenge will be to control their future growth lest they also are transformed into the big cities that Chinese policy is intended to avoid. Concern for excessive growth of medium-sized as well as big cities would seem particularly relevant in view of China's efforts, begun in 1984, to develop 14 coastal cities as centers for foreign economic activities. Interviews with Chinese officials indicate their awareness of the serious challenge posed by undertaking economic development of these cities while continuing to control their population growth.[2]

Unlike big and medium-sized cities, generally seen as the locations of heavy and light industry, smaller cities and towns are viewed as potential locations for handicraft and workshop activities, with workers supplied largely from the rural surplus labor force. Such places, it is argued, require less government investment while also serving as catalysts for "changing" rural populations into urban ones. What is perhaps most interesting about the stress on development of small places as the proper course of urbanization is the magnitude of the transformation it would involve. Although planners recognize that any one town can absorb only a limited number of people, they point out that the large number of such places in China allows the aggregate effect to be great: If each of the nation's 2,100 county seats increased its population to 50,000 people, 39 million more people would be absorbed over the current 61 million county town residents (Ye, 1982). And if each of the 54,000 commune seats increased to an average of only 5,000 persons, some 270 million would reside in these centers.

Comparative costs and difficulties of urbanizing in this way, rather than through allowing increased migration to big or even medium-size cities, have been considered. Counter-arguments that bigger cities are more efficient because of locational considerations and availability of infrastructure are discounted, however, because China's social system is seen as able to compensate for these advantages by providing in smaller places adequate job opportunities, housing, schools, commercial activities, and recreational facilities. Whether, in fact, all of these amenities will be available in smaller places or whether persons living in smaller places and rural areas will be provided with easier access to city amenities remains to be demonstrated. What is clear is that existing policy is premised on the validity of this argument.

Despite growing attention in China to urbanization and effects of migration, serious obstacles have hampered efforts to assess urban growth patterns and migration's role in this growth and in the alleviation of rural-urban disparities in the quality of life. Among major obstacles have been (1) limitations inherent in existing legal and conceptual views of what constitutes migration and urban places, and (2) lack of adequate data to assess the nature of urbanization and the extent and type of population movement that characterize the urban and rural scenes. The 1982 Chinese Census marks a major advance in coping with some of these problems in the attention it has given to definitional concerns and in the wealth of data it has collected.

In exploiting the data available in early census publications (SSB, 1982, 1983b), I attempt to gain insights into China's urbanization patterns in relation to the Chinese policy of controlled urban growth. In doing so, I also make limited use of material and insights gained as part of three months of field research on the ways in which the agricultural responsibility system adopted in rural locations has affected population movement and urbanization in China.

Defining Urban Places and the Urban Population

Considerable confusion has characterized reports on the size of China's urban population since 1949, in large part because of the varying definitions used for *urban place* and *urban population.* Unlike a number of other countries, China has relied not only on criteria related to minimal numbers in a given location or to the percentage of individuals engaged in nonagricultural activity. In China, the definition of *urban* at times has been affected by a unique perspective for viewing the urban population—a perspective based largely on a combination of where people live and who is responsible for providing their grain needs. But this has not always been the case, nor even consistently so at a particular time.

Defining Urban and Rural Places

According to Ernest Ni (1960), no definition of rural and urban was given in Chinese sources until 1955. However, in data on urban places by size class and in the absence of any sharp changes in the urban population in annual statistics covering the period 1949-56, indicators suggest that the definition employed by the State Council in 1955 was quite similar to that used by the 1953 Census. According to the 1955 definition, a place was urban if it (1) had a municipal people's committee or was the seat for a people's committee of the *xien* (county) level or above; (2) had a permanent population of 2,000 or more, of which at least 50 percent were not in agriculture; or (3) had a permanent population of between 1,000 and 2,000 of which at least 75 percent were nonagricultural and concurrently was also a commercial, industrial, education, health, or communication center (Ni, 1960:3). Evidently, in minority areas a town could also qualify as an urban place if it had a total population of less than 2,000 inhabitants, a considerable number of whom were engaged in industry or commerce (Ullman, 1961:4). This combination of criteria yielded a total of 5,568 urban places in the 1953 Census, of which 164 were municipalities of 20,000 population or more. Of the 5,404 towns, 256 had between 20,000 and 100,000 residents, and the remaining 5,148 had fewer than 20,000, including 193 towns with less than 1,000 people.

The total population living in the 164 cities and 5,404 other urban places was not the equivalent of the urban population. Of the 52.3 million persons in cities, 43.5 million, or 83 percent, were classified as urban, the rest presumably being classified as rural on the basis of economic activity. In other urban places (towns), 33.7 million, or 95 percent of the 35.3 million residents, were urban, suggesting that the larger places were more overbounded than the smaller ones and therefore included more rural persons. In all, therefore, the 1953 Census identified 77.3 million, or 13.3 percent, of China's 582.6 million population as urban.

For the 1964 Census, the criteria were changed. In December 1963, the State Council specified that: (1) an area could be designated a town if industries, commerce, and handicraft trades were relatively concentrated, and it had a population of more than 3,000 with 70 percent being nonagricultural; (2) an area could be designated a town if it had a population between 2,500 and 3,000 with more than 85 percent being nonagricultural and requiring direct administration of the county government; (3) however, places where minority nationality people lived could also be designated as towns if industries, commerce, and handicraft trades were concentrated there and direct administration was needed, even though the population was under 3,000 and under 70 percent was engaged in nonagricultural production; (4) cities were places with populations over 100,000 or places of fewer than 100,000 that were provincial capitals, heavy industrial bases, fairly big centers for gathering and distributing goods and materials, or important towns in border areas requiring direct administration of provincial or regional authorities (Ma, 1984:12-13).[3]

When the new criteria were applied, towns not meeting the minimum standard were transferred to the jurisdiction of communes, resulting in their classification as rural places. Small cities not meeting the State Council's criteria were reclassified as towns and put under the jurisdiction of county governments. The net result was a reduction in the number of towns from 5,404 in the 1953 Census to 3,148 in the 1964 Census. Since the reclassification evidently had not been completed by the time of the census, some places were still officially identified as towns and included in the urban population in the 1964 Census even though they did not meet the new criteria. The completed reclassification after 1964 yielded fewer than 3,000 towns; at the time of the 1982 Census only 2,664 towns met the official criteria. Thus, between the 1953 and 1982 censuses, the number of designated towns was reduced by over 50 percent, from 5,404 to 2,664 places. However, since most of these places were small, the aggregate number living in them probably did not exceed more than a few million and so did not seriously affect comparability with later censuses in the total size of the urban population.

The criteria adopted in 1963 also applied to cities; as a result, some cities were reclassified as towns. The change in number of cities between 1964 and 1982 was not sharp, however, because the number of cities downgraded according to the 1963 criteria was offset by the number of new cities developed in the post-1963 period in response to national planning. By the end of 1982, according to the *1983 Statistical Yearbook* (SSB, 1983a), China had already increased the number of cities from the 236 identified in the 1982 Census to 245, and at least 14 new locations were reclassified as cities in 1984 (*Renmin Ribao* [People's Daily], 20 January 1984). The 1964 and 1982 Censuses then used the same basic criteria for designating places as urban, although the 1964 Census included 484 towns later shifted to rural status; it is not possible to determine how many people this involved.

Defining the Urban Population

Understanding the various criteria used to define the urban population requires recognition of the central role the household registration system plays in defining an individual's residence status and in controlling permanent migration.

The Registration System and Migration

Each individual in China has an official place of residence, the record of which is maintained at the brigade level in rural areas and at the neighborhood level in urban places by the Public Security Bureau. To effect a permanent change in residence, permission must be granted by the appropriate authorities in the places of origin and destination. Peasants can generally obtain an urban household register in only a limited number of ways. The most prominent are: (1) university enrollment, which carries with it urban household registration, which is then retained; (2) city or industrial expansion into farmlands, which may entitle peasants displaced in the process to urban household registration; (3) permanent employment in an urban place, which leads to urban registration. In such cases family members of the employee must generally retain their rural household register, even if they in fact live in the city.

Given the close interrelations between the registry system and population movement in China, an individual is considered a permanent migrant only if the move involved a change in household registration. Thus, persons living in cities who are not de jure residents of those cities are not counted as part of the city population in any enumeration based on household registers. Nonetheless, a considerable amount of "temporary" movement to urban places exists in China, and such movement is often officially sanctioned. Since de facto residents may be substantial in number and selective in their socioeconomic and demographic characteristics, their omission from urban registers and statistics distorts data on the size and composition of urban places, and data on the rural populations at places of origin.

1982 Census Qualifications

In partial recognition of the situation just described, the 1982 Chinese Census identified separately and counted as residents those people who had lived in a given locality for over one year even though they were registered elsewhere and those who had resided in the locality less than one year, but were absent from their place of registration for more than a year.[4] In this respect, the 1982 Census results are not and should not be identical with register enumerations, despite the heavy reliance on the latter to ensure complete coverage in the census.

Particular attention was given in the census to the problems inherent in identifying the permanent residence of mobile individuals. For example, people in rural communes and production brigades often went temporarily to another city or country as peddlers or construction workers, or for other activities, but came back frequently. They were not considered as being away for more than one year and were enumerated in the places of their household registration. Persons involved in activities that required geographic mobility—such as prospecting, transportation and communication, mobile sales of handicraft and sideline production, and construction—were also enumerated at the place of their household registration to facilitate enumeration and avoid duplication or omission. An indeterminate number of persons living in cities without any kind of official sanction may also have reported themselves as temporary urban residents (for less than one year) to avoid bureaucratic difficulties; they, too, would have been enumerated by the census as living at their place of registration.

Among the 1.002 billion people enumerated (excluding Taiwan, Hong Kong, Macao, and Tibet), preliminary tabulations of census results indicate that 98.9 percent (990.6 million) lived in their place of registration (SSB, 1982). Of the remainder, 6.3 million people resided in places in which they had lived more than one year without permanent registration; and 4.8 million reported residence with registration still to be settled. Only a small number (210 thousand) were reported as residing in the particular location at which they were enumerated for less than one year but absent continuously from their legal place of registration for more than one year. Still fewer (57 thousand) were reported as living overseas. Since it is likely that the large majority of "temporary" migrants, including those away for at least a year, are rural-to-urban movers, the volume and composition of this segment of the Chinese population has particular significance for the nation's urban development.

Temporary Movement

In addition to the population officially registered in urban places and those without registration but counted by the 1982 Census as living in cities and towns, a growing number of "temporary" migrants also swells the urban population. This segment of the population has increased dramatically since 1979 as a response to the growing amount of surplus rural labor created by implementation of the agricultural responsibility system.

At least four different forms of temporary mobility can be identified: (1) temporary residents may be construction workers (who sometimes number in the thousands) recruited from rural areas to provide labor needed to build major projects, including satellite towns, factory complexes, and university expansions; (2) communes may send groups of their residents to the city to operate collective shops or engage in other commune-sponsored enterprises; (3) peasants come to cities for days or even weeks to sell their produce and sideline products; (4) growing numbers of individuals move from rural areas into cities where they are hired privately for their services and skills in such jobs as child care, housekeeping, and carpentry. All continue to be officially registered in their rural places of origin and hold only temporary residence permits in the city.

Unfortunately, no body of statistics is readily available to document the volume and characteristics of these types of temporary movement. This documentation difficulty reflects the nature of the registration system. Although a "permanent," legal transfer of household registration is carefully controlled and documented, temporary movement is not. Temporary residents can easily obtain temporary registration from the Public Security Station if they are living with relatives or friends. Residence in a hotel or in work unit facilities requires no temporary registration at all. Temporary residents involved in free markets are registered with the Industrial and Commercial Bureau. Many others do not register at all, since the regulations are often not rigorously applied. The net result is a complex system that does not lend itself to centralized statistics or at least to their ready availability.

Not only does the system's deficiency frustrate city planners who are unable to assess the volume and character of temporary movement for proper planning of urban services, it also distorts estimates of the size and characteristics of the urban population. Temporary migrants are undoubtedly selective in terms of a number of socioeconomic characteristics, including age, sex, and occupation. Many remain in the urban location for extended periods of time and in terms of their economic activity and residential needs would under most definitions be included in the urban population.

The Criterion of Grain Sources

That many temporary migrants are not classified as urban relates to the nation's regulations controlling rural-urban migration and to another criterion often used to define the urban and rural population—the individual's source of grain supply. As long as a person holds a rural registration, regardless of occupation or de facto residence, the commune of official residence is responsible for supplying the individual's annual grain allotment. In urban places, grain is purchased from state outlets, and the state is responsible for providing adequate supplies.

Reliance on the source of grain supply as the basis of classification as urban or rural therefore creates a major problem of comparability in Chinese data on urban/rural distribution. This definition was widely used after the 1964 Census, at least into 1984, with resulting enumerations far different from those revealed by the censuses (compare for example, Aird, 1982:279-82; see also Banister, 1984:264-66). The magnitude of the problem is illustrated by a news account in the *Beijing Review* (1983) that compared the 1982 Census results, showing an urban population of 206 million, with the 138 million reported as urban in the previous year by registry statistics, which indicated an urban growth of 68 million in one year. The report hastened to explain that in addition to the effects of natural increase, permanent migration, and some increase in number of cities between 1981 and 1982, an important factor accounting for the increase in urban population was the different method of defining the urban population. The 1981 statistics, based on registry data, count as urban only those residing in urban places whose grain was supplied by the state (termed *commercial grain*), a group defined operationally as nonagricultural. Excluded were those living in urban places who were dependent on their own (or on their commune's) production for their grain supply and were operationally classified as agricultural, even though they may have been engaged in nonagricultural activities.

What particularly complicates the dichotomy is that under this system, individuals who have moved into an urban location, and have lived there more than one year but have not changed their registration, continue to be classified as agricultural, even if engaged in nonagricultural work. The original designation, under which they are registered, determines their status, and the key criterion is grain source. Similarly, members of communes located within officially designated city boundaries are also defined as "agricultural" because they obtain their grain from their communes. Under this system, the urban population more closely resembles a de jure rather than a de facto count, with only those registered in urban places and receiving commercial grain counted as urban residents.

The effect of the differences in definition is illustrated by data available from the *1983 Statistical Yearbook* (SSB, 1983a:107-08). Excluding counties (and the county towns) under the cities' administration (referring only to inner city populations) but including all residents of the city regardless of how they obtained their grain, 145.2 million persons were reported as living in China's 239 cities at the end of 1982. Of these cities, 85 had more than 500,000 population and accounted for 74.9 percent of the total city population. Only 17 cities had populations of fewer than 100,000 people; these cities accounted for less than 1 percent of all people living in Chinese cities.

But note how the size and distribution changes if the city population is restricted, by definition, to residents obtaining their grain through state outlets. The aggregate number shrinks by almost one-third to 97.1 million, implying that almost 50 million people living in cities are directly dependent on communes for their grain. The number who actually are members of communes located within city limits and the number who live in the city while remaining registered commune members cannot be ascertained from available data. Moreover, the effect on the city hierarchy is also dramatic. The number of big cities with more than 500,000 population declines from 85 to 48, and the percentage of total city population living within big cities drops from three-fourths to below two-thirds (63.8 percent). For small cities, the change in definition has the opposite effect: the number of cities of less than 100,000 increases from 17 to 55; and the population of such places almost triples, while its percentage of the total rises from less than 1 percent to almost 4 percent. If places of 100,000 to 300,000 are included in the small city category, the corresponding changes in number are from 108 to 160 and the percentage of total city population rises from 12.8 to 23.9 (SSB, 1983a:107).

The effect of criteria employed in identifying the urban population can also be seen by comparing the size of the urban population of individual inner cities using the criterion of residence and agricultural/nonagricultural status. The ratio of the latter statistic to the former varies significantly from city to city (SSB, 1983a:108). Shanghai, for example, at the end of 1982 had 6.27 million urbanites (excluding the population of the rural counties controlled by the municipality) if based on residence and 6.22 million if based on source of grain supply, so there was minimal difference as indexed by a ratio of 99.2 nonagricultural population per 100.0 total residents. Beijing's ratio was noticeably lower (85.9), reflecting the inclusion of many more residents supplied by grain directly from communes. For big cities such as Chengdu (57.1), Xian (73.5), and Guangzhou (76.3), the ratios were even lower. For all of the 20 cities with more than one million nonagricultural residents, the ratio was 80.8; this contrasted with a ratio of 70.4 for the 20 cities with populations between 500,000 and one million. The contrast suggests that overall size affects the extent to which the different criteria "produce" different urban populations. In general, larger cities are characterized by a high ratio, suggesting that they generally encompass fewer rural areas within their *inner* city boundaries; when the cities expand to incorporate rural areas, these become more rapidly urbanized than newly incorporated rural areas of somewhat smaller cities. Whether this relation extends to medium- and small-sized cities remains to be determined when comparable data on a city-by-city basis become available.

Suburban Counties: Urban or Rural?

Another complicating factor in defining urban population size in 1982 is added, however, because of the state policy (p. 189) that has allowed selected cities to place a number of adjoining counties under their jurisdiction. According to the 1982 Census (SSB, 1983b:table 11), 58 cities of all size classes have suburban counties disproportionally concentrated among big cities. Of these, 45 percent had suburban counties, compared with only 16 percent of medium-sized cities and 6 percent of small cities. For all these places, the city boundaries were effectively extended to encompass inner city districts and the officially designated suburban counties. If no attention is given to this distinction, as in the case of some United Nations reports, China's urban population is greatly expanded. For example, in 1982, the combined inner city and suburban county population numbered 227.1 million, a 50 percent increase over the 145 million urban population of the cities' inner districts, which raises the percentage of the population in cities from 14.5 to 22.7 percent of China's total population.

Some towns classified as urban are within these suburban counties. Since currently available data do not allow separation of these towns from the suburban counties, one cannot ascertain how many of the 61 million reported by the census as living in all towns in China are encompassed within suburban counties. Clearly, if the nation's total urban population were to include not only all the cities and towns, but also the balance of the suburban counties over which cities have jurisdiction, then China's level of urbanization would be well beyond the 20.6 percent reported by the 1982 Census. The choice of data used also affects the nature of China's urban hierarchy. The possibility of using the combined inner city and suburban county population instead of only residents of the inner city as the total city population, argues for careful attention first to which data set is being used.

The Challenge of Comparability

Overall, my assessment suggests that depending on which criteria are adopted, very different views emerge of the character of overall urbanization in China. Insights are difficult to draw, however, on how the nature of urban places changes depending on who is included or excluded under differing definitions. Nor is it possible to judge completely the role of urbanization in the development/modernization process without such information.

The problem of how to define urban populations and the magnitude of the differences are exacerbated by the continued reliance of registry statistics and other data sources on defining *urban* in terms of grain source. This definition introduces major discrepancies that create severe problems of comparability over time and between China and other nations. One can only hope that as the Chinese continue to focus efforts on improving their statistical systems, the problems associated with the use of widely different definitional criteria will be resolved. The 1982 Census decision to include as urban all who had resided for a year or more within cities and towns designated urban by the State Council, regardless of their grain ration source or place of registration, is a major forward step. It warrants replication in other statistical systems to ensure comparability.

As a step toward enhancing standard definitions and comparability, the *1983 Statistical Yearbook* (SSB, 1983a;103) has issued a time series covering 1949-82 in which annual estimates beginning in 1964 have been adjusted to make them comparable with the 1982 criteria. Pre-1964 data have not yet been adjusted to ensure comparability with 1982, creating a break in the continuity of the series between the 1963 and 1964 year-end statistics (Banister, 1984). These differences in comparability must be recognized in the ensuing analysis for data referring to the unadjusted pre-1964 population figures and for data sets thereafter using urban definitions different from those of the 1982 Census. It must, at the same time, be recognized that the quality of the various data sets—including the 1953 and 1964 censuses—varies, which may also affect comparability. As Ansley Coale (1984) has noted, however, various quantitative comparisons among numbers derived from the censuses of 1953, 1964, and 1982 and from the large-scale fertility survey conducted in 1982 show a surprising degree of consistency. (See also Banister, 1984:241-43.)

Levels and Patterns of Urbanization in China

Changes in Urbanization Levels

In 1949, China's total population, according to State Statistical Bureau (SSB, 1983a) estimates, was just more than half of what it was at the time of the 1982 Census: 541,670,000 people in 1949 compared to 1,003,790,450 in 1982. By the end of 1982, the State Statistical Bureau estimated that China's population was 87 percent greater than it had been at the end of 1949. This growth, equivalent to an annual rate of 1.87 percent, did not apply uniformly to urban and rural places.

In 1949, only 10.6 percent of China's population lived in urban places, yet their number (57.6 million people) was more than the total populations of all but 11 countries of the world as recently as 1983. By the 1982 Census, the urbanization level had doubled to 20.6 percent but was still well below the level in developing countries as a whole, about 32 percent in 1982 (United Nations, 1982). (See figure 7.1.) The 206.6 million living in urban places in China were exceeded in number by the total populations of only three countries of the world—the United States, the Soviet Union, and India. By the end of 1982, according to SSB estimates, the number had already risen to 211.5 million, making China's urban population 3.7 times greater than it had been at the end of 1949. By contrast, and despite an overwhelming estimated size of 803.9 million at the end of 1982, China's rural population was only 1.7 times greater than it had been 33 years earlier. These 800 million rural people, a growing number of whom are becoming surplus labor, constitute a huge reservoir of potential migrants to cities, and present China with a major challenge of how to control its urbanization rates while it proceeds to develop and modernize the countryside.

Between 1949 and 1982, the pace of urbanization in China was uneven. Changing policies related to control of urban growth and location of industry, political factors associated with the Great Leap Forward and the Cultural Revolution, and natural catastrophes have together affected the speed of changes in the levels of urbanization. Moreover, because of efforts at different times since the 1950s to reduce the birthrate—success varied over time and between urban and rural places—annual growth rates in urban and rural places have varied considerably from year to year, even while the long-run pattern has been one of rising urbanization levels, concurrent with growing numbers of persons in urban and rural places.

China's first census in 1953 followed a period of reconstruction, and marked the beginning of the first Five-Year Plan (1953-58). At that time, China's urban population was enumerated at 77.3 million and its end-of-year population is estimated by the State Statistical Bureau to have been 78.3 million persons, equivalent to an urbanization level of 13.3 percent (table 7.1). These figures suggest that the initial years following establishment of the People's Republic in 1949 resulted in substantial urban growth. In absolute terms, the number of persons living in cities increased during this four-year interval by 21.6 million, almost as much as did the number living in rural areas, 25.7 million. Given the different base populations, the urban growth rate was much higher—36 percent for the four years, compared to only 5 percent in rural areas.

The first Five-Year Plan had a clear bias toward heavy industry and building large urban centers. It also coincided with the establishment of the people's communes, which caused considerable rural dislocation and massive

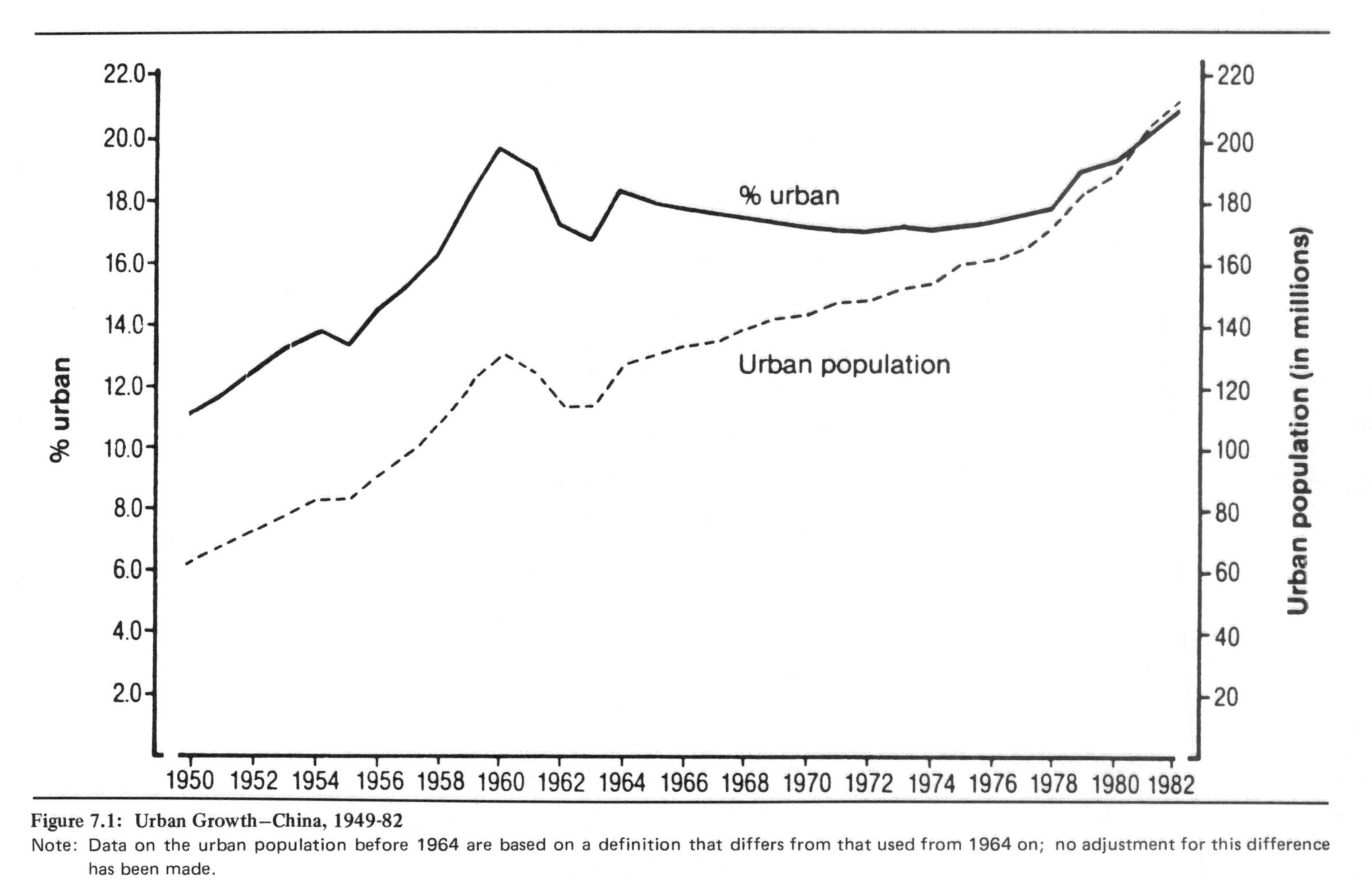

Figure 7.1: Urban Growth–China, 1949-82

Note: Data on the urban population before 1964 are based on a definition that differs from that used from 1964 on; no adjustment for this difference has been made.

TABLE 7.1
Urban and Rural Population—China, 1949-1982
(in thousands at year end)

Year	*Total*	*Urban*[a]	*Rural*[b]	*Percentage Urban*
1949	541,670	57,650	484,020	10.6
1953	587,960	78,260	509,700	13.3
1960	662,070	130,730	531,340	19.7
1964	704,990	129,500	575,490	18.4
1970	829,820	144,240	685,680	17.4
1974	908,590	155,950	752,640	17.2
1980	987,050	191,400	795,650	19.4
1982	1,015,410	211,540	803,870	20.8

SOURCE: SSB (1983a: 103-4).
a. Population of inner cities and towns, including military population.
b. Counties, including suburban counties of cities, exclusive of towns.

rural-to-urban migration (Orleans, 1982:278-79). The outflow of peasants was so great that the resulting pressures on cities led, during 1956-57, to mobilization of millions of people, especially youths, to move from the cities to the villages and also to attempts to institute controls on rural-urban movement. But cities continued to grow. The controls, which met with only limited success, were abandoned altogether in 1958 when the Great Leap Forward began. For the next two years—until natural calamities and mismanagement brought a halt to the Great Leap Forward—the urban population was swollen by in-migrants.

Reflecting the effects of these various policies, the SSB data on urban and rural populations indicate a continuing growth of the urban population from 77 million people at the time of the 1953 Census to a peak of 130.7 million at the end of 1960; the 1960 level of urbanization was thus 19.7 percent, well above that reported in the 1953 Census and almost double that in 1949. With the collapse of the Great Leap Forward, millions of urban residents were relocated to rural areas (Orleans, 1982:279); the urban population declined to a low of 116.5 million people at the end of 1963, accounting for only 16.8 percent of the total population. A growth spurt followed (the reasons are not clearly explained by the data) so that by 1964, the urban population had risen to 129.5 million, or 18.4 percent of the total population.

Since 1964, according to SSB estimates, the urban and the rural populations have been characterized by continuous increases. The unrest associated with the Cultural Revolution between 1966 and 1976 and the return of rusticated youth and others to cities have contributed to urban growth. Despite restrictions on rural-urban migration, such movement has also contributed to the growth of big cities, and even more to the growth of medium- and small-sized places. The result is that between the end of 1964 and the end of 1982, the urban population grew from 129.5 million to 211.5 million, averaging an annual increase of 2.7 percent. The rural population also experienced substantial growth at an average rate of 1.9 percent, increasing from 575.5 million to 803.9 million by the end of 1982.

As earlier discussion of China's definition of *urban* indicated, the urban population includes those living in cities and towns. At the time of the 1982 Census, China had 236 cities and 2,664 towns.[5] The number of people living in cities far exceeded the town population; just over 70 percent of all of the urban population was living in cities, equalling 14.5 percent of China's total population.[6] The 30 percent of the urban population living in towns constituted only 6.1 percent of China's people; the remaining 79.4 percent live in rural areas. Clearly, urbanization in China has been largely a city phenomenon; whether the recent emphasis on towns as a potential location for increasing urban settlement will change this situation remains to be seen.

Regional Distributions

China's population is quite unequally distributed in relation to the geographic size of the various regions.[7] Most extreme are the East region which in 1982 encompassed 29 percent of China's population but only 8.5 percent of its land area, and the Northwest, which contained only 7 percent of the population but one-third of the nation's territory. Other regions also showed imbalances between their proportion of the country's population and the land area. Thus, the three regions encompassing most of China's coastal provinces (East, Northeast, and Central-South) contained two-thirds of the total population but only one-third of the land; and the two regions containing only land-locked provinces (Northwest and Southwest) accounted for 59 percent of the land area but only 23 percent of the population. The North region, which includes the coastal municipality of Tianjin as well as sparsely populated Inner Mongolia, had a closer balance between population and land area: 11.4 percent and 8.3 percent, respectively.

Urban Population Distribution

Data by region (center panel of table 7.2) indicate that China's urban population is also unevenly distributed. Just more than 60 percent of the urban population resides in three regions—the North, Northeast, and East—whereas only 47 percent of the rural population lives in these sections of the country.

The Northwest is sparsely populated and accounts for only 7 percent of China's urban and rural populations. Reflecting their much greater density, but lower levels of urbanization, the Central-South and the Southwest regions account for 32 percent of the urban population but 46 percent of the rural. Sichuan Province alone (located in the Southwest), with its 100 million people and highly rural character, is a major factor in this distribution.

Of interest, too, is the regional distribution of the city and town population. The differences between these two distributions within regions is much less than the differential regional distribution of the urban and rural populations. Only in the North, and to a lesser extent in the Central-South, does the percentage of city population located in the region differ substantially

TABLE 7.2
Population of Cities, Towns, and Rural Areas by Region—China, 1982

Region	*Cities*	*Towns*	*Total Urban*	*Rural Areas*	*Total*
Number of Persons (millions)					
North	24.2	5.2	29.4	85.1	114.6
Northeast	25.9	11.4	37.3	53.6	90.9
East	41.0	18.7	59.7	234.7	294.4
Central-South	27.1	14.9	42.0	230.0	272.0
Southwest	16.5	7.8	24.3	138.5	162.7
Northwest	10.0	3.9	13.9	55.5	69.3
Total PRC	144.7	61.9	206.6	797.4	1,003.9
Percentage Distribution by Region					
North	16.7	8.4	14.3	10.7	11.4
Northeast	17.9	18.4	18.1	6.7	9.1
East	28.4	30.2	28.9	29.4	29.3
Central-South	18.7	24.0	20.3	28.8	27.1
Southwest	11.4	12.6	11.7	17.4	16.2
Northwest	6.9	6.4	6.7	7.0	6.9
Total PRC	100.0	100.0	100.0	100.0	100.0
Percentage Distribution by Urban/Rural Category					
North	21.1	4.6	25.7	74.3	100.0
Northeast	28.5	12.5	41.0	59.0	100.0
East	13.9	6.4	20.3	79.7	100.0
Central-South	9.9	5.5	15.4	84.6	100.0
Southwest	10.1	4.8	14.9	85.1	100.0
Northwest	14.3	5.7	20.0	80.0	100.0
Total PRC	14.4	6.2	20.6	79.4	100.0

SOURCE: SSB (1982: Table 5).

from the percentage of town population. Relatively twice as many of China's city residents live in the North than do its town residents (16.7 versus 8.4 percent), whereas the Central-South region contains substantially more of China's town than city population. The East region, however, clearly contains the largest proportion of China's urban population, judged either by those living in cities or towns. The index of dissimilarity indicates that differences between the regional distribution of city and town populations are less than differences between urban and rural; only 4.4 percent of the population would need to be redistributed for the city and town distributions to resemble each other among regions, whereas 7.5 percent would have to be redistributed before percentages of the urban and rural populations were similar.

Although these regional variations were still pronounced in 1982, comparison with 1953 Census data indicate that they had lessened considerably during the 30-year interval. In 1953, the Northeast had the highest levels of urbanization of any region, with more than one-third of its population urban (table 7.3). The North and East had the next highest percentages urban, but at

TABLE 7.3
Number and Distribution of Urban Population by Region and Percentage Change—China, 1953 and 1982 (in thousands)

Region	Population		Percentage Urban	Percentage Distribution of Urban	Percentage Change 1953-82	
	Total	Urban			Total	Urban
1953[a]						
North	65,000	12,705	19.5	16.4	–	–
Northeast	43,753	15,619	35.7	20.2	–	–
East	179,459	24,221	13.5	31.4	–	–
Central-South	159,563	12,954	8.2	16.8	–	–
Southwest	99,469	8,435	8.4	10.9	–	–
Northwest	35,359	3,323	9.4	4.3	–	–
Total	582,603	77,257	13.3	100.0	–	–
1982[b]						
North	114,566	29,452	25.7	14.3	76.3	131.8
Northeast	90,947	37,315	41.0	18.1	107.9	138.9
East	294,408	59,750	20.3	28.9	64.1	146.7
Central-South	271,956	41,940	15.4	20.3	70.4	223.8
Southwest	162,713	24,249	14.9	11.7	63.6	187.5
Northwest	69,347	13,882	20.0	6.7	96.2	317.8
Total	1,003,937	206,588	20.6	100.0	72.3	167.4

SOURCES: (a) Orleans (1982); (b) SSB (1982).

substantially lower levels. Minimum differences characterized the other three regions, each being 8-9 percent urban.

As previously discussed, this pattern had changed substantially, testifying to some success in China's efforts to decentralize its population and cities. While the Northeast continued to be the most urbanized region, the Central-South, Southwest, and Northwest greatly increased their urban percentages to between 15 and 20 percent. As a result, while the Northeast in 1953 was more than four times as urbanized as the least urbanized regions, in 1982 it was not even three times as urban as the least urbanized, and more often the ratio was 2:1 or 1.5:1. Urbanization, broadly defined, had become considerably more widespread.

The change in overall urbanization levels is also reflected in the changing distribution of the urban population. The urban populations of the three regions in the north and east declined from 68 percent of the country's total to 61 percent, while that of the Central-South increased from 17 to 20 percent, and that of the two western regions from 15 to 18 percent of the total. This redistribution reflects effects of differential growth rates among the regions. As the percentage of change in table 7.3 shows, in all regions the urban population grew at faster rates in this 29-year period than did China's total population, but the extent of the differential varied. For the total population, the 29-year period witnessed regional growth ranging between 64 and 108 percent, with the Northeast being highest and the East and Southwest the lowest. During this same time, however, the urban population grew by between 132 and 318 percent, and growth was especially great in the Central-South and the West.[8] Thus, while the urban population increased by somewhat more than 100 percent in the North and East, it grew by as much as 224 percent in the Central-South and by as much as 318 percent in the Northwest. Policies designed to relocate industry and develop urban places in inland areas obviously had an impact.

These regional data show that distribution of the urban population between cities and towns varies as well (table 7.2, lower panel).[9] The Northeast region, considerably developed during the Japanese occupation, has undergone further substantial industrial development since then. The Northeast is China's most urbanized region with more than four of every ten persons living in urban places and about three of these four living in cities. Not surprisingly, next most urbanized is the North, which includes two of China's three municipalities—Beijing and Tianjin; one-fourth of its population is urban and by far the greatest proportion live in cities.

Urbanization levels in China's remaining four regions cluster within the range of 15-20 percent. Somewhat surprisingly, the East, which includes Shanghai, and the Northwest, encompassing the provinces along the Soviet border, have equal urbanization levels and virtually the same distribution pattern of urban population between cities and towns, a ratio of over 2:1. Reasons for these similar patterns are, however, quite different. In addition to the major metropolis of Shanghai, the East includes a number of older coastal cities; its urban population totals 59 million persons, but it also has very densely populated rural areas. By contrast, urban places in the Northwest consist much more of recently developed industrial cities and a relatively sparsely settled terrain. The total urban population (about 14 million people) is only about 23 percent as large as that of the East region; in fact, with its 59 million urbanites, the East's urban population is not far below the total population of the Norhtwest. The lowest levels of urbanization characterize the Central-South and Southwest regions of China, with 15 percent of their populations in urban places, divided in a ratio of about 2:1 between cities and towns.

Distribution of Urban Places

Still another perspective for evaluating regional differences in urbanization is to examine the location of the cities and towns themselves rather than the populations living in them. The data in table 7.4 present statistics for the 236 cities and 2,664 towns that constitute urban units of the 1982 Census. The results differ somewhat from the patterns indicated by the distribution of urban population. Judging by the number of cities, regardless of size, the Central-South region is the most urbanized, with 27 percent of China's cities located there, followed very closely by the East, which accounts for just more than one-fourth of the cities. No other region contains more than 15 percent, and the Northwest has less than 10 percent of China's cities. Again, influence of the coastal areas is clear. By contrast, towns are much more widely dispersed. The East and Central/South reverse the order, but the difference between the two is not great; together they account for 52 percent of all of China's towns, just as they accounted for slightly more than 52 percent of its cities. However, the Southwest is also prominent among the regions for a substantial percentage of towns, probably reflecting a relatively high population density coupled with a relatively low proportion of cities, so that towns are relied on more heavily for commercial functions.

Considerable regional variation exists for the average population size of cities and towns in the respective regions (table 7.4). For China as a whole, the 236 cities averaged a population of 613 thousand in 1982, ranging from a high of 807.8 thousand for cities in the North to averages of 432.4 thousand in the Northwest and 422.9 in the Central-South. While the high average sizes of the North and East are not surprising, given the location of many of the major metropolises in these regions, the unusually high average for the Northeast, reflecting the considerable industrial/urban development, is noteworthy.

Towns in China obviously represent a category in themselves, averaging only about 23 thousand persons. The Northeast is outstanding; its average exceeds that of China as a whole by 15.4 percentage points. The North, East, and Central/South regions vary minimally with respect to average size of towns and are close to the all-China average. The more remote regions in the country have averages below 20 thousand. This pattern, too, may be related to the smaller number of cities in these comparatively large geographic areas. Towns in these regions must serve multiple functions, accomplished through many small towns scattered over a wide area. Further insights into these patterns may be gained through examining statistics on the distribution of cities by size.

Distribution by Size Class of Cities

National Patterns

Between 1953 and 1982, the number of cities in China increased from 173 to 236. Comparative distribution of these cities by size class[10] (table 7.5) clearly indicates a substantial upward shift to larger size categories. In 1953, four of every ten cities had under 100,000 inhabitants and an additional 30 percent had between 100,000 and 200,000 (Ullman, 1961:10). Compared with 1953, when there were 71 cities under 100,000 only 18 cities under that size were identified in 1982 (7.6 percent of all cities). Even if cities in the next largest size class (100,000-200,000) are included, the percentage of small cities remained below one-third of the total, compared to two-thirds 30 years earlier. In fact, except for the size group under 200,000, every size category had a greater absolute number of cities in 1982 than in 1953. All but one category increased about threefold in number; cities ranging in the one million to 2.5 million category were five times more numerous in 1982 than in 1953. China had only

TABLE 7.4
Number of Cities and Towns by Region and Average Population Size—China, 1982

Region	*Number* Cities	Towns	*Percentage Distribution* Cities	Towns	*Average Size (thousands)* Cities	Towns
North	30	221	12.7	8.3	807.8	23.6
Northeast	34	295	14.4	11.1	762.5	38.6
East	60	752	25.5	28.2	683.9	24.9
Central-South	64	646	27.1	24.2	422.9	23.0
Southwest	25	539	10.6	20.3	659.1	14.4
Northwest	23	211	9.7	7.9	432.4	18.7
Total PRC	236	2,664	100.0	100.0	613.0	23.2

SOURCE: SSB (1982: Table 5).

9 million-plus cities in 1953, but 38 in 1982, 8 of them with 2.5 million or more inhabitants. Although the three municipalities—Shanghai, Beijing, and Tianjin—dominated the urban hierarchy of China, the greatest proportion of Chinese cities was in the medium-sized category; more than one-third of the cities had between 200,000 and 500,000 residents.

The distribution using cities as units of analysis does not fully reveal the size category in which the urban population is concentrated. Examination of the distribution in 1982 of the city population by city size shows that the metropolises dominated the urban hierarchy even more so than their numbers would indicate (table 7.5). While cities of a million or more constituted only 16 percent of all cities, they accounted for just more than 50 percent of all of China's population living in cities, and almost half of these people lived in cities of 2.5 million persons or more.[11] Medium-sized cities, by contrast, while accounting for just more than one-third of the total number of cities in China, encompassed only 19 percent of the population living in China's cities. An even greater discrepancy characterizes the smaller cities which, while accounting for 28 percent of all cities, contained only 6 percent of the city population. Clearly, China's city population is unequally distributed by size category and helps to explain why government policy invokes strict control on further growth of big cities while encouraging growth of small cities and towns to serve as alternative centers of settlement and linkages between larger cities and the countryside (Pannell, 1984:210-13).

Restricting the assessment of distribution by size category to cities omits the populations living in towns, also part of the urban hierarchy. The 2,664 places classified as towns accounted for over 90 percent of all urban places in China and for almost 30 percent of the population in urban places. Although published census statistics do not include data on the size of towns, such information is provided by Ding (1984), who describes the urban/rural distribution of China's population using 1982 Census results. These data show considerable variation in the size of places designated towns. Of the 2,664 towns, 26 had more than 100,000 residents and an additional 231 contained 50,000-100,000 people. Such places are similar in size to the 66 small

TABLE 7.5
Distribution of Cities and City Population by Size, Class and Percentage Change—China, 1953 and 1982

Size Class	1953[a] Number	1953[a] Percentage Distribution	1982[b] Number	1982[b] Percentage Distribution	Percentage Change 1953-82
Cities					
5 million and over	1	0.6	3	1.3	+200.0
2.5-4.99 million	2	1.1	5	2.1	+150.0
1.0-2.49 million	6	3.5	30	12.7	+400.0
500,000-999,999	16	9.2	46	19.5	+187.5
200,000-499,999	28	16.2	85	36.0	+203.6
100,000-199,999	49	28.3	49	20.8	0
Under 100,000	71	41.1	18	7.6	–74.7
Total PRC	173	100.0	236	100.0	+36.4
City Population (millions)					
5 million and over	6.2	11.8	17.0	11.7	+174.2
2.5-4.99 million	5.5	10.5	15.6	10.8	+183.6
1.0-2.49 million	9.3	17.8	42.6	29.4	+358.1
500,000-999,999	11.3	21.6	32.8	22.7	+190.3
200,000-499,999	8.5	16.2	28.0	19.3	+229.4
100,000-199,999	7.2	13.7	7.4	5.1	+2.8
Under 100,000	4.4	8.4	1.4	1.0	–69.2
Total PRC	52.4	100.0	144.8	100.0	+176.3

SOURCE: (a) Ullman (1961); (b) SSB (1983b).

cities of less than 200,000. On the other hand, over 90 percent of China's towns had under 50,000 population, and 58 percent were reported as having under 20,000. Should China succeed in developing a significant number of new towns as urban growth centers and increasing the population living in existing towns so that they can absorb a substantial proportion of the agricultural surplus labor, the predominance of these smaller places in the urban hierarchy, judged by their numbers and the percentage of urban population living in them, should increase significantly in the years ahead.

Regional Distributions

The distribution of cities by size for all regions also points to considerable regional variation in the urban hierarchy. Two of China's three municipalities are in the North, separated by only a short distance. The third, Shanghai, is in the East. Neither region has a city in the next category, 2.5 to 5 million people. The five cities in this class are distributed among three other regions—the Northeast, Central-South, and Southwest. Thus, all regions but the Northwest have at least one city of 2.5 million or more population. The Northwest's largest city, Xian, had only 2.2 million people in 1982. Cities in each of the other classes are distributed among all regions.

When comparisons are made in terms of big, medium, and small cities (table 7.6), for China as a whole, the big- and medium-sized cities are nearly

TABLE 7.6
Percentage Distribution of Cities and City Population by Size Class and Region—China, 1982

Size Class	*North*	*Northeast*	*East*	*Central-South*	*Southwest*	*Northwest*	*Total*
Cities							
Big	33.3	58.8	43.4	21.8	32.0	26.1	35.6
Medium	40.0	26.5	33.3	37.5	44.0	39.1	36.0
Small	26.7	14.7	23.3	40.7	24.0	34.8	28.4
Total	30	34	60	64	25	23	236
City Population							
Big	79.2	86.6	76.8	59.3	76.1	62.3	74.6
Medium	16.5	11.5	17.7	28.8	19.2	27.7	19.3
Small	4.3	1.9	5.5	11.9	4.7	10.0	6.1

SOURCE: SSB (1983b).
NOTE: Percentages sum to 100.

equal, with small cities not far behind. However, this pattern does not extend to each region. The North and Southwest are quite similar while the Northeast has a conspicuously larger proportion of big cities than any other region and the smallest of small cities. The East also has a relatively larger proportion of big cities. The reverse characterizes the Central-South region, where more cities are small and fewer are big. Like the Central-South region, the Northwest also has a relatively smaller proportion of big cities, but the remainder are almost equally divided between the other two categories.

The population distributes very differently among size categories by region than do the cities. In all regions, 60 percent or more of the city population resides in big cities. For all but the Central-South and Northwest regions, the percentage is above 75 and for the Northeast it is 87 percent. All regions have far more of their city population living in medium- rather than small-sized cities, but in only the Central-South and Northwest is it as high as one-fourth. These two regions are also the only ones to have at least 10 percent of their city populations in small cities. Despite the stress in the 1980s on small places, in China as a whole and in most regions, small cities still account for only a low percentage of total city population.

Changes in Big City Distribution, 1953-82

Given China's concern with controlling growth of big cities, particular attention to this size category (500,000 and over) seems in order. Such a focus provides further insights into changing urbanization patterns between the 1953 and 1982 Censuses. In 1953 China had 25 big cities (Ullman, 1961), compared to the 85 identified in the 1982 Census. Although there had been a sharp rise in the number of big cities, their geographic distribution did not change substantially. In 1953, 17 of the 25 cities (68 percent) were located in the North, Northeast, and East regions and only four (16 percent) were in the West; the remaining four were in the Central-South region. By 1982, the number of big cities in the three northern and eastern regions had more than tripled to 57, but still constituted 67 percent of all the big cities in China. Moreover, the West and the Central-South each still contained 16 percent of the big cities. Overall, therefore, while these three decades witnessed significant increases in the number of big cities in each region, these data point to minimal change in regional distribution, suggesting that development of new big cities inland has been balanced by big city growth in the earlier developed coastal areas.

If only million-plus big cities are considered, some change in regional distribution becomes apparent. In 1953, two-thirds of the million-plus cities were in the North, Northeast, and East; by 1982, these regions still accounted for 71 percent. But while the West had only 3 such cities in 1953 (11 percent of all million-plus cities), it had 11 in 1982, 18 percent of all such cities in China. The Central-South region gained only two million-plus cities in these three decades, halving its percentage of China's total from 22 to 10 percent. The overall impression from these comparative data for 1953 and 1982 is that both eastern and western China have been characterized by big city growth, with the western region showing more change in the relative number of very big cities.

The changing distribution by size category of China's big cities has involved considerable shifting in size classification of individual cities—shifts marked by net gains and losses to individual categories. Of the 85 cities of 500,000 and over, only 3 were in the same size class in 1982 as they had been in 1953: Shanghai in the 5 million-plus group, Nanjing in the 1-2.5 million class, and Wuxi in the 500,000-1 million group. The other 82 cities shifted upward, including Beijing and Tianjin, which moved from the 2.5-5 million class to the 5 million-plus class. Of the 60 big cities in 1982 that had fewer than 500,000 residents in 1953, 14 grew to over a million. Most of the newcomers to big-city status—40 of the 60 additions—were located in the North, Northeast, and East. However, ten big cities were added to the four in the two western regions in 1953, reflecting concerted efforts to spread major urban centers inland.

Further insights into the extent of change in China's urban hierarchy can be obtained by comparing the 25 largest cities and their rank size in 1953 with the 25 largest in 1982. Each of the 25 largest cities in 1953 had at least 500,000 population; in 1982, each had over 1.2 million inhabitants; 18 of these were among the 25 most populated in 1953, and their comparative rank had changed minimally. In fact, the Spearman rank correlation between the two sets was .978. All seven of the 1953 largest cities not qualifying in 1982 were located in the eastern half of China; of the seven newcomers to the list, three were situated in the western half of the nation. These data again point to a limited shift westward in the location of big cities.

Primacy

Concern with urbanization in developing countries has focused not only on the overall rate of urbanization and urban growth, but also on the nature of the urban hierarchy and the extent of spatial polarization as evidenced by conditions of primacy. In a number of developing countries, including some in Asia, the major city—usually the national capital—has far outpaced other cities in its growth rate; it encompasses a disproportional share of the nation's economic, political, cultural, and educational activities, and receives a large share of total investment from the national government and overseas. The same locations have very often been characterized by high degrees of social and economic inequality among their residents, even while remaining the goals of a large number of prospective migrants from rural areas and small towns because these cities are viewed as providing far greater opportunities than other locations.

Chinese planners and policymakers have been cognizant of this potential situation and concerned with existing conditions in China's big cities and the possibility that their uncontrolled growth would lead to problems similar to those characterizing many other primate cities in Asia and elsewhere. In response, during the 1960s and 1970s, the Chinese government attempted to develop policies designed to promote wider income distribution, reduce reginal inequalities, and create a more balanced urban hierarchy, which would lead to a greater decentralization of economic activities. In doing so, the intention was to slow population growth in the largest cities, while allowing continued increases in medium-sized and smaller urban centers. Primacy of the larger cities would thereby be reduced, and growth of towns and the smallest cities would lead to more effective linkages to rural areas and expedite modernization and development of the countryside. Given this general program, one would expect primacy levels characterizing China in 1953 would have been considerably reduced by 1982.

Because China is so vast and activities are still largely organized in terms of provincial and regional networks, a fair assessment of the extent of primacy should be undertaken on regional and provincial levels rather than on the national level.

Measurement of the degree of primacy in 1982 relies on census data on size of cities (using inner city data only) to construct a four-city primacy index. This index compares the size of the leading city in the country, region, and individual provinces, respectively, with the size of the next three largest cities in the same geographic grouping. If the cities conform to the rank-size rule, then the resulting index will be approximately 100 since, according to the rule, the second city should approximate half of the first, the third city one-third of the first, and the fourth city one-fourth of the first; that is, the second to fourth cities together should approximate the size of the first city. An index above 100 points to primacy and one below 100 suggests that the next few cities are well above what one would expect according to a rank-size normal distribution.

The expectation, based on the earlier review of the urban hierarchy, is that China as a whole will show no indication of a primate condition, but that primacy will be characteristic of certain regions and even more so in selected provinces (see also Yeh and Xu, 1984). The data confirm this expectation (table 7.7). The index for China as a whole is only 43, suggesting that the largest cities in China do not differ much from each other in size and that no one city dominates the urban scene, as indexed by population size. This is not surprising, given the urban history of China as well as the total size of the country. This conclusion is confirmed even if primacy is measured in terms of an 11-city index (38) instead of a 4-city index (43).[12]

Somewhat surprisingly, the four-city primacy index for each of the six regions also points to a general absence of primacy conditions, with the sole exception of the East region where Shanghai is so predominant. When the index for the East region is based on 11 cities rather than 4, however, it drops to 90.5, suggesting that in relation to the larger urban structure, even Shanghai is not exceptionally large. For all other regions, the four-city primacy index is well below 100. The same conclusions are reached with the 11-city index (table 7-7). Overall, these indexes therefore suggest that on a regional and national level, little evidence of primacy exists, with the possible exception of the East region.

When the level of comparison is reduced to provinces and autonomous regions, conditions of primacy manifest themselves in selected areas. (The primacy index could not be calculated for Beijing, Tianjin, and Shanghai since each city is equivalent to a province and therefore has no smaller cities within the same political unit. Nor was the index constructed for Tibet, Qinghai, and Ningxia because the nature of the urban structure in these frontier provinces did not lend itself to such calculations.) No evidence of primacy emerges from the data for provinces in the North, East, or Northeast; in the latter, only Liaoning Province shows a slight indication of primacy, but the index of 102.2 does not point to strong primacy. In fact, in most of the provinces in these three regions, the index is considerably below 100. In the eastern provinces of China, including those which are most densely settled, the larger cities apparently do not dominate the urban hierarchy, as judged by size alone. Indeed, the data suggest that these three regions of China may be characterized by an urban hierarchy conforming closely to the goals national policies advocate.

In contrast, in the Central-South and two western regions, 6 of the 11 provinces for which primacy indexeds were calculated have indexes above 100, and usually considerably so. Hubei's high index reflects the dominance of the industrial city and transport junction of Wuhan, which numbered 3.3 million people in 1982. In Guangdong, Guangzhou continues to dominate the urban scene, probably reflecting its importance vis-à-vis Hong Kong and as a port of entry for China. The high primacy characterizing the four other provinces is undoubtedly related to their less developed conditions. In the Northwest, Lanzhou, in Gansu Province, is particularly noteworthy among cities whose population has grown rapidly as a result of efforts to relocate industry inland from the coastal regions. And in Shaanxi Province, the ancient city of Xian has taken on increased importance as a provincial and regional center as reflected in its primacy index.

TABLE 7.7
Primacy Indexes by Region and Province—China, 1982

Region and Province	*4-City Primacy Index*	*11-City Primacy Index*
China	43.0	38.0
North Region	66.6	77.9
Beijing	–	
Tianjin	–	
Hebei	53.8	
Shanxi	92.4	
Mongolia	81.9	
Northeast Region	68.6	61.9
Liaoning	102.2	
Jilin	68.4	
Heilongjiang	91.6	
East Region	111.4	90.5
Shanghai	–	
Jiangsu	93.2	
Zhejiang	38.7	
Anhui	57.3	
Fujian	84.3	
Jiangxi	58.0	
Shandong	56.7	
Central-South Region	58.2	58.6
Henan	67.5	
Hubei	309.1	
Hunan	53.4	
Guangdong	160.3	
Guangxi	70.6	
Southwest Region	44.4	47.9
Sichuan	61.8	
Guizhou	110.8	
Yunnan	170.5	
Tibet	–	
Northwest Region	75.5	77.0
Shaanxi	177.7	
Gansu	279.3	
Qinghai	–	
Ningxia	–	
Xinjiang	89.2	

SOURCE: SSB (1983b).

Previous research on the changing system of Chinese cities (Chang, 1976) has documented that one of the most remarkable features of China's urban development since 1949 has been rapid growth of provincial capitals. Based on comparisons of the census enumeration of 1953 and estimates of provincial capital populations in 1970, Chang found that 10 of 25 capital cities of provinces and autonomous regions had more than doubled their population, and 5 had been tripled. A number of these rapidly growing cities are located in frontier regions; many of the other, older capitals had undergone substantial growth as the result of improved linkages to national transportation networks, improvements in infrastructure, and development of heavy and light industries. The net effect of these improvements, Chang observed, was an increase in the urban primacy of many provincial-level capital cities in relation to other cities within the province.

In his analysis of provincial capitals, Chang calculated a two-city primacy index for 1970, based on estimates of city size for that year, and made some comparisons with the index of 1949. His evaluation indicated that "primacy declined in well-developed provinces and it increased in less developed provinces" (Chang, 1976:404). In particular, he found that in 1970 three of the four capital cities with a primacy index greater than 10 (more than ten times the size of the second city) were located in inland provinces of western regions. The three—Lanzhou (Gansu), Xining (Qinghai), and Kunming (Yunnan)—were all newly emerging industrial or transport centers as was the fourth city, Wuhan, which had become a key industrial center at the juncture of the Yangtze and Han Rivers in Hubei Province, Central-South region. Of these four, only Wuhan had had an index greater than 10 in 1953; by 1970, it was 12.8. Lanzhou's index had increased from 4.7 to 19.3, and Kunming's from 4.4 to 14.7. Overall, for the 23 provinces and autonomous regions for which comparisons could be made between 1953 and 1970, 15 had higher indexes in 1970 than in 1953, and only 8 had lower ones. Xining's index was not measurable in 1953 since no second municipality existed in Qinghai Province at that time.

If the 2-city index is calculated for the 25 provincial capitals in 1982 (Tibet was not included in the 1970 analysis), results show that no city had an index greater than 10, suggesting the higher growth rates of smaller cities in China between 1970 and 1982. In fact, the index is lower in 20 of the 25 provinces and autonomous regions that can be compared between 1982 and 1970 and often the differences are considerable. For example, for Lanzhou, which had the highest index in 1970, it had declined from 19.3 to 7.0; for Wuhan from 12.8 to 8.7; for Guangzhou from 8.3 to 3.3. For Xining the index decreased from 16.7 to 9.6; its index was the highest of all capitals in 1982. Further attesting to lower primacy levels is that the number of capital cities with indexes less than 2 increased from 9 in 1970 to 15 in 1982. Whereas four capital cities in 1970 were estimated to have indexes below 1, indicating they were exceeded in size by another city in the province, this was true of seven capitals in 1982.

While primacy had declined over these 12 years, the newly industrializing inland provinces generally continued to display the highest indexes. In fact, the rank correlation between primacy indexes in 1970 and 1982 was .70, suggesting comparatively little change in ranking; the major changes, therefore, seem to be the general reduction in absolute levels of primacy rather than changes in comparative rankings in primacy conditions among provinces. Interestingly, a similar conclusion emerges from a comparison between rankings in 1982 with those in 1953; for the 23 provinces and autonomous regions between which comparisons can be made, the rank correlation is .67.

Discussion

The concerted efforts of the Chinese government to control population growth through its one-child family policy has understandably received worldwide attention. Less well known outside China is the considerable attention given by Chinese government officials at all levels to problems related to rural-urban population distribution, rates of urban growth, and relations between employment opportunities and rural and urban development. These concerns have led to the emergence of a clearly and firmly articulated policy regarding population movement and distribution of population between rural and urban places and among urban places of different size.

China's officials consider efforts in these areas, like those in fertility control, to be of critical significance for future national development and modernization. From an international perspective, efforts to control population distribution and achieve orderly urban growth merit close monitoring and evaluation for the lessons they may provide for other developing countries. To the extent that many developing countries have rated problems of population distribution even higher than those of population growth, China's experiences should be of particular interest. They may provide insights on ways to avoid some of the negative consequences of too rapid urban growth at the same time that benefits of urbanization for the overall development process and the absorption of surplus rural labor can be realized.

Availability of data from the 1982 Chinese Census has provided a unique opportunity to assess in some depth urbanization patterns in China and changes since the Census of 1953. This analysis has documented that in the three and one-half decades since establishment of the People's Republic, the urbanization level has risen slowly, from 10.6 percent in 1949 to 20.6 percent in 1982. Moreover, despite some changes in the urban hierarchy and some evidence of success in control of big city growth, almost two-thirds of the total urban population remains concentrated in big cities (500,000 and more) and almost 40 percent live in metropolises of more than a million. In fact, with 38 such metropolises, China has more million plus cities than any other country of the world, just as its 1982 urban population of 206 million exceeds that of all other countries of the world; indeed, despite their low percentage of China's total population, China's urban residents are more numerous than the total populations living in any country of the world except India, the Soviet Union, and the United States. The Chinese situation is particularly challenging because, concurrently, about 800 million people still live in rural areas and are largely engaged in agricultural activities. This means that the concerted efforts now in process to modernize and develop the country must, sooner or later, involve the absorption of several hundred million more rural people into nonagricultural activities and possibly into urban places. China's urban population also remains imbalanced in its geographic distribution, although the extent of this imbalance has diminished some through efforts to achieve higher rates of industrialization and urbanization and city development in the inland provinces.

For many of the 37 years since establishment of the People's Republic, smaller cities and towns did not fare well; inadequate production of rural commercial products meant that functions of small urban places as bridges between larger urban centers and rural areas were generally curtailed. Residents of many smaller places moved to big or medium-sized cities; other town residents joined communes. Since 1980, however, Chinese policymakers see these smaller places as playing a key role in the future urbanization of China as well as in overall rural development (Fei, 1984). These areas are being increasingly looked to as potentially absorbing the vast surplus rural labor that is developing, and as incipient urban centers that will provide alternative locations for industry and commerce. As such, these smaller places are thereby expected to relieve the pressure on larger cities by providing urban amenities to rural residents who might otherwise seek these attractions by migrating to cities, and by also providing job opportunities to increasing numbers of rural residents who will continue to live in rural areas, sometimes even functioning partly as peasants while commuting to small towns and cities to work in private, collective, or state-owned industry and commerce.

The limited evidence already available from the 1982 Census on characteristics of the population living in China's 2,664 towns suggests that these places have already assumed an urban character and are likely to play a key role in China's future urbanization. The basic urbanization policy, adopted in 1980, calls for strict control of the expansion of big cities, rational development of medium-sized ones, and vigorous efforts to build up small cities and towns. Such a policy is premised on the belief that the close linkage of industry and agriculture in smaller cities and towns will allow fuller use of local natural resources, raw materials, and manpower. Thus they are expected to absorb the surplus rural labor force resulting from the combined effects of population growth and introduction of the responsibility system in agriculture.

Local industry and workshops operated by communes and brigades and small private enterprises are seen as providing the operative mechanism in small town development. In some areas, the government gives direct help to peasants to enable them to make the transition from agricultural to nonagricultural activities. Such help includes financial assistance, provision of raw materials, and technical training. Peasants may obtain permits to become specialists in such service activities as tailoring, carpentry, and blacksmithing, and then to hire apprentices. Others may be given permission to open shops or operate small factories, or to purchase tractors, carts, or boats for transportation. Technically qualified peasants may be allowed to operate nurseries, bookshops, or clinics and are offered technical guidance to enhance the quality of their activities. These private economic activities are a drastic break with earlier restrictions. Accordingly, reports indicate, the government is seeking to provide legal protection and adequate publicity to the new policy as a way of ensuring that peasants are not deterred from engaging in such activities out of fear of being stigmatized by fellow peasants as being "out-of-line."

The policy stress placed on the potential value of stronger linkages between rural and small urban places and the even broader networks into which such places fit are evidenced in the Circular of the Communist Party of China Central Committee, issued early in 1984 as Document no. 1 (*Renmin Ribao* [People's Daily], 22 January 1985) on the topic of rural work. It called for various measures to improve the infrastructure for commodity circulation, including better provision of storage facilities, warehouses, transportation, and communication; it also recognized that big and medium-sized cities play a key role in rural development by providing free markets for peasants; offering sites for wholesale markets for farm produce and sideline products, and becoming sites where trade centers might be created.

Concurrently, this key document recognizes that with development of greater labor divisions in rural areas, many more people will withdraw from cultivation and farming to engage in sideline activities, such as forestry, fishing, and animal husbandry, and will also transfer into industry and commerce in small towns. As the document phrased it, "If we could not change the situation of 800 million people engaged in farming, peasants would not be well off, the country could not prosper and be strong, and the four modernizations would not be realized."

The document also recognizes existing commune- and brigade-run enterprises as pillars of the rural economy, often closely linked to large factories in cities. In particular, the document envisages rural industry as centralized in towns to take advantage of economies with respect to energy use, storage costs, transportation, water needs, sewage disposal, and other infrastructure. Such concentration is assumed also to serve as an impetus for providing the rural population with improved cultural, educational, and service undertakings, by making towns the economic and cultural centers of rural areas.

Perhaps the strongest evidence of the emphasis placed on development of small towns as alternatives to further growth in large cities and of their role in rural development is the provision in the document that peasants will be allowed, on an experimental basis, to settle in towns to engage in industry, business, and service trade providing they make their own grain arrangements (presumably mainly with the communes where they have been members) and do not become dependent on state supplies. The policy implies that peasants would be allowed to build homes in the towns, although they would not have permanent registration there; urban residence might be terminated if conditions should change such that continued peasant residence burdened the town's ability to provide jobs, housing, and grain supplies. Anticipating such developments, the document urges towns to plan for short- and long-term construction and, in doing so, to use land sparingly. Clearly, recent developments in rural areas and the very substantial exodus by peasants out of agriculture into industrial, commercial, and service work has given impetus to this emphasis on the town as a place of residence and a link in the rural-urban network.

These links have been strengthened even further by a document released by the State Council in 1984 (*Renmin Ribao* [People's Daily], 22 October 1984) stipulating that peasants engaged in nonagricultural work may obtain registration in towns at the commune level. In doing so, peasants must return their assigned land to the production brigade, but they reserve the right to return to their place of origin. The same document urges authorities in commune towns to make building materials and rentals conveniently available.

Policymakers anticipate that linkages to larger urban places will be reenforced by these developments. One such link is created through the tie-in of small rural factories with large urban plants. The small workshops produce component parts for the larger factories, thereby obviating the need for more urban construction and movement of workers into cities. Commodity production for the urban market also helps relieve the pressures resulting from the inability of urban industry to meet market demand. Such interaction between urban and rural industries and the involvement of urban experts as advisers to the newly developing rural industrial and commercial activities provides further opportunities for urban ideas and knowledge to spread to rural areas. Rural industry is seen as contributing to the creation of urban facilities in rural locations, and the transformation of some commune seats into small towns. Employment in small factories and commercial establishments, coupled with use of commune income to improve schools, medical, recreation, and business facilities, and road and other infrastructure, is envisaged as leading to a higher quality of rural life and reduction in the desire to move to cities. These developments are not without their problems: considerable concern has already arisen about (1) the extent to which materials produced in rural workshops match the quality of those produced in the large urban factories, (2) availability of adequate energy in rural areas, and (3) pollution of the rural environment.

Evidencing the rapid pace of policy implementation since 1980 and elaborated on in Document no. 1 at the beginning of 1984 is the report in *Renmin Ribao* ([People's Daily], 9 September 1984) that by the end of June 1984 the total number of towns in China had risen to 5,698, setting, as they put it, "a record figure since liberation." The report indicates that since release of Document no. 1, 2,900 new towns were established in different places in the country and a total of 10 thousand towns is expected by the end of the year. While the report does not indicate the exact process by which this dramatic increase was achieved, the very fact that it is reported testifies to the weight given to small town development as the direction of future urbanization.

News reports in China indicate that the growing prosperity characterizing many rural areas is revitalizing old Chinese towns. Enterprises in such towns are allowing substantial numbers of the rural labor force to shift from agricultural to nonagricultural employment; many of them work in factories and workshops established in small towns while continuing to live in their villages. Some are able to take a factory or service job in the town while concurrently continuing to engage in agriculture under the contract system that is part of the responsibility system. Some work only at their town jobs or in nonagricultural activities in the village, while other household members engage in farming or sideline activities. A growing number of rural households throughout China consist of such mixes, giving rise to the label "half worker/half peasant households" (*China Daily*, 1984c:1). By the end of 1983, industrial township enterprises are reported to have absorbed 32 million farm workers (*China Daily*, 1984b).

As a result of the diversification of economic activities stimulated by small town development and the responsibility system, considerable temporary movement from rural to urban places has evolved. Such circulation has provided still another mechanism for absorbing surplus labor, transferring capital from urban to rural places, and satisfying the desires of many peasants to participate in urban life. From the point of view of the rural area, such migration allows surplus labor to engage in productive activity, and avoid becoming a liability to the brigade or commune; by going to the city or to another rural area to earn income, the individual is able to contribute to the support of agricultural development through remittances used by the brigade for purchase of machinery, fertilizer, and other items. At the same time, since such individuals usually go to the city only if they know they can find employment, they are not seen as putting pressure on urban facilities, but rather as providing a desirable function. Nor do they place a burden on the state, since responsibility for their basic food needs and social welfare continues to rest with the commune. All these temporary urban dwellers, as well as the rural labor force engaged in nonagricultural work, are thus considered to be contributing to modernization of the countryside and are also seen as forging links between countryside and city and integrating farming, industry, and service activities.

Together, Chinese policymakers see such temporary movement and the job opportunities expected to be provided by rural-based industry and sideline activities in small towns as preventing surplus rural labor from flooding into the large cities on a permanent basis. Such development is thus regarded as a way of avoiding the experiences of other developing countries which have been characterized as having led to the mass movement of population into big cities and the transfer of rural poverty to urban places. In contrast, the Chinese argue that their new agricultural policy and the responsibility system provide the best approach to raising the rural income level, creating more job opportunities for the anticipated massive increase in surplus rural labor, and fostering the eventual concentration of commodity production and specialized activities in selected rural areas which have the potential of becoming small urban centers.

While the immediate available evidence points to success in meeting these goals, it is premature to judge whether these changes can operate efficiently on a large enough scale to absorb the millions of surplus agricultural laborers who will need to be absorbed. Moreover, it is not at all clear whether the urban amenities provided in this way will be sufficient to meet the preferences many persons will develop for participation in the consumer society represented by life in the medium-sized or big city. Rising standards of living may well lead to continuously rising levels of aspirations that can be met fully only by permanent residence in a big city. The extent to which changes now in process can substitute for such urban residence need careful monitoring and evaluation; the experiences with current policies and efforts to implement them have obvious implications not only for China's urban structure, but for urbanization in developing countries.

Strongly favoring growth of small cities and towns as alternates to bigger city growth, while also determined to control growth of big cities, Chinese planners do seem increasingly to recognize the important role larger cities can play in China's overall development and development of smaller urban places and the rural hinterlands of big cities. Since the early 1980s, big cities, such as Chongqin, Wuhan, and Shanghai have been encouraged to help smaller places by providing technical expertise and financial support, and creating local industrial establishments, which may be branch operations of larger firms centered in the cities. Unified planning, designed to exploit and integrate the advantages offered by each of the localities within the region, is intended to rely on the key city as the center and break the barriers among urban and rural regions, provinces, and departments by creating intertrade and transregional economic zones and networks (Lin and *Beijing Review* Staff, 1984). In this way, it is hoped that the large cities will "carry" along the smaller, less-developed places in the modernization process. Creating these networks is not envisaged as increasing migration to the big city; rather the hope is that the greater integration achieved will stimulate growth rates of smaller urban places and encourage movement to them.

Still one other development testifies to the continuing importance attached to large urban places in China. Reflecting the needs associated with its modernization efforts and the need for technology and foreign exchange, 14 coastal cities in China have been designated economic development zones to join the four special economic zones (SEZs) as centers of foreign business investment. Like the SEZs, the port cities are to be allowed to practice flexible economic policies, including greater control over foreign economic relations and trade and greater power to offer preferential treatment to overseas investors (*China Daily*, 1984a). All of the cities are located in economically developed areas, and have relatively solid foundations of industry, science, and technology. Plans call for establishment within the cities of a number of economic and technical development districts where manufacturing, commercial, and scientific facilities will be located, away from existing areas of high urban density. It is not anticipated, however, that this will lead to rural-urban migration; rather, these centers are expected to rely on their own labor supply or on migrants from other urban places in the same general area.

Recent policies giving big cities leadership roles in the development of smaller urban places and rural hinterlands and policies designating the 14 coastal cities as special centers of economic development clearly indicate that big cities will continue to play a key role in China's modernization process. Inevitably, these policies bring into question whether China can concurrently succeed in its goal of strictly controlling the demographic growth of big cities. Are these policies in contradiction to each other? Success in efforts to achieve greater development and leadership at the center, especially when this will involve expansion of manufacturing, commercial, and technological activities, would be expected to lead to a demand for more manpower than is available from the resident population. Especially if current migration restrictions are relaxed in the interests of more efficient matching of skills with needs, migration to these cities may take on increased importance. Since both policies involving new roles for selected cities have been introduced since the early 1980s, it is still premature to assess either their success or demographic impact.

It is fortunate that the introduction in 1979 of the responsibility system and the urbanization policies adopted in the 1980s preceded or followed by only a short time the 1982 Census. As a result, census data can provide baseline information on the structure of the urban hierarchy. Such data allow comparisons of urban and rural populations at a time when efforts were initiated to foster small city and town growth and to absorb surplus rural labor in nonagricultural activities. Using the 1982 data, later surveys and censuses will be able more easily to assess changes resulting from these policies and evaluate their success in achieving greater equity in spatial population distribution and its access to a better quality of life.

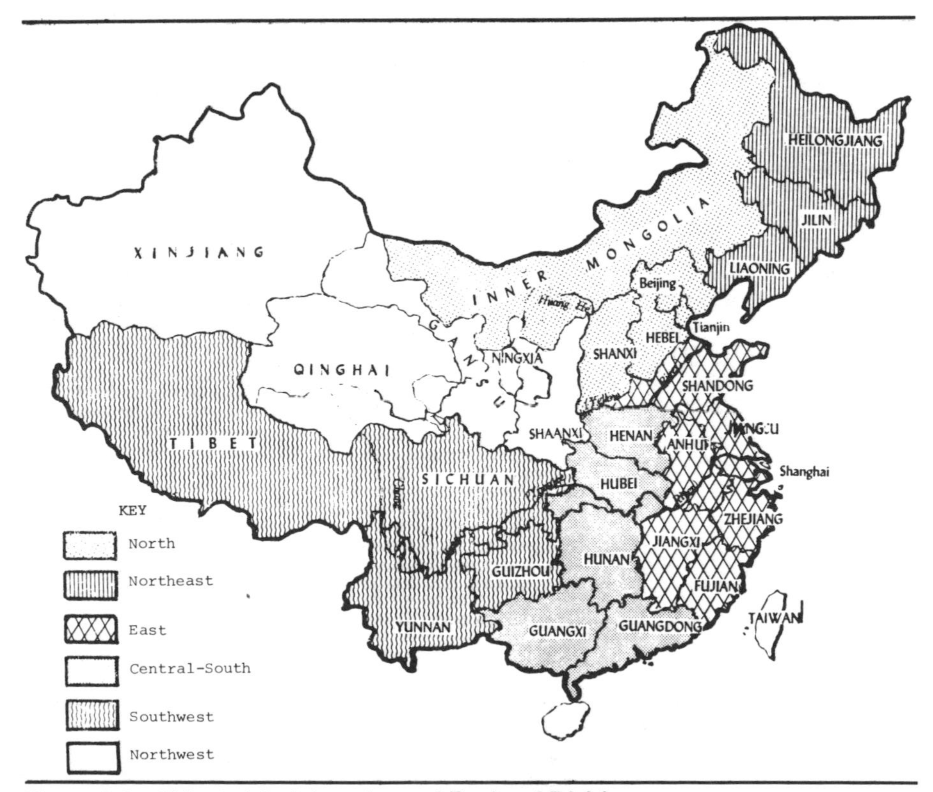

Figure 7.2: China's Administrative and Regional Divisions

Notes

1. Cities may be classified in three categories: (a) *big cities* have 500,000 or more residents; those above a million are called *metropolises*; (b) *medium-sized cities* range between 200,000 and 499,999; (c) *small cities* range from 50,000 to 199,999.

2. See also pp. 219-21.

3. In China, one important way of ranking cities is by their administrative status. The highest ranking cities—municipalities—are administratively placed under direct control of the central government. Only three big cities—Beijing, Shanghai, and Tianjin—rank as municipalities. Each has status equal to that of a province and has under its jurisdiction inner city districts and rural counties. In each municipality a majority of the total city population lives in the inner city, ranging from a high of 66 percent in Beijing to only 53 percent in Shanghai.

Ranking below municipalities are cities directly under provincial government leadership: the 26 capital cities of the provinces and autonomous regions that serve as provincial political, cultural, economic centers. In 19 provinces, the capital is the largest city; of the 26 provincial capitals, 19 exceeded one million in population in 1982; of these, 7 had more than two million. The remaining cities in China have usually gained size or importance through economic function, but occasionally because they serve some other key function in the province.

4. In the census, household members were classified as: (a) residing and registered in the locality, including individuals who had been away from place of registration for less than one year; (b) residing in a place more than one year, but registered elsewhere; (c) residing here less than a year, but absent continuously from the place of registration for more than one year; (d) living in the locality at time of census, with registration status not yet settled, such as demobilized soldiers waiting for job assignments, students, and ex-criminals released from institutions; (e) originally living in the locality, but abroad at census time for work or study and thus without registration.

5. By the end of June 1984, reports released by the National Conference on the Development of Small Cities and Towns indicated the number of towns in China had risen to 5,698. This impressive increase was attributed to effects of the burgeoning commodity economy and new emphasis on town development (*Renmin Ribao* [People's Daily], 9 September 1984).

6. Unless otherwise indicated, city population is restricted in inner city districts; suburban counties which include a considerable rural population are omitted, although administratively part of the city (*shi*).

7. Regions delineated in this chapter correspond to the administrative areas established in 1949-50 (Ginsburg, 1952). Designed initially for political-administrative purposes, they were not and are not designed as coherent economic planning units. Nevertheless, these regional distinctions are still used in China for statistical and other purposes even though their administrative role has fallen into disuse. Here they serve as a way in which to examine geographic urbanization variations, in large part consistent with regional distinctions used by other analysts of Chinese spatial development (cf. Chiu, 1980:99-107). The same regions have also been used recently by other evaluators of China's population situation (e.g., Orleans, 1982:table 2; Taylor, 1984).

8. *West* refers to the combined Southwest and Northwest regions.

9. Unfortunately, published results of the 1982 Chinese Census (the 10-percent tabulations), the most comprehensive set of information yet available from the census, do not provide a breakdown of the urban population by cities and towns for each of the 29 provinces, municipalities, and autonomous regions. This information was published in the earlier manual tabulations from the 100-percent count (SSB, 1982), but these differ slightly from results of the 10-percent sample. The manual count, for example, shows 14.4 percent of the total population living in cities and 6.2 percent in towns; sample data show this distribution to be 14.5 and 6.1 percent, respectively.

10. Statistics by city size class are based on table 11 of the 10-percent sample of the 1982 Census (SSB, 1983b:28-211). In table 11, data on individual cities and counties are presented by province, with subdivisions for inner cities and suburban counties where applicable. These city data provide the basis for aggregations by city size class. Table 11 data have been corrected according to information received after publication. Three Jiangi Province counties were incorrectly

identified as cities and one city each in Hunan, Yunnan, and Shaanxi Provinces was classed as a county in the published data. My tables 6.5 to 6.7 incorporate the corrected data. Despite the corrections, a discrepancy (597,990) still exists between total inner city population based on aggregate data from table 11 and totals in other census tables (e.g., table 20).

11. See note 1.

12. The 11-city primacy index compares the largest city in a given area with the 10 next largest; in China it can only be used nationally or regionally since many provinces do not have 11 cities.

Bibliography

Aird, John S. 1982. "Population Studies and Population Policy in China." *Population and Development Review* 8:267-97.

Banister, Judith, 1984. "An Analysis of Recent Data on the Population of China." *Population and Development Review* 10:241-71.

Beijing Review Staff. 1983. "Why the Jump in Urban Population between 1981 and 1982?" *Beijing Review* 26(14 Feb.):28.

Chandler, Tertius, and Gerald Fox. 1974. *3000 Years of Urban Growth.* New York: Academic Press.

Chang, Sen-Dou. 1976. "The Changing System of Chinese Cities." *Annals of the Association of American Geographers* 66:398-415.

China Daily Staff. 1984a. "Port Cities Encourage New Foreign Business." *China Daily* (14 July):1.

———. 1984b. "Peasants Flock to Join Town Enterprises." *China Daily* (10 Aug.):1.

———. 1984c. "Rural Prosperity Rejuvenates Towns." *China Daily* (18 Aug.):4.

———. 1984d. "More Rural Markets Are Urged for Industries." *China Daily* (12 Sept.):3.

Chiu, T. N. 1980. "Urbanization Processes and National Development." In *China.* See Leung and Ginsburg, 1980.

Coale, Ansley J. 1984. *Rapid Population Change in China, 1952-1982.* Report/Committee on Population and Demography, no. 27. Washington, DC: National Academy Press.

Ding, Yisheng. 1984. "The Urban and Rural Distribution of China's Population." *Renkou Yanjiu* (Population Research) no. 4(July):14-17.

Fei, Xiaotong. 1984. "Planning Population Growth and Distribution." *China Reconstructs* 33(May):24-27.

Ginsburg, Norton S. 1952. "China's Changing Political Geography." *Geographical Review* 42:102-17.

Goldstein, Sidney, and Alice Goldstein. 1984. "Population Movement, Labor Force Absorption, and Urbanization in China." *Annals of the American Academy of Social and Political Science* 476:90-110.

Koshizawa, Akira. 1978. "China's Urban Planning: Toward Development without Urbanization." *Developing Economics* 16(1):3-33.

Leung, C. K., and Norton S. Ginsburg, eds. 1980. *China: Urbanization and National Development.* Research paper no. 196. Chicago: Department of Geography, University of Chicago.

Li, Mengbai. 1983. "Planned City Growth." *China Reconstructs* 32(Nov.):7-9.

Lin, Wang, and Beijing Review Staff. 1984. "Economic Reform: Facts Behind the Shanghai Economic Zone." *Beijing Review* 27 (16 Apr.):16-23.

Ma, An. 1984. "An Evaluation on the Quality of the Data of the 1982 Population Census of China." Paper presented at the International Seminar on China's 1982 Population Census, Beijing, 26-31 March.

Mao Tse-tung. 1977. *On the Ten Major Relationships.* Beijing: Foreign Languages Press.

Ni, Ernest. 1960. *Distribution of Urban and Rural Population of Mainland China, 1953 and 1958.* Series P-95, no. 56. Washington, DC: U.S. Bureau of the Census.

Orleans, Leo A. 1982. "China's Urban Population: Concepts, Conglomerations, and Concerns." In *China under the Four Modernizations,* Part 1. Selected Papers submitted to the Joint Economic Committee, 97th Cong. of the United States, 2d Sess. Washington, DC: U.S. Government Printing Office.

Orr, Ann C. 1984. "Labor Force Participation, Employment and Unemployment of Women in China." Paper presented at the annual meeting of the Population Association of America, Minneapolis, 3-5 May.

Pannell, Clifton, E. 1984. "China's Changing Cities: Urban View of the Past, Present, and Future." In *China: The 80s Era,* edited by Norton S. Ginsburg and Bernard A. Lalor. Westview Special Studies on East Asia. Boulder, CO: Westview Press.

People's Republic of China. Population Census Leading Group. n.d. *Instructions for Filling Out the Questionnaire of the Third National Population Census.* Beijing: State Council.

People's Republic of China. State Statistical Bureau. 1982. *Major Figures on the Third Census of Population in China.* Beijing: China Statistical Publishing House.

———. 1983a. *Statistical Yearbook of China 1983.* English ed. Hong Kong: Economic Information & Agency.

———. 1983b. *Ten Percent Sampling Tabulation on the 1982 Population Census of the People's Republic of China.* Beijing: China Statistical Publishing House.

PRC. *See* People's Republic of China, Population Census Leading Group.

SSB. *See* People's Republic of China, State Statistical Bureau.

Taylor, Jeffrey R. 1984. "Employment and Unemployment in China: Results from the Ten Percent Sample Tabulation of the 1982 Population Census, People's Republic of China." Paper presented at the Workshop on China's 1982 Population Census, East-West Center, Honolulu, 3-8 December.

Ullman, Morris B. 1961. *Cities of Mainland China, 1953 and 1958.* Series P-95, no. 59. Washington, DC: U.S. Bureau of the Census.

United Nations. Department of International Economic and Social Affairs. 1980. *Patterns of Urban and Rural Population Growth.* Population Studies no. 68. ST/ESA/SER.A/68. New York.

———. 1982. *Estimates and Projections of Urban, Rural and City Populations, 1950-2025: The 1980 Assessment.* ST/ESA/SER.R/45. New York.

Ye, Suzan. 1982. "Urbanization and Housing in China." *Asian Geographer* 1:1-11.

Yeh, Anthony Gar-On, and Xueqiang Xu. 1984. "Provincial Variation of Urbanization and Urban Primacy in China." *Annals of Regional Science* 18(3):1-20.

Zhuo, Zhu. 1981. "Rationalization of Population Distribution." In *China's Population: Problems and Prospects,* edited by Liu Zheng, Song Jian, and others. Beijing: New World Press.

8

Giant Cities and the Urban Hierarchy in China

Xiangming Chen

With over a billion citizens, China is the largest nation in the world, but only 20.8 percent of her people live in cities and towns of 3,000 or more. Interestingly, half of China's total urban population resides in 38 cities of a million people or more. What problems are associated with a relatively low level of overall urbanization and concentrations of urban people in a substantial number of very large cities? In this chapter I consider this issue by (1) examining the structure and integration of the Chinese urban system, (2) identifying some major problems associated with growth of very large cities that analysis reveals lack integrated urban development, (3) assessing the potential of China's policy responses to these problems, and (4) presenting a prospective view of the largest Chinese cities.

AUTHOR'S NOTE: Portions of this research were presented at the annual meeting of the Southern Sociological Society, Charlotte, NC, 11-14 April 1985. I am indebted to Professor Joel Smith, whose comments and criticisms, both substantive and stylistic, have considerably facilitated the completion of this work. I am grateful to Marcia Spray for her kind and competent assistance in preparing the tables and references for the final draft. I would also like to thank the two editors of this volume for their useful suggestions on the final revision.

The 1981-82 data on the 16 cities used in this chapter have been updated for 1984-85 and expanded to cover 33 Chinese cities. These newer and more comprehensive data, currently being analyzed in my dissertation and related research, will let one draw more general conclusions about the giant cities and the urban hierarchy in China.

Urbanization Research on China

Because of a paucity of statistical information on the Chinese population in the last 30 years, little has been known about population distribution and urban growth in the world's largest nation. Limited descriptive and historical studies (Chan, 1981; Gernet, 1977; Kapp, 1974; Murphey, 1974) have provided detailed chronological accounts of individual Chinese cities with special features, such as the treaty port of Shanghai or the former largest city of Hangzhou (around the 13th century). These studies tend to emphasize unique sociopolitical antecedents as influences on the growth of large Chinese cities. Although they provide a historical context that helps place the postrevolutionary phase of urbanization in perspective, they offer little guidance for a structural analysis of either very large Chinese cities or the urban system as a whole.

Another major body of research focuses on changes in policies toward urban system planning in postrevolutionary China (Buck, 1981; Kwok, 1981; Murphey, 1980). These policy-oriented studies reveal how the post-1949 Chinese government, driven by a Marxian view that such cities as pre-1949 Shanghai represent decadence, power, corruption, and lust for capital, adopted a counterurbanization policy. Buck (1981), tracing the evolution of China's urban policy from 1949 to 1979, presents a detailed discussion of China's efforts, especially after 1958, to produce spatially balanced and decentralized development. Kwok (1981) suggests there have been four periods in the transformation of China's urban development and identifies the rationales for spatial considerations and locational emphases for each. Although such analyses are largely descriptive, they catch a thematic thread that runs through these urban planning strategies. One of the major goals of this discussion is to determine whether there is a Chinese model of urban development by dissecting and elaborating the underlying rationales and structural components of recent responses to urbanization.

Definition and Structure of Urban China

I begin with some definitions and classifications of the Chinese urban system and review some statistics on distribution of cities in China.

Urban places in China are defined in terms of population size. However, there have been changes in the definitions used since the early 1950s. *Urban places* were then defined as settlements with 2,000 inhabitants or more, at least half of whom were engaged in nonagricultural pursuits. Places of 1,000 to 2,000 population might also be classified as urban if 75 percent or more of the population were nonagricultural (Ullman, 1969:89). Since 1964 (the second census), *urban places* have been defined as towns with 3,000 inhabitants, 70 percent or more of whom were nonagricultural. Populations of 2,500-3,000, 85 percent or more of whom were nonagricultural, also are urban. Settlements of more than 20,000 people were defined as *cities*. More recently, cities have been grouped into three categories based on size: (1) large (500,000 and over); (2) medium-sized (200,000 to 500,000); and (3) small (below 200,000). A recent study (Ding, 1984) defined a new category of *very large cities* which includes those with 1,000,000 people or more.

See table 8.1 for a description of the urban population of China in terms of locality categories and several measures of concentration and distribution. The very high level of population concentration at the top of the urban hierarchy is not primarily accounted for by one or two primate cities (Chen, 1986). Although the proportion of the total population in urban areas of a million and over (these include the rural counties) is only 20 percent (column 2), 52 percent of the urban population is located in such places (columns 3 and 4). Concentration in very large cities also is indicated by the considerable gap between the largest and next largest places on two other measures (columns 5 and 6). The proportion of the urban population in the second category (500,000-999,999) is less than half that in the largest category, while the proportion of the total population in the same category is twice (41 percent) that in places of a million and over. This phenomenon is mainly because of the presence of a sizeable rural population in the smaller size group. Thus, the high concentration in very large cities is exemplified by the urban/rural ratio for the top category (.339) being six times that of the second largest category (column 6). Still another indication of the same phenomenon is that although the number of million-plus-population cities (38) constitutes 16 percent of the 239[1] cities, their combined population accounts for more than half the total urban population, showing the top-heavy concentration in the urban system. Nevertheless, concentration of total population for China in 1982 (1.05) (see table 8.1) was much less than similar levels for 1960 in Brazil (2.92), Denmark (4.25), Ghana (2.03), the United States (4.06), United Kingdom (4.97), and Yugoslavia (2.36) (Gibbs, 1966:173). This concentration suggests that population growth at the top will increase the structural imbalance in the urban hierarchy.

Although the level of urbanization in China only increased by about 5 percent from 1950 to 1980, the number of million-plus cities grew from 6 in 1949 to 44 in 1983, an astronomical rate of 633 percent. Against this backdrop, we now consider the group of cities with populations of two millon or more.[2] They constitute 35.2 percent of all cities of such size in developing countries. The 16 cities of two million and over constitute 13 percent of the 124 cities of over one million in the less developed world; however, their combined population is 25 percent of the 330 million people in such cities.

Because of lack of information in the past, little has been known about population concentration, industrial complexity, and the municipal infrastructure of very large cities in China.

Recent information on these 16 cities, however, provides an opportunity to examine some heretofore unexplored issues: (1) relationships among population size, density, and municipal functions; (2) variations in the association between central cities and their suburban areas; and (3) differences between the socioeconomic status of the populations of Shanghai and Beijing and those of

TABLE 8.1
Distribution of Urban Localities—China, 1982

Locality Size	(1) Number of Cities	(2) Percentage per Thousand of Total Population in Each Size Class and Over	(3) Cumulated Percentage per Thousand of Urban Population to Size Class	(4) Percentage per Thousand of Total Urban Population in Size Class	(5) Percentage per Thousand of Urban to Total Population in Size Class	(6) Ratio of Urban to Rural Population in Size Class
> 1,000,000	38	.197	.518	.518	.230	.339
500,000-999,999	47	.409	.749	.231	.050	.052
300,000-499,999	46	.226	.872	.123	.054	.085
< 299,999	125	.219	1.000	.128	.133	.153
Total	256	1.051[a]		1.000		

SOURCE: Computed from People's Republic of China (1983).
a. Scale of population concentration = Σ X, where X is the proportion of the total population in each size category and over. For interpretation and explanation of this measure, see Gibbs (1966).

other very large cities. Moreover, these data permit comparisons between the ecological patterns and processes of industrial, highly suburbanized Western cities and those in China.

Information for our study has been taken from the 1982 Statistical Yearbook of China (People's Republic of China, 1983). All 16 cities for which information is provided have populations of more than two million. If we follow the definition that very large cities are those with more than a million people (Ding, 1984), these 16 cities deserve to be called giant cities, standing on the top layer of the urban hierarchy. These cities do not constitute either a random or representative sample of the Chinese city population; they have been designated by the state as nationally important "key" cities on the basis of their size or political and administrative positions as national or provincial capitals, strategic locations, or industrial centers. Their historical experiences and geographical distributions make comparisons among them and between them and other cities meaningful.

We have information on each city and its two components: (1) the urban district and (2) suburban (rural) counties under municipal jurisdiction. This distinction is best understood in relation to conventional spatial ecological concepts. There is no isomorphism between the Chinese definitions and concepts and those for Western metropolises because urban growth patterns and municipal administrative systems differ greatly in the two parts of the world. However, we may draw some parallels for the sake of subsequent analysis. The *urban district* in Chinese usage is a broad concept that includes city districts. The number and size of urban districts in very large cities vary greatly. While Shanghai has 12 urban districts, there are 9 in Tianjin. Qingdao's urban districts cover only 118 square kilometers, the smallest of the 16, whereas Shanghai's urban districts amount to an area of 627 square kilometers, about five times the size of Qingdao. The *city district* is a much larger area than the Central Business District (CBD) of large Western cities and approximates the central city of the United States. The urban district usually includes the city's outskirts and is an appendage to the city district; it is structurally similar to the urban fringe: the subzone of the rural-urban fringe that is contiguous to the central city (Pryor, 1968:206). In reality, the urban district is analogous to the urbanized area including the central city in the United States. These urban districts vary considerably in their degree of urbanism, since they cut across the city edge, with portions located within the city limits and parts stretching into adjacent rural areas.

Suburban (rural) counties are included in these cities' boundaries for administrative reasons. There also is considerable variation in the number of rural counties administered by municipal governments of the 16 cities. For example, Nanjing has five such counties, while nine are under Beijing's jurisdiction. Shanghai, largest of the 16 cities, embraces ten surrounding counties. These counties bear some resemblance to the rural fringe—that subzone of the rural-urban fringe contiguous with the urban fringe—and have a lower density of occupied dwellings than the median density of the total rural-urban fringe, a higher proportion of farm than nonfarm and vacant land, and a lower rate of increase in population density, land use conversion, and commuting (Pryor, 1968:206). To a certain extent, these suburban (rural) counties are comparable to portions of Standard Metropolitan Statistical Areas (SMSAs) outside urbanized areas. While possessing predominantly rural characteristics, these counties are diverse in their level of development. Some, bordering on the urban districts of Shanghai and Beijing, contain a considerable number of industries and have been selected as sites for satellite towns to absorb industry and population from the central cities. Others, located mostly on the edges of municipal boundaries, are completely agricultural and reflect little urban influence because only weak transportation and communications links are available.

Information on the two urban sections allows us to examine their differences and the functional relationships between urban districts and their affiliated counties. This exercise will help provide a more specific and realistic assessment of the population size and industrial capacity of cities, which despite their being very large, have very diverse functional and spatial segments within their administrative boundaries. Therefore, regardless of their inherent limitations, this material should reveal much more than previously has been known about the general profiles and structural attributes of very large Chinese cities.

Measures of a variety of characteristics of these 16 cities and their respective urban districts and suburban (rural) counties are available. There is demographic information on population size, density per square kilometer, and rate of natural increase. Municipal infrastructure and industrial complexity are indicated by the number of industrial enterprises, heavy and light industrial output values, and numbers of units providing retail and food services. I have standardized these measures into rates and per capita values, thereby reducing the skewness of frequency distributions affected by the considerable size variation among the cities.

Indicators of industrial capacity and economic concentration are standardized as production enterprises per 10,000 population and industrial output value per capita. Municipal infrastructure is expressed by food services, colleges and high schools, hospital beds, and so forth, per 10,000 persons. Standard of living indicators in these very large cities include consumption sales per capita, housing space per capita, and so on. These measures are calculated separately for cities as a whole, urban districts, and attached counties, and are comparable across units of place.

Population Growth and Municipal Characteristics

Patterns of difference in urban structure are examined in the context of the dynamics of change in size of those cities during the last 150 years. See table 8.2.

Because of its special status as the imperial capital from the mid-1660s to 1911, Beijing was the largest city in China and one of the largest in the world around 1825. Shanghai's impressive growth from the late 1900s to 1949 reflected its role as the country's leading port and the rapid development of its commercial and transportation facilities. Shanghai's average annual growth rate of 3.5 percent[3] from 1900 to 1949 surpassed that of almost all other cities. Increasingly, extensive overseas trade in the first half of the 20th century contributed to the continued high growth of the old port cities of Tianjin and Guangzhou (Canton). The Manchurian cities of Shenyang (Mukden) and Changchun grew rapidly in the 20th century in response to the development of heavy industry under the Japanese rule in the 1930s and concentrated Soviet aid after 1949. Favored by transport advantages, cities along the Yangtze River in Central China, such as Wuhan and Nanjing, experienced rapid population growth. Nanjing also benefited from being the Nationalists' wartime capital during the 1940s. As the largest military and financial center for the Nationalists in southwest China, Chongqin added 753 thousand people to its population up until 1949.

Growth patterns took a different turn after the Revolution of 1949. Some of these 16 cities, even though they may only have been small towns 100 years earlier, exceeded the one million mark after 1949 and have grown rapidly as a result of governmental policy to build them into major industrial centers. This policy not only produced substantial controlled migration of technical personnel and workers from large coastal cities to interior cities, but also speeded up natural increase of their populations, particularly in the 1950s when there were few effective population control policies. Lanzhou's population soared from approximately 100 thousand in 1949 to 2.4 million in

TABLE 8.2
Population of Large Cities—China, 1825-1982 (in thousands)

City, Province	*1825*[a]	*1900*[a]	*1937*[b]	*1949*[b]	*1953*[b]	*1970*[b]	*1977*[c]	*1982*[d]
Shanghai	–	37	3,480	4,447	6,204	7,000	10,810	11,810
Beijing	1,350	1,100	1,565	1,672	2,768	5,000	8,300	9,190
Tianjin	175	700	1,067	1,708	2,694	3,600	6,280[e]	7,780
Chongqing, Sichuan	–	250	281	1,003	1,773	2,400	–	6,510
Changchun, Jilin	–	–	205	605	855	1,200	1,050[f]	5,750
Guangzhou, Guangdong	900	670	1,157	1,413	1,599	2,500	4,970	5,610
Shenyang, Liaoning	180	–	527	1,121	2,300	2,800	4,200[e]	5,140
Dalian, Shenyang	–	–	–	–	766	1,650	–	4,720
Qingdao, Shandong	–	–	527	759	917	1,300	–	4,260
Wuhan, Hubei	–	450	1,353	1,062	1,427	2,560	3,670	4,180
Chengdu, Sichuan	175	475	481	620	857	1,250	–	4,020
Nanjing, Jiangsu	200	–	1,013	1,137	1,092	1,750	3,200	3,740
Jinan, Shandong	–	–	437	591	680	1,100	1,100+[g]	3,350
Xi'an, Shaanxi	259	–	–	400	787	1,600	1,300[g]	2,940
Lanzhou, Gansu	–	–	–	100	397	1,450	2,000+	2,400
Taiyuan, Shanxi	–	–	–	–	721	1,350	–	2,200

SOURCES: (a) Chandler and Fox (1974: 372); (b) Pannell (1984: 99-103); (c) Ma (1981b: 224); (d) People's Republic of China (1983).
e. 1975.
f. 1978.
g. 1974.

1982 (10.1 percent per annum) attributed, in a certain degree, to the stimulus of a booming oil-refining industry. Both Chengdu and Chongqin grew at 5.8 percent annually for the last 33 years, as both received a number of military-related industries and other organizations from northern and eastern provinces during the Cultural Revolution. On the east coast, as a result of the efforts to control population growth in very large cities since the 1960s, smaller port cities like Dalian (6.5 percent) and Qingdao (5.4 percent) grew faster than the older and larger port cities of Guangzhou (4.3 percent), Tianjin (4.7 percent), and Wuhan (4.2 percent). This policy had a particularly strong effect on China's largest city, Shanghai, which at 3.0 percent per annum had the slowest average growth rate from 1949 to 1982. In general, these very large cities have grown very rapidly both absolutely and proportionally in comparison with small cities; 115 cities of 200 thousand or less in 1953 dropped to only 65 in 1982 (Ding, 1984). However, it should be noted that the high growth rates of these very large cities also include changes resulting from the annexation of surrounding rural counties.

Table 8.3 contains basic population statistics for the urban districts and suburban (rural) counties of the 16 cities. The two differ substantially in population density and size. The three largest cities (Shanghai, Beijing, and Tianjin) all have larger urban district populations, while the other cities vary considerably in distribution of population between the two sections. Rather than reflecting widely varied settlement patterns, however, this situation reflects the absence of uniform standards for dividing urban districts and suburban (rural) counties and original differences in the size and system of municipal administrations. That urban district populations grow faster is the opposite of the national population growth pattern—1.3 percent per annum for cities and 1.5 percent per annum for counties. This unusual demographic characteristic of very large cities in China is considered in more detail later.

Table 8.4 contains descriptive statistics for selected major demographic and socioeconomic attributes of the 16 cities. Mean values for almost all indicators are higher for urban districts than for suburban rural counties. This fact suggests a higher concentration of population, industrial capacity, and municipal functions (including recreational facilities and social services) in urban districts. Compared with national averages, the 16 cities are already more developed than the country as a whole in terms of industrialization, educational opportunity, and standard of living. Their higher standards of living and better municipal infrastructures are reflected in per capita measures of educational, recreational, medical, and food service facilities. For example, the number of college students per 10,000 population (71) in these cities is about seven times the national average (11). While 10,000 people share accessibility to only 13 doctors as a national average, 30 doctors are available for 10,000 residents in these cities, on average.

However, development of municipal utilities and economic functions in these cities is uneven. Demographically, the 16 cities have very high residential density, especially in the urban districts. This great density is also displayed by their poor public housing conditions. Although their gross

TABLE 8.3
Population, Density, and Rate of Natural Change of Large Cities for Urban Districts and Suburban Counties—China, 1981-1982 (in thousands)

	Urban Districtss			*Suburban (Rural) Counties*		
City	*Population*[a]	*Density*[b]	*Percentage Change*	*Population*[a]	*Density*[b]	*Percentage Change*
Shanghai	6,270	27,261	2.2	5,540	930	0.8
Beijing	5,550	2,055	2.2	3,640	258	1.4
Tianjin	5,130	1,200	2.1	2,650	377	1.7
Chongqing	2,650	1,742	2.0	3,860	464	0.6
Changchun	1,740	1,559	2.5	4,010	226	0.4
Guangzhou	3,120	2,320	1.3	2,490	239	1.2
Shenyang	4,020	1,476	2.6	1,120	223	0.8
Dalian	1,480	1,150	1.9	3,240	280	1.4
Qingdao	1,180	4,836	1.9	3,080	538	1.0
Wuhan	3,220	2,075	1.9	950	325	0.6
Chengdu	2,470	1,707	1.7	1,550	642	1.3
Nanjing	2,130	2,457	2.0	1,610	418	1.0
Jinan	1,320	2,733	1.9	2,030	462	1.0
Xi'an	2,180	2,532	1.6	760	481	2.3
Lanzhou	1,430	674	3.5	970	79	2.2
Taiyuan	1,750	516	2.8	450	114	0.7

SOURCE: People's Republic of China (1983).
a. In thousands.
b. Measured as number of persons per km^2.

industrial output per capita is three times the national average, these 16 cities possess less agricultural capacity, on average, notwithstanding that their rural counties are more productive than the national average in grain output. On another score, 5.8 percent of the total labor force in these cities is engaged in retail services, only half that of the nation as a whole. In comparison with large urban centers in developed societies and even some metropolises in developing countries, several public services in these 16 cities are worse in per capita terms. For example, while there is an average of 3.87 public buses for 10,000 residents, 500 inhabitants have access to one bus on average in Bogotá, Colombia (Mohan, 1984). The mean statistics in table 8.4 reveal the general characteristics of the 16 cities. However, they do not capture the extreme discrepancies in distributions of some of the measures.[4]

Within these municipalities is a considerable gap between urban districts and rural counties. That the urban districts are much more crowded, industrialized, and functionally integrated has two implications. First, there is a major difference between patterns of city structure in developed and developing societies. In Western societies, population density in central cities rises, then falls with distance from the center; in non-Western societies it remains high throughout the city's area. Compactness and crowding diminish over time in Western cities as populations disperse into surrounding areas; they tend to remain constant in non-Western cities with less expansion at the

TABLE 8.4
Means of Selected Demographic and Socioeconomic Characteristics of 16 Large Cities, Their Urban Districts, and Suburban Counties—China, 1982

Characteristics	*National*	*City*	*Urban District*	*Suburban (Rural) Counties*
Population (thousands)	1,015,410	5,225	2,853	2,372
Population density (km^2)	106.00	714.00	3,518.00	379.00
Natural increase rate (1981-82) (percentage)	1.40	1.72	2.13	1.15
Industrial enterprises/ 10,000 population	3.83	5.57	7.18	3.58
Gross industrial output/ capita[a]	542.24	1,579.58	2,735.79	224.41
Farmland/capita (acres)	0.25	0.16	0.05	0.30
Grain output/capita (lbs)	626.52	518.23	157.51	979.70
Nonproduction investment/capita[a]	24.77	107.08	176.86	15.32
Completed living space/ capita (year end in m^2)	5.60[b]	0.36	3.47	0.07
Food services/10,000 population	6.18	8.57	11.17	4.51
Public buses/10,000 population	—[c]	—[c]	3.87	—[c]
Percentage of employed labor force in retail service	11.06	5.82	4.94	14.63
Consumption sales/ capita[a]	214.84	473.58	706.67	193.15
Cinemas/100,000 population	14.15	15.39	14.30	17.68
Public libraries/ 1,000,000 population	1.90	1.89	2.30	2.02
College students/ 10,000 population	11.36	71.00	122.64	—[c]
Vocational schools/ 100,000 population	0.03	8.97	14.39	4.05
High schools/ 100,000 population	0.21	9.49	7.65	12.62
Hospitals/10,000 population	0.65	3.30	5.20	1.00
Hospital beds/ 10,000 population	20.30	35.40	52.42	14.48
Doctors/10,000 population	12.90	29.34	46.21	8.39

SOURCE: People's Republic of China (1982: 35-102).
a. In Chinese yuan–1 yuan = 0.4 U.S. dollars.
b. This measure, called urban housing space per capita, is calculated on the basis of existing housing space per urbanite, instead of the annually completed living space used to compute the indicator of housing conditions for the cities as a whole and their suburb and rural county sections.
c. Cannot be computed for absence of information.

urban periphery (Berry and Kasarda, 1977), though limited decentralization of industries and services has occurred in some more developed and better-planned urban centers in Third World countries. In contrast to the centrifugal forces in the large metropolises of Western industrial societies, a centripetal drift still operates in large urban centers of developing countries generally.

The data show that very large Chinese cities share the growth pattern of large cities in other non-Western societies. Faster population growth in urban districts (see table 8.3) suggests that they draw people from suburban counties, rather than having necessarily high rates of natural increase. This internal demographic growth of urban districts is matched with little external infrastructure and socioeconomic development in the farther outskirts and rural counties. Until the materialization of this development, residents in surrounding rural counties are likely to be halted from drifting into the urban districts of these cities.

Second, these intracity discrepancies evidence a lack of functional integration and coordination in the administrative boundaries of the 16 cities, also suggesting uneven socioeconomic development across the two sections. Possibly there is within the two sections some degree of functional interdependence which cannot be integrated because of highly limited city-suburban interaction. This structural segmentation largely results from (1) an overconcentration of resources in the urban districts and (2) underdeveloped transportation links between the two sections. Inadequate municipal planning has stalled more balanced socioeconomic development. Just as the continued sprawl of Western metropolises is stimulated by rapid changes in land use promoted by widespread mass transportation and communications, the absence of dynamic suburbanization in China is attributable to weak intra- and intercity transport and communications systems that hamper attempts to redistribute resources of urban districts to rural hinterlands and maintain a smooth information flow between the two sections.

Distribution of population and industrial strength of cities depends on (1) how each one's size and functions fit into the whole demographic and economic system and (2) how each layer of the urban system is internally structured. Focusing on the latter, I have done correlational analyses on the 16 cities to explore the relationships among population concentration, industrial capacity, and municipal development in the top rung of China's urban system. The ecological and industrialization models postulate positive relationships among city size, industrial capacity, and municipal functions (Bean et al., 1972; Berry and Kasarda, 1977; Eberstein and Frisbie, 1982). (For a brief review of these theoretical models, see Chen, 1985a.) Correlations of three demographic measures (size, density, and rate of natural increase) with various indicators of industrial complexity and service functions reveal that population size (natural log[5]) is strongly related to light industrial output per capita (.64)[6], industrial profit and tax revenue per capita (.56), and production investment per capita (.48). Population density (natural log) is positively associated with industrial enterprises per 10,000 people (.58) and light industrial output per capita (.66).

The relationships within the two sections are more interesting: the magnitudes of the coefficients are generally larger for each of the two sections than for the city as a whole, e.g., the correlation between density and heavy industrial output per capita is .24 (city) in contrast to .55 (suburban county), and .71 (urban district); the correlation among density, industrial profit, and tax revenue per capita is .43 for the city and .91 for the urban district. The relationship is less strong for the rural counties, though size and density are correlated with industrial output per capita at .61 and .69, respectively. These positive coefficients indicate that the larger and more densely populated the urban districts and suburban counties are, the higher is their level of industrialization, especially light industry.

The associations between population and nonproduction investment are in opposite directions for the two sections of the 16 cities. While size and density are weakly associated with nonproduction investment per capita at .24 and .30 (urban district), the comparable coefficients are –.84 and –.64 for the rural counties, indicating that larger and more crowded rural counties generally have less nonproduction investment. On the other hand, size and density are associated weakly (.33 and .35) with housing investment per capita, whereas the same coefficients are .49 and .67 for the rural counties. This association reflects the less strong relationship between population size and density and housing improvement in the urban districts of the 16 cities. The negative relationship (–.47) between population size and number of food services per 10,000 people suggests that the larger and more crammed the cities are, the fewer are food services such as restaurants and snack bars on a per capita basis. From the perspective of economies of scale, this finding may be a lack of indication that there has been much increase in the size of these food service establishments to compensate for their small number. However, this situation may be improved quickly since an increasing number of restaurants and eateries, many privately-owned, have been appearing in these very large cities in response to the new economic reforms. Rate of natural increase has a negative relationship with grain output per capita (–.68), implying that population grows faster than the availability of grain in these cities. In fact, much of the food supply for these very large cities must come from other provinces.

The relationship between consumption sales per capita and population size and density is positive, as expected. However, it is positive (.57) only in urban districts, being –.59 for rural counties. On average, larger crowded counties have fewer high and vocational schools, although population growth and establishment of more secondary educational institutions tend to be associated in the urban districts. Differences in the magnitudes of the coefficients indicate a closer association between urbanization and such service facilities as hospitals and cinemas per capita in urban districts than in rural counties. These analyses provide only limited support for the proposition that city size and density are positively related to industrial capacity and municipal infrastructure development. They alert us to useful directions for further lines of inquiry as information for more units in the city system becomes available. Improvements in conceptualization and measurement of various indicators will permit more rigorous tests of more stringent hypotheses with better data.

A major premise of ecological theory is that high levels of industrialization and economic growth in large urban centers tend to be accompanied by more services and a better municipal infrastructure. Correlations between indicators of industrial capacity and municipal functions and standards of living show that five of six industrialization characteristics have strong positive correlations (.70 and over) with consumption sales per capita and hospitals per 10,000 people. Their positive relationships with housing opportunity measures and development of higher education (college students per 10,000 people) are much more moderate. The same five indicators of industrial capacity (heavy and light industrial output per capita) have weak negative associations with retail services per 10,000 population, suggesting that business and commercial networks in these very large cities grow behind the pace of industrialization. This finding contradicts suggestions that large cities with high levels of industrialization have extensive trade linkages (Berry and Kasarda, 1977).

To check the possibility that these associations are unduly influenced by extreme values for Shanghai, the analysis was repeated excluding that city. A comparison of coefficients is revealing: for example, the correlation between consumption sales per capita and gross industrial output per capita only drops from .74 to .68, while the correlation of gross industrial output with completed living space per capita increases sharply from .33 to .75. Other industrialization indicators behave similarly in relation to consumption sales and housing. These comparisons indicate the tremendous influence on overall patterns of Shanghai's high consumption level and bad housing conditions. The negative and weak positive relationships between levels of industrialization and the development of cultural and educational facilities become strongly positive when Shanghai is deleted. This phenomenon suggests that Shanghai's overwhelming industrial capacity is not being translated into recreational and educational services per capita, possibly because of overcrowding. The two employment characteristics (percentage of labor force employed in state and collective enterprises) whose associations with urban conditions are in opposite directions must be interpreted carefully. Size of the proportion of labor force in state enterprises (most of which is in large organizations and government agencies) has close positive relationships with indexes of the standard of living (.47 with housing space and .51 with availability of hospital service). On the other hand, higher proportions of employees in collective enterprises are associated with lower levels of recreational and service functions, the correlations with housing space and hospitals over 10,000 people being –.50 and –.47, respectively. Nevertheless, retail services, many of which are collectively-owned, have only a small positive relationship with percentage of employees in such enterprises.

The hypothesis that industrial growth and municipal service functions are associated is largely validated in very large Chinese cities: the more industrialized the cities, the more developed their municipal service functions and standard of living. Higher agricultural activities, in contrast, are associated with lower levels of municipal functions and lower standards of living. Moreover, results indicate an unbalanced and less systematic integration between industrial growth and development of municipal services at present (e.g., consistently weak relationships between indicators of industrial capacity and such municipal functions as retail trade and cinemas). The strong impact of Shanghai on stability of the coefficients also attests to this fact.

Having examined the general pattern of relationships among various structural dimensions of these cities, I now consider how each stands in relation to the others on a set of representative characteristics. Standard scores for nine indicators were computed to summarize divergences from the average free of the underlying metrics. (See table 8.5.) I then ranked the cities in terms of their standard scores, that is, the city ranking highest receives a score of 1, the lowest a score of 16.

Consistent with the generally known demographic dominance and high socioeconomic status of Beijing and Shanghai in the urban hierarchy, the two cities rank over all other cities. Beijing's higher rank reflects its more balanced development as the nation's capital. Older ports like Tainjin and Guangzhou (Canton) and newer ports along the Yangtze (e.g., Wuhan and Nanjing) have achieved more balanced development than inland cities like Lanzhou and Taiyuan, except for Xi'an and smaller port cities (e.g., Dalian and Qingdao). The old industrial city of Changchun, which ranks lowest on industrial capacity, the housing indicator, and medical facilities, but highest on grain output exemplifies a lack of integrated development. The big gap in standard scores between Shanghai and Beijing and some other cities reflects the difficulties of creating a more balanced and homogeneous top layer of the urban structure in China.

Persisting Problems and Policy Implications

In this initial attempt to evaluate the various dimensions and features of a set of very large Chinese cities, the analysis suggests that they exhibit some of the characteristics of large Western metropolises. Population size, level of industrialization, and municipal service functions are positively related, as they tend to be in most Western cities. These observations, however, should be interpreted cautiously, for data limitations do not permit systematic and rigorous statistical analyses. Nevertheless, even the descriptive measures clearly indicate uneven development between urban districts and rural counties with a concentration of population and resources in the former. Urban districts are functionally more integrated; their demographic measures are more strongly related to economic and service function indicators than are those for rural counties. This finding suggests that very large Chinese cities have the features of large Western urban centers in their earlier stages of urbanization. All of this has implications for designing and implementing urban policies.

TABLE 8.5

Ranks of Cities of Over Two Million on Selected Socioeconomic Characteristics[a]—China, 1982

Cities	*Population Size*	*Gross Industrial Output*	*Grain Output*	*Food Services*	*Percentage Labor Force in State Enterprises*	*Sales for Consumption*	*Completed Living Space*	*Colleges and Universities*	*Hospitals and Clinics*	*Overall Rank*
Shanghai	1	1	11	10	4	1	10	5	1	2
Beijing	2	4	10	15	2	2	3	1	2	1
Tianjin	3	2	14	14	7	6	2	10	3	3
Chongqing	4	14	5	9	9	16	4	15	15	12
Changchun	5	16	1	13	15	15	16	13	16	15
Guangzhou	6	10	6	2	8	3	13	12	8	6
Shenyang	7	8	2	4	16	5	11	7	13	9
Dalian	8	11	7	16	14	11	15	14	14	15
Qingdao	9	12	8	12	11	14	14	16	12	14
Wuhan	10	3	13	7	10	4	1	3	10	4
Chengdu	11	13	3	5	3	13	6	9	6	7
Nanjing	12	7	4	8	13	7	8	4	7	8
Jinan	13	15	9	6	12	12	12	11	11	13
Xi'an	14	9	12	1	5	8	7	2	4	5
Lanzhou	15	5	16	3	1	10	9	6	9	10
Taiyuan	16	6	15	11	6	9	5	8	5	11

SOURCE: Computed from People's Republic of China (1983: 35-102).

a. Figures are per capita unless otherwise noted.

The primary implication concerns policy responses to a persisting dilemma associated with growth processes and functional roles of these very large cities. On one hand, these cities have higher standards of living and more urban amenities than the rest of the nation (see table 8.4), although they still lag substantially behind large Western cities on a per capita basis. However, they also share many of the serious problems of very large or primate cities in the Third World. Shanghai's population density in the urban district is higher than that of Tokyo, Mexico City, New York, and São Paulo. In 1982, population per square kilometer measured 43 thousand in the central area of Shanghai which covered only 141.7 square kilometers. Of the 121 community blocks in the urban district, 20 had registered residents of over 100 thousand per square kilometer; 5 had residential densities of over 150 thousand per square kilometer (Gui and Zhu, 1984).

High density means extremely crowded housing, with many Shanghai households still sheltering three generations under one roof, separated only by hanging curtains. Living space per head is estimated to be less than two square meters. While defying the imagination of Westerners used to spacious and private living, this shortage of housing space has been a chronic problem, as population growth has outpaced housing construction (see Chen, 1985c for the case of Beijing). Public transportation is becoming increasingly clogged. With about two million people in Beijing going to work every day by public transport, buses and trolley-buses have turned into boxes choked with passengers. From 1982 to 1983, bus riders increased by 4.5 percent and the volume of subway transportation grew by 9.0 percent, though the capital city's population only showed a 1.7 percent growth. During peak rush hours in some cities, passengers per square meter on public buses are as many as 10 to 13 people, exceeding the maximum limit of 9 persons per square meter set by the state. Being late for work is not uncommon because overcrowded buses move very slowly in the narrow streets. In urban districts of these cities, motor vehicles now average 12 kilometers per hour, dropping by 40.4 percent from 20 kilometers per hour before 1966. A recent survey showed that the average speed of motor vehicles on Shanghai's main streets does not exceed five kilometers per hour. Torrents of bicycles (estimated at one per three persons in some cities) rush for space with and even outrun motor vehicles in the streets, causing nightmarish rush hour traffic. An increasing number of mopeds, driven mostly by fashion-oriented youths, has added additional hazards and is in part responsible for the recent rapid increase in road accidents.

Industrial pollution is a constant menace, especially in cities of North China.[7] Water shortage also has become a threat to daily life in these very large cities. The two reservoirs providing Beijing's water supply have recently fallen below their water margins. To insure industries' access to water, the municipal government designed a plan of conservation that would supply water during limited hours of the day and rotate the supply across different sections of the city. Quite a few enterprises in Tianjin once had to shut down temporarily because of a water shortage. Telephone and telegraph systems in some of these cities are dated and overloaded, with very few telephones per capita and scarce telegraph services. Lines frequently are tied up and customers queue up to send telegrams, resulting in delayed communications.

Unemployment has risen in many cities as a result of the joint impact of youths returning from the countryside and large increases in the labor force from the second baby boom (1962-70). However, the early introduction of successful family planning in some cities has kept their unemployment from rising unmanageably. In Shanghai, for example, the rapid fertility decline of the mid-1960s reduced labor force entry (at age 16) in 1982 to 61 percent of its 1979 level for the whole city and 41 percent of the 1979 level for the urban district (Gui and Zhu, 1984). At the same time, those urban youths whom state enterprises are unable to absorb through planned employment procedures have been hired by the growing tertiary sector of collectively- and individually-owned enterprises on either a tenure-track or contractual basis.

Despite these problems, these giant cities have always played critical roles in the national and international economies for China. Together they account for 28.0 percent of the annual national gross industrial output, although they contain only 8.2 percent of the total population and occupy only 1.5 percent of the total land area. Their industrial dominance and locational advantages are fully recognized by the Chinese government. In 1984, 5 of the 16 (Tianjin, Shanghai, Qingdao, Dalian, and Guangzhou, all located on the coast), together with 9 other smaller coastal cities, were opened to direct foreign investment (Chen, 1985b). They also have been encouraged to create within their administrative boundaries Special Economic and Technology Development Districts which offer low-tax environments and special institutional facilities to attract capital by setting up joint ventures with overseas companies. Chongqin, Wuhan, Shenyang, and Guangzhou have been selected as pioneering cities in urban economic reforms and given provincial autonomy in economic planning and business management. All 16 have been designated as national or regional urban centers to coordinate economic growth in surrounding areas that have been expanded to cross provincial boundaries.[8] Collectively, these cities will remain the backbone of China's future economic growth.

Government factors for these cities will stimulate their further industrial growth but also may worsen some of their problems. More investments in these cities from the state and abroad could induce an acceleration of their population growth, as the demand for manpower would attract more migrants. As a result, residential housing could become more crowded. Creating more industries could lead to heavier pollution of the urban environment. A larger proportion of investment going into industrial expansion could reduce nonproduction investment and increase strain on municipal infrastructure. In other words, increasing such advantages as market access, better-developed educational systems, skilled labor forces, rich commercial information, and a strong technological base may offset the efficient accommodation of more demographic growth and needed infrastructure improvements. For example, under the new agricultural policy and government effort to promote consumption, a large number of peasants come into these cities either on bicycle or by their own trucks or tractors to sell privately-grown foodstuffs or self-made handicrafts on the free market. The initial trickle of peasants has turned into a constant flow, creating more pressure on the overburdened roads, hotels, and other services. The freer commercial exchanges brought about by economic reforms have attracted more business people from other provinces to these cities, especially Beijing and Shanghai. More and more peasants who have become prosperous are now eager and able to afford to visit and experience the "bright lights" of Beijing and Shanghai. At the same time, limited child care in cities like Beijing has attracted rural women from the South, young or middle-aged, to come as babysitters (in some cases also housekeepers) for double-career families. Although each stimulus for temporary migration may not be significant in and of itself, collectively they contribute much to the floating population (unregistered) in these cities. The combined floating population in 10 of these 16 cities is estimated to have recently reached a record of 4,833,000, as much as 14 percent of their total legal residents.[9]

These dilemmas, some of which are unintended consequences of policy clashes, may be avoided if China is successful in pursuing its broad and integrated urban planning strategy. In 1980, the Chinese state instituted a comprehensive urban planning policy, officially codified as "controlling the expansion of big cities, rationally developing medium-sized ones, and rigorously building up the small cities" (M. Li, 1983:7). What led to formulation of this policy was the Chinese government's rising concern about various problems associated with increased urbanization occurring hand in hand with rapid modernization. In the early 1980s the state established the Ministry of Urban Planning and Environmental Protection, which has been in charge of designing and coordinating urban development activities. National conferences of mayors have been held to discuss what strategies should be employed to deal with problems facing their cities. The overarching national policy mentioned above is based on theoretical principles that hold promise of a unique model of urban development. By disaggregating the abstract and general principles underlying these policies, we may clarify the rationales designed to guide urban growth in China.

First, the policy instrument aims to forestall the emergence of primate cities at the top level of the urban hierarchy, for Shanghai and Beijing are very likely to attain such dominance without rigorous population control. The demographic explosion of primate cities in the developing countries is a common problem resulting from joint pressures of high natural increase and inmigration. This threat is less severe in China where rural migration into the very large cities has been regulated relatively efficiently because success of family planning in Shanghai and Beijing has greatly reduced their fertility levels (Chen, 1986).

Simultaneously, suburban growth of very large cities is being facilitated through recent government efforts to overcome the urban-rural barrier by building satellite towns.[10] The objective is gradually to make the city proper the administrative, financial, and information center, while satellite towns and rural counties become industrial, science and education, and tourist centers. Rural residents are allowed to run various service businesses in these satellite towns. Export-oriented factories in Shanghai's vicinity have grown from 47 in 1978 to 130 in 1984. The very large cities also have begun to channel and disperse some population and industry to the new satellite towns in their suburbs and governed rural counties, which, in comparison to the suburban areas in Western metropolises, are presently underdeveloped and have great potential for growth. Probably these very large cities will become more metropolitanized by building and absorbing satellite towns in much the same fashion that large urban centers in the West have engulfed incorporated cities in their suburban areas. This suburbanization will connect cities proper more closely to the rural counties beyond.

Second, with respect to the middle level of the urban system, the Chinese strategy can be thought of as regionalization of central (secondary) cities to facilitate integration across geographically bounded areas. The historical experience of urban development along regional lines in China (Skinner, 1977) has provided structural conditions for this pattern to occur. The sizable number of secondary cities with disproportionally small populations (the middle two categories in table 8.1) indicates a potential pool of candidates to play out that function. The absence of such a middle tier in the urban systems of many Third World countries lends support to the dependency perspective of urbanization: one or two primate cities, created and perpetuated through and after colonialism (Chen, 1985a), have become dominant over small cities and rural areas.

The underlying strategy of strengthening secondary cities on a regional basis can be thought of as a synthesis of central place theory and the countermagnet approach (Richardson, 1984). These cities have been allowed to extend their influence into surrounding areas and beyond via technical assistance and commodity export and by coordinating overall development of the regional economy. They also have been encouraged to compete with Shanghai and Beijing at the national level by expanding their economic and commercial exchanges across administrative boundaries. Polarization of primate cities at one end of the urban system and small towns at the other has created and sustained uneven development in Third World nations. China's vast rural areas and varied geographic conditions imply that the coordinating and stimulating functions of secondary central cities are crucial for achieving an integrated and balanced development.

Third, this policy package proposes what can be conceptualized as the citifying of rural towns. Several purposes and functions of this citification process can be identified: (1) connecting secondary and small cities with the villages, (2) absorbing unemployed rural labor by providing local industrial and service jobs, and (3) diversifying and improving the infrastructure (e.g., residential housing, educational institutions, service and recreational facilities) and thereby increasing rural access to urban life. This policy is expected to have two consequences: (1) peasants' departure from the land without leaving their home township and (2) peasants' entrance into local factories without entering the big cities. If the projection holds true that China's urban population will grow 31.0 percent from 1980 to 2000 (versus Latin America's 17.3 percent, North America's 5.7, Asia's 34.2, or Africa's 47 percent),[11] the large number of rural residents becoming urbanized can only be absorbed by these citified towns. If this policy is smoothly implemented, we will witness an unprecedented major population redistribution in China that will involve total redefinitions of both *urbanization* and *urban places.*

The second wave of economic reforms, aiming at more autonomous and decentralized decision making for administrative authorities in urban areas and individual cities, tends to reinforce uneven growth of cities of varying sizes and locations. For example, the 14 well-endowed coastal cities recently opened to encourage direct foreign investment and joint ventures will develop more rapidly, thereby widening the long-standing gap between coastal and inland cities (Pannell, 1981, 1984). Shanghai is also likely to experience more rapid growth on the basis of its unique economic conditions, although it is not now demographically a primate (Chen, 1986; Ginsburg, 1980; Pannell, 1984). Despite their being distributed in an uneven manner geographically with the very large and more industrialized cities located on or near the east coast, large Chinese cities conform more to the rank-size distribution than do those in many developing countries. Because of its slower growth, the gap between Shanghai's population and those of other very large cities has narrowed over the last 30 years. However, Shanghai has remained the largest economic center in the country. In 1983 its industrial output accounted for one-ninth, its foreign trade one-sixth, and its port capacity one-third of the national total (Z. Li, 1985). Its recorded annual industrial productivity per worker (28,684 yuan or about $11,474) was approximately 46 percent higher than the second most productive city of Tianjin (19,608 yuan or about $7,843).

Second, the labor demand generated by new industrial enterprises in favored cities may become a strong pull for rural population, depending on the structural changes that will take place in their labor markets. Thus, the strategy of selecting some cities for faster development may produce size-class jumping—some secondary cities may move into the top category of large cities. On the other hand, the low aggregate population growth in urban areas and continued control on rural-urban migration tend to forestall upward shifts by larger cities.

We have noted a continued commitment of the Chinese government to intervene to modify and redirect the dynamics of urban growth under intensive industrialization. Although data presented earlier (table 8.1) indicate it is probably too late to avoid concentration at the top level of the urban structure, current policies are preventing overconcentration of population and resources in very large cities. They offer a potentially promising solution to the dilemma presented by the structure and functions of these cities—tradeoffs between allowing their continued growth at the risk of primacy problems and stopping their growth at the cost of suppressing their economic functions. At the core of these urban-system planning strategies is the desire to avoid the path of urbanization experienced by most developing countries by pioneering a new urban growth pattern. Although the outcome of these policies remains to be seen, their general implications may be assessed and more recent measures to remedy, if not cure, the existing urban malaise should be highlighted.

This comprehensive picture of very large cities in the context of China's urban policies suggests that an effective Chinese model of urban development has yet to crystalize. As a late-comer to urbanization and economic development, China may have the advantage of observing and learning from the experiences of other developed and developing countries. It will continue to face the crucial question of vertical and horizontal functional integration within its very large cities and between them and lower strata of the urban system. The key to integration is to improve the transport and communications systems.

A host of developments in this respect has occurred recently. Some cities have begun operating mini-bus services as an alternate transportation mode to lessen pressure on existing public bus systems. Working schedules of some factories and government organizations have been rearranged to reduce the amount of traffic during rush hours. The second line of the subway in Beijing (which first opened in 1969) went into operation in 1984. A planned five-year construction project will add a new subway line and extend its service into the city's outskirts. A rigid licensing system has been introduced to reduce the number of mopeds on the road. Tractors are not allowed to cruise into the central city any more. A ceiling has been imposed to limit the number of trucks entering the city during daylight and driving on the major and busy roads. More circumferential bus routes have been created to avoid through-city traffic. More efficient arrangements have replaced some of the older two-way street intersections. Some pedestrian bridges and tunnels have been constructed to alleviate traffic tie-ups. Tianjin has constructed an 18-kilometer beltway around the city to ease traffic congestion. A short highway is to be built around Shanghai, linking the central city with Baoshan and Jinshan, its two booming industrial satellite towns. Expressways are planned to be built between Shenyang and Dalian, Beijing and Tianjin, Shanghai and Nanjing, and Guangzhou and Shenzhen (the largest of the four Special Economic Zones on China's southern coast) (Chen, 1985c). The last is scheduled to be completed by 1990.

Digital-controlled exchange telephone systems purchased from abroad are to be installed in several cities, for example, Shanghai, Nanjing, Guangzhou, Dalian, and Wuhan. Ongoing priority projects include (1) a coaxial 1800-line carrier communication trunk linking Beijing, Wuhan, and Guangzhou, and (2) multiple-microwave communication cables connecting Nanjing with Shanghai, Shenyang with Dalian, and Jinan with Qingdao.

The city of Wuhan has founded the country's first regional airline—Wuhan Airline Company. Shanghai is building a new container dock, which, when completed, will handle 200 thousand standard containers a year, the largest facility of this kind in the country.[12] These measures should help to improve the limited transportation and communication within and among these very large cities, as well as between them and the outside world.

Although these planned and implemented solutions are unlikely to overcome the problems overnight, each can function as a quick pay-off intervention and will have a remedial effect on this complex gamut of urban pathologies. However, it is more difficult for China to foster the transfer and flow of resources from coastal to inland cities. In a sense, China is trying to follow the model of the United States, which successfully developed westward from the industrial and financial centers on or near its east coast. Nevertheless, in the United States there was a second coast in the West that, with its rich natural resources and convenient access to maritime transportation, stood not only as a target but as a magnet for development. Although Shanghai can be perceived as functioning like a "New York in China" in a broad sense by offering both financial and industrial assistance to inland areas, the horizontal transmission of resources and technology from east to west across the wide national urban landscape is predicated on building an extensive network of transportation and communications. Realistically, China has a long way to go to succeed.

The urban-system planning strategies essentially are conditioned by China's sociocultural past and unique structural characteristics, including geographic diversity, a large number of port cities on a long shore, a massive rural population, and incomplete colonial penetration. Furthermore, the critical role of a strong state in shaping and pushing these policies is peculiar to China's political system. For all these reasons, these policies are not automatically transferable to other developing countries.

Nevertheless, some components of the Chinese model may be applicable cross-nationally. Regionalizing the structural and functional status of secondary cities as economic coordinating centers is a promising move for balancing the whole urban structure. Increasing the influence of these cities across their tributary areas helps strengthen their positions vis-à-vis primate cities. This strategy may retrieve the missing middle link in many rapidly urbanizing countries whose provincial and regional centers often are bypassed or serve only as stepping-stones for migrants to the capital or other primate cities. The comprehensive citification of rural towns also may be appropriate for some developing countries, especially those in Asia and Africa which are still largely rural. It is conducive to breaking down the isolation of very large rural sections by generating local economic dynamism and transmitting it to cities. More important, the emergence of a network of rural towns, which have efficient agricultural marketing and flourishing local industries and services, helps retain many potential peasant migrants at the lower levels of the urban system. Otherwise the very limited opportunities provided by underdeveloped towns and small cities make them only transitional stops for migrants on their way to larger and primate cities. China's strategy of strengthening and building rural towns and expanding some into full-fledged small cities is part and parcel of the more general and multidimensional rural economic reforms, which predated current urban reforms in cities. This experience could become a crucial ingredient in an integrated urban-planning policy for some developing countries.

Balanced patterns of urban growth approximating the rank-size pattern are difficult to achieve in many Third World countries where urbanization has already evolved into entrenched primacy. Partial strategies only have a minimal impact and a complete restructuring of the urban system requires centralized governmental effort and huge unavailable resources. The Chinese experience is a typical mixed bag of continued problems besetting the nation's giant cities and constantly emerging measures for dealing with them. Clearly China's modernization relies to a large extent on the industrial capacity and economic functions of these very large cities; on the other hand, it is imperative for China to minimize the price to be paid for favoring these cities through trial-and-error urban-planning policies.

Given that varied political, cultural, and economic factors have shaped the courses and patterns of urban growth and large cities in developing countries, it seems more desirable for each to adopt selective urban-planning programs which target its own problems. The potential diversity of governmental and private responses to the pressing problems associated with rapid growth of large cities in Third World countries cannot be covered in a single case-oriented analysis of China. However, the implications of Chinese policies for some general and theoretically relevant issues in comparative urbanization research can and should be seriously considered.

A Final Look Down the Road

Returning to the basic concern implied by the title of this chapter, one finds troublesome questions arising when speculating about the demographic and socioeconomic futures of China's giant cities. Is China's comprehensive urban-planning strategy, coupled with its piecemeal and ad hoc approaches, sufficient for monitoring the pulse of these very large cities and ensuring their healthy development? Will the strict population control in the past be effective enough from now on in light of socioeconomic changes so that urban primacy in a demographic sense and its accompanying problems can be avoided? Available demographic projections have presented a rather pessimistic scenario: Shanghai and Beijing's populations are predicted to swell to 23.7 and 20.9 million (including population in their metropolitan areas), respectively, by the year 2000. This will rank them third and sixth on the list of the 35 largest cities in the world (United Nations, 1981). To achieve these gigantic sizes, the cities will have to grow at 4.0 percent and 4.7 percent per annum, far exceeding their current growth rates (see table 8.3). Realization of this scenario may bring a high degree of urban primacy if cities right below Shanghai and Beijing and those lower on the urban hierarchy grow at a slower pace in the comparable span of time. Although the ongoing tendency is that these other cities have been growing faster than Shanghai and Beijing (Chen, 1986), continued control of the latter two is highly necessary, particularly in most of the 1980s when rapid increase in marriages of baby boomers from the early 1960s has begun generating an upswing in birth rates of these two largest Chinese cities.

A few other factors also could contribute to a faster growth of very large cities in China: (1) the reformed rural economy, which has moved away from the traditional mode of labor-intensive and low-yielding grain production, will continue to drive peasants off the land, thereby amassing a pool of potential rural-urban migrants; (2) recently loosened control on between-city and rural-urban migrations may thicken and channel the volume of migrants in the direction of very large cities, especially Shanghai and Beijing, which remain by far China's most alluring urban centers. Given the retrospective evidence that rigorous population control policies have reduced the growth rates of Shanghai and Beijing to almost the lowest in comparison with cities of similar sizes in the developing world today, we cannot take a blind faith stance toward the rapid prospective growth rates suggested by pure demographic projections. On the other hand, any major policy shifts concerning population control could add considerable force to the inherent demographic momentum and large base of giant cities in China. Therefore, it is not unreasonable to assume that a close interaction between the two aforesaid determinants could reverse the growth trend of Shanghai, Beijing, and other giant cities in the years ahead.

Aside from speculating about the counterfactual conditionals because of the unpredictability of policy swings in China, the actual conditions very large cities in China face today present problems unlikely to be solved simultaneously. The fundamental one seems to be a lack of parallel development of a largely labor-intensive urban economy and these cities' ability to absorb more laborers and provide them with adequate service facilities. For example, although Shanghai will probably never repeat the present situation of Mexico City with massive numbers of urban migrants jamming its peripheral shanty towns, Shanghai has not been very successful in dispersing its population from its overcrowded central city because of its suburbs' inferior and unattractive infrastructure. People who have taken jobs in Shanghai's satellite towns cling to their residence in the central city at the cost of commuting for two hours every day on congested public transportation. So long as infrastructure development in satellite towns of very large cities continues to lag, there will be little attraction for industrial firms and service establishments to decentralize. Therefore, these large cities will continue to bear the pressure of concentrated industry and population on their existing municipal functions. This continued pressure will, in turn, offset efforts to improve the urban environment and living conditions in China's giant cities.

Notes

1. During 1983, 44 new cities were established, raising the total number of cities in China to 289 (6 are left out of the 1982 data in table 8.1) by the year's end in 1983 (*Renmin Ribao* [People's Daily] March 4, 1984, p. 5).

2. Comprehensive information is available for only 16 Chinese cities with more than two million people. Data are missing for four other cities of such size, except population and growth rate; therefore, they have been excluded from the analysis. (For a study of urban primacy in China involving the size and growth rate of the 44 largest cities in China, see Chen, 1986).

3. The annual growth rate is calculated with the formula for obtaining compounded increases.

4. For example, population density for the urban district of Shanghai is 27,261 per square kilometer, while there are only 516 residents per square kilometer in the urban district of Taiyuan, smallest of the 16 cities. Another abnormal distribution is reflected in the gross industrial output per capita measure for suburban and rural sections of the city: Shanghai claims 1,813.90 yuan in gross industrial output for each of its inhabitants, whereas the city of Chongqin only has 32.64 yuan.

5. Population size and density have been transformed to logarithms to normalize the underlying skewedness caused by one or two very large and highly densely populated cities.

6. Almost all product-moment correlation coefficients presented are statistically significant at least at the .05 level. Since the number of cases is small and not a random sample, we avoid emphasizing statistical inference and comment only on size and sign of the correlation.

7. A recent survey of air pollution in 52 Chinese cities revealed that the average amount of soot per cubic meter in the atmosphere varied from 427 to 1,358 micrograms. The higher number is four times worse than the limit of 300 micrograms recommended by the state (*China Daily*, 1985b).

8. For example, Shanghai will expand the economic integration of urban and rural areas in a newly established Shanghai Economic Zone which includes Zhejiang, Jiangsu, Auhui, and Jiangxi provinces (*China Daily*, 1985a).

9. This figure was revealed by a recent survey conducted by the Ministry of Public Security. This rapid increase of transient population has imposed additional strain on public transport and lodging facilities in these very large cities. A good number of these passers-by can't find hotels to stay in and have to lodge in public bath-houses, railroad stations, or even on the streets (*People's Daily*, 1985).

10. Shanghai's dynamic economy has spilled into its seven satellite towns, particularly Banshan, Jinshan, and Minhang. The massive Banshan Iron and Steel Complex, which has imported 300 thousand tons of equipment and technology from Japan and West Germany, is being built at Wusong, north of the city. Jinshan, south of Shanghai, was converted from a deserted shore 12 years ago into a "petrochemical city" when the Shanghai General Petrochemical Works went into operation (*China Daily*, 1985a).

11. These rates are calculated forward from the most recent United Nations estimates and projections (United Nations, 1985).

12. Boasting one of the biggest harbors in the world, ice-free all year, the shipping frontage and handling capacity of the Shanghai Port will be further enlarged by this new container dock (*China Daily*, 1985c).

Bibliography

Bean, Frank D., Dudley L. Poston, Jr., and Halliman H. Winsborough. 1972. "Size, Functional Specialization, and the Classification of Cities." *Social Science Quarterly* 53:20-32.

Berry, Brian J. L. 1981. *Comparative Urbanization: Divergent Paths in the Twentieth Century*. Rev. and enl. 2d ed. New York: St. Martin's Press.

Berry, Brian J. L., and John D. Kasarda. 1977. *Contemporary Urban Ecology*. New York: Macmillan.

Buck, David D. 1981. "Policies Favoring the Growth of Smaller Urban Places in the People's Republic of China, 1949-1979." In *Urban Development in Modern China*, edited by Lawrence J. C. Ma and Edward W. Hanten. Westview Special Studies on China and East Asia. Boulder, CO: Westview Press.

———. 1983. "New Municipal Plan for Beijing." *Urbanism Past and Present* 8(2):14-22.

Chan, James. 1981. "The Central Places of Gungdong Province, China." In *Urban Development in Modern China. See* Buck, 1981.
Chandler, Tertius, and Gerald Fox. 1974. *3000 Years of Urban Growth.* New York: Academic Press.
Chen, Xiangming. 1985a. "The Challenge of Global Urbanization and City Growth: A Chinese Response." Paper presented at the annual meeting of the Southern Sociological Society, Charlotte, NC, 10-13 April.
———. 1985b. "Comparing Development Strategies in East-Asia: A Preliminary Examination of China's Special Economic Zones (SEZs)." Paper presented at the annual meeting of the American Sociological Association, Washington, DC, 26-30 August.
———. 1985c. "The One-Child Population Policy, Modernization, and the Extended Chinese Family." *Journal of Marriage and the Family* 47:193-202.
———. 1986. "A Comparison of Urban Primacy: The Special Case of China." Paper presented at the annual meeting of the Population Association of America, San Francisco, 3-5 March.
China Daily. 1985a. "Shanghai Embraces Grand Design." (25 Apr.):2.
———. 1985b. "Urban Pollution at 'Critical Levels.' " (25 Apr.):3.
———. 1985c. "Urban Giant Makes Big Strides." (10 July):2.
Davis, Kingsley. 1972. *World Urbanization 1950-1970.* Vol. 2. *Analysis of Trends, Relationships, and Developments.* Population Monograph Series no. 9. Berkeley: Institute of International Studies, University of California.
———. 1974. "Asia's Cities: Problems and Options." *Population and Development Review* 1:71-86.
Ding, Yisheng. 1984. "The Urban and Rural Distribution of China's Population." *Renkou Yanjiu* (Population Research), no. 4(July):14-17 (in Chinese).
Eberstein, Isaac W., and W. Parker Frisbie, 1982. "Metropolitan Function and Interdependence in the U.S. Urban System." *Social Forces* 60:676-700.
Gernet, Jacques. 1977. "Daily Life in China on the Eve of the Mongol Invasion 1250-1276." In *Third World Urbanization*, edited by Janet Abu-Lughod and Richard Hay, Jr. Chicago: Maaroufa Press.
Gibbs, Jack P. 1966. "Measures of Urbanization." *Social Forces* 45:170-77.
Ginsburg, Norton. 1980. "Urbanization and Development: Processes, Policies, and Contradictions." In *China: Urbanization and National Development,* edited by Chi-Keung Leung and Norton Ginsburg. Research Paper no. 196. Chicago: Department of Geography, University of Chicago.
Golden, Hilda H. 1981. *Urbanization and Cities: Historical and Comparative Perspectives on Our Urbanizing World.* Lexington, MA: D.C. Heath.
Goldstein, Sidney, and David F. Sly. 1975? *The Measurement of Urbanization and Projection of Urban Population.* International Union for the Scientific Study of Population Working Paper no. 2. Dolhain, Belgium: Ordina Editions.
Gui, Shixun, and Baoshu Zhu. 1984. "The New Technical Revolution Faces the Population Problem in Shanghai." *She Hui* (Society) 4 (2):8-12 (in Chinese).
Hamer, Andrew. 1985. "Urbanization Patterns in the Third World." *Finance & Development. A Quarterly Publication of the International Monetary Fund and the World Bank* 22 (Mar.):39-42.
Hay, Richard, Jr. 1977. "Patterns of Urbanization and Socio-Economic Development in the Third World: An Overview." In *Third World Urbanization. See* Gernet, 1977.
Kapp, Robert A. 1974. "Chungking as a Center of Warlord Power, 1926-1937." In *The Chinese City between Two Worlds,* edited by Mark Elvin and G. William Skinner. Studies in Chinese Society. Palo Alto, CA: Stanford University Press.
Kearsley, G. W. 1984. "Urbanisation and Development in the Third World." *New Zealand Journal of Geography* no. 76 (Apr.):16-18.
Kwok, R. Yin-Wang. 1981. "Trends of Urban Planning and Development in China." In *Urban Development in Modern China. See* Buck, 1981.
Li, Mengbai. 1983. "Planned City Growth." *China Reconstructs* 23(Nov.):7-9.

Li, Zhaoji. 1984. "New Postures of China's Open-Door Policies." *Liaowang* (Outlook Weekly) 8 (Feb.):9-11 (in Chinese).

Lo, Chor-Pang, Clifton W. Pannell, and Roy Welch. 1977. "Land Use Changes and City Planning in Shenyang and Canton." *Geographical Review* 67:268-83.

Ma, Lawrence J.C. 1981a. "Introduction: The City in Modern China." In *Urban Development in Modern China. See* Buck, 1981.

———. 1981b. "Urban Housing Supply in the People's Republic of China." In *Urban Development in Modern China. See* Buck, 1981.

Mohan, Rakesh. 1984. " 'The City Study': Understanding the Developing Metropolis." *World Bank Research News* 5(3):3-14.

Mote, F. W. 1977. "The Transformation of Nanking, 1350-1400." In *The City in Late Imperial China,* edited by G. William Skinner. Studies in Chinese Society. Palo Alto, CA: Stanford University Press.

Murphey, Rhodes. 1974. "The Treaty Ports and China's Modernization." In *The Chinese City between Two Worlds." See* Kapp, 1974.

———. 1980. *The Fading of the Maoist Vision: City and Country in China's Development.* New York: Methuen.

Pannell, Clifton W. 1981. "Recent Growth and Change in China's Urban System. In *Urban Development in Modern China. See* Buck, 1981.

———. 1984. "China's Changing Cities: Urban View of the Past, Present, and Future." In *China: The 80's Era,* edited by Norton S. Ginsburg and Bernard A. Lalor. Westview Special Studies on East Asia. Boulder, CO: Westview Press.

People's Daily. 1985. "Rapid Increase of Transitory Population in Ten Large Cities." (3 July): 8 (overseas ed.; in Chinese).

People's Republic of China. State Statistical Bureau. 1982. *Statistical Yearbook of China 1981.* English ed. Hong Kong: Economic Information & Agency.

———. 1983. *Statistical Yearbook of China 1982.* English ed. Hong Kong: Economic Information & Agency.

Pryor, Robin J. 1968. "Defining the Rural-Urban Fringe." *Social Forces* 47:202-15.

Richardson, Harry W. 1984. "The Goals of Population Distribution Policy." In *Population Distribution Policies in Development Planning. Papers of the United Nations/UNFPA Workshop on Population Distribution Policies in Development Planning. Bangkok, 4-13 September 1979.* Population Studies no. 75. ST/ESA/SER.A/75. New York: Department of International Economic and Social Affairs, United Nations.

Schnore, Leo F. 1957. "The Growth of Metropolitan Suburbs." *American Sociological Review* 22:165-73.

———. 1963. "The Socio-Economic Status of Cities and Suburbs." *American Sociological Review* 28:76-85.

Skinner, G. William. 1977. "Regional Urbanization in Nineteenth-Century China." In *The City in Late Imperial China. See* Mote, 1977.

Ullman, Morris B. 1969. "Cities of Mainland China: 1953-1959." In *The City in Newly Developing Countries: Readings on Urbanism and Urbanization,* edited by Gerald W. Breese. Modernization of Traditional Societies Series. Englewood Cliffs, NJ: Prentice-Hall.

United Nations. Department of Economic and Social Affairs. 1981. *World Population Trends and Policies, 1981 Monitoring Report.* Vol. 1. *Population Trends.* Population Studies no. 79. ST/ESA/SER.A/79. New York.

———. 1985. *World Population Prospects. Estimates and Projections as Assessed in 1982.* Population Studies no. 86. ST/ESA/SER.A/86. New York.

———. Statistical Office. 1983. *Demographic Yearbook 1981.* 33d ed. ST/ESA/STAT/SER.R/11. New York.

9

India's Giant Cities

Hans Nagpaul

India is known as a land of villages. About 75 percent of its population still lives in about 600 thousand small, scattered settlements, mainly dependent on agriculture for their livelihood (India, Ministry of Agriculture, 1976). The predominant rural form of life often overshadows the growing proportion of her population residing in urban areas as well. The total urban population of the present Indian territory (excluding Jammu and Kashmir) was 26 million (11 percent of the total population) in 1901, 108 million in 1971, and is now reported to be about 160 million—approximately 23 percent of the total population according to the 1981 Census. Thus, during the last 80 years, the urban population has increased by almost 135 million.[1]

A notable feature of urban development in India has been the rapid growth of cities exceeding 100,000 people. In 1931, 31 such cities in undivided India had a total population of about 10 million; by 1951 the Indian Union alone had 71 such cities whose population totalled 26 million; and in 1961 the number increased to 109 with a 36 million population containing 48 percent of the urban population and 8 percent of the country's total population.

By 1971, cities of 100 thousand population had increased to 142 and the number had reached 215 by 1981 (India, Registrar General, 1982; United Nations, Department of International Economic and Social Affairs, 1983). On the whole, growth of cities of all sizes has accelerated. Among the cities there has also been a marked concentration of population in the very large urban centers and as much as half the total city population lives in 12 cities of more than one million each. In this context table 9.1 provides data about urban population growth from 1901 to 1981; table 9.2 traces growth of cities with 100

TABLE 9.1
Growth of Urban Population—India, 1901-1981

Census Year	*General Population (millions)*	*Total Urban Population (millions)*	*Urban Population as Percentage of Total Population*
1901	235.5	25.7	10.9
1911	249.0	26.3	10.3
1921	248.1	28.2	11.2
1931	275.5	33.4	12.0
1941	314.8	43.8	13.9
1951	356.9	61.9	17.3
1961	436.5	79.2	17.9
1971	547.0	108.8	19.8
1981	685.1	159.7	23.2

SOURCE: Census of India (1981); United Nations, Department of International Economic and Social Affairs (1983).
NOTE: Figures have been rounded. All relate to present territory of India (excluding Jammu and Kashmir). Since the 1961 Census, definition of *urban* has been more rigorous than the definition adopted in earlier censuses.

thousand plus population from 1901 to 1981; and table 9.3 shows population growth in the ten largest cities. In 1971, the city of Nagpur exceeded one million population; in 1981, Jaipur joined this group making a total of 12 million plus cities.

Historical Growth of Cities

Urbanization in India originated long before the Christian era. Abundant historical evidence is available to show that urban centers existed as early as 2500-1500 B.C. Though the ancient Hindu cities are irrevocably gone, we still find relics of the past in some parts of the country. During the Muslim period, most of the ancient cities fell; some underwent changes and only a few were able to preserve their traditional nature; some new cities also took shape under the Muslim regime (Ghurye, 1963; India, Ministry of Education, 1967; Piggot, 1945). During the 18th century, European colonial rule brought about sweeping economic and political changes which affected the process and growth of urbanization considerably and led to establishment of new commercial and administrative cities. At the same time, a number of traditional cities dwindled and some even disappeared. New cities rose mainly as centers of raw materials export and foreign goods import. Administrative cities, educational centers, and military stations carrying out political functions of the foreign rulers grew rapidly. The rise of these cities, though mainly nonindustrial in character, facilitated the movement of labor and capital and set the stage for the industrial revolution.

The beginnings of the factory system from 1880-95, as reflected through establishment of the textile and jute industries along with the introduction of a network of railroad systems, gave birth to several new cities, and also

TABLE 9.2
Growth of 100,000 Plus Cities–India, 1901-1981

Year	*Number*	*Population (millions)*	*Percentage of Total Urban Population*	*Percentage of Total Population*
1901	24	6.5	23.0	2.7
1911	23	6.9	24.2	2.7
1921	28	7.9	25.3	3.2
1931	31	9.6	27.4	3.4
1941	46	15.7	36.0	5.0
1951	71	25.8	41.7	7.2
1961	109	36.0	48.4	8.3
1971	142	57.0	52.4	10.0
1981	215	95.1	60.0	14.0

SOURCE: Census of India (1981); United Nations, Department of Economic and Social Affairs (1983).
NOTE: Figures have been rounded.

TABLE 9.3
Population Growth in Ten Largest Cities–India, 1901-1981 (in millions)

City	*1901*	*1931*	*1941*	*1951*	*1961*	*1971*	*1981*
Calcutta	1.5	2.0	3.5	4.6	5.5	7.0	9.1
Bombay	0.8	1.3	1.7	2.8	4.1	6.0	8.2
Delhi	0.2	0.4	0.7	1.4	2.3	3.6	5.1
Madras	0.5	0.6	0.8	1.4	1.7	2.5	4.2
Bangalore	0.1	0.3	0.4	0.8	0.9	1.6	2.9
Ahmedabad	0.2	0.3	0.6	0.8	1.5	1.6	2.5
Hyderabad	0.4	0.4	0.7	1.8	1.2	1.8	2.5
Kanpur	0.2	0.2	0.5	0.7	0.9	1.3	1.7
Poona	0.8	0.2	0.3	0.6	0.7	0.8	1.7
Lucknow	0.2	0.3	0.4	0.5	0.7	0.8	1.0

SOURCE: Census of India and United Nations (1981); Department of Economic and Social Affairs (1983).
NOTE: Figures have been rounded. For Calcutta, Bombay, Delhi, and Madras metropolitan population is indicated.

brought an industrial orientation to some existing cities. The pace of industrialization, however, remained sluggish during British rule, considering India's vast potential and there was no attempt to lay sound foundations for industrial expansion by developing basic and capital goods-manufacturing industries.

During the Second World War, the British government became more conscious of the necessity for industrial expansion and substantial progress was achieved in the postwar period. But industrialization, on the whole, played a limited role in the growth of urbanization, and consequently only a few cities had a decidedly industrial character. Even in many industrial cities, a large proportion of the labor force in manufacturing continued to be engaged only in small-scale establishments (Buchanan, 1934; Gadgil, 1971; Lokanathan, 1935).

Since attaining independence, the Indian national government has enunciated a comprehensive industrial policy for systematic and thorough expansion of organized modern industries. Inauguration of the first five-year plan in 1951 paved the way for establishment of large-scale industries and subsequent plans for overall economic development promoted a bold program of industrial development. Several new industrial towns were established which grew into cities in a single decade. The policy of decentralizing industries to bring about balanced regional development and population dispersal was also encouraged to some extent. The impact of industrialization on urban growth has not been dominant because multiple push factors in the rural economy have been much more responsible for the in-flow of rural migration to urban places. Nevertheless, the ambitious programs of industrialization introduced during the last three decades have intensified urbanization and promoted further rural migration to towns, cities, and metropolitan centers in recent years (India, Planning Commission, 1958).

The Exploding Giant Cities

Bombay, Calcutta, Delhi, and Madras have become the exploding giants, not only of India but of South and Southeast Asia. Almost a fourth of India's urban population of cities of a hundred thousand plus live in these giant cities. Of the 41 million population in million-plus cities, about 27 million—65 percent of the total residing in large cities—live in the four giant cities (Misra, 1978; United Nations, Department of International Economic and Social Affairs, 1983). See table 9.4 for current and projected populations of these cities.

No scholar or writer from any field wants to predict future needs and basic requirements of food, shelter, clothing, education, and health for a population of about 50 million likely to inhabit these cities in the year 2000. The task is staggering with bewildering implications. These cities are centers of almost every human activity ranging from industry, commerce, finance, education, health, science and technology to production of consumer goods. The cities are also the very life blood of the Indian society and provide most of the revenue to their respective states and the national government. Almost all forms of change radiate from these cities to other parts of the country. The giant cities present contrasting styles of life; old and new, oriental and occidental, and wealth and poverty exist side by side. Rural enclaves within these cities are widespread and an estimated one-fourth to one-third of the residents live under impoverished conditions. Rates of crime and other forms of disorganization have increased slowly but steadily. In comparison to major cities in the United States, these rates are still low. (See table 9.5.)

Calcutta originated in 1690 as a trading center. Susequently, the British East India Company made it the capital city of its empire in 1772, a political

TABLE 9.4
Projected Population for Giant Cities–India, 1981 and 2000
(in millions)

	1981	*2000*
Calcutta	9.1	19.6
Bombay	8.2	19.0
Delhi	5.1	13.2
Madras	4.2	10.3

SOURCE: United Nations, Department of International Economic and Social Affairs (DESA) (1976).
NOTE: The United Nations DESA issued similar estimates for the year 2000 in 1985.

status it held until 1912 when the capital was moved to Delhi. Today Calcutta serves as India's chief port for international trade and is also considered the gateway to the most heavily populated region of northeast India. People from almost every other part of India and South Asia are settled here. Calcutta is the world center of jute production and has many other important industries including cotton mills, tanneries, iron works, railway coach and automobile assembly plants, metal goods and manufacturing, electrical equipment, shoe factories, cement making, and glass works. The city is considered dynamic because demonstrations, strikes, and riots take place in its streets quite frequently. Although several hundred thousand Calcuttans sleep in the streets every night, the city shows relatively low levels of crime that, by conventional Western social theory, calls for new lines of explanation (N. K. Bose, 1968; Dutt, 1977; Nagpaul, 1969; Sen, 1960).

Next in population size is Bombay, an island and India's most Westernized city. Portuguese traders founded Bombay in the mid-16th century. In 1661 King Charles II of England received the city as a wedding gift and subsequently rented it to the East India Company. By 1850 Bombay had become a metropolis with a population of half a million. Vying with Calcutta industrially and commercially, Bombay handles almost one-third of India's trade.

All types of industries ranging from cotton textiles, pharmaceuticals, chemicals, plastic wares, and petroleum distribution to consumer goods are found in Bombay. India's large motion picture industry flourishes in the city. People from almost every Indian state live here, and multiple languages are spoken, though most people are Maharashtrians.

Bombay has exclusive neighborhoods for the middle and upper classes while almost a million people live in flimsy shanties and at least several hundred thousand sleep in the streets. High rises are numerous; overcrowding is widespread; traffic jams are frequent; and general living conditions are poor. In 1980, more than five thousand people were killed in accidents, 202 committed suicide, and 160 were murdered. All forms of deviance are reported to have increased in recent years (Arunachalam, 1978; India, Ministry of Home Affairs, 1983; Lakdawala, 1963).

TABLE 9.5
Murders, Suicides, and Accidental Deaths in Giant Cities—India, 1980

City	*Total Population (millions)*	*Murders*[a]	*Suicides*	*Accidental Deaths*
Calcutta[b]	3.3	93	23	620
Bombay	8.3	187	202	5113
Dehli	5.4	160	255	1941
Madras[b]	3.4	34	263	574

SOURCE: India, Ministry of Home Affairs (1983, 1984).
NOTE: These figures are taken from the cited reports; there are minor variations from the data given in Table 9.4 based on India Census and the United Nations statistics.
a. 1979.
b. City only.

Delhi lies on the western bank of the Yamuna River in north India. The city and area surrounding it are believed to have been the site of many ancient and medieval cities. Mogul emperors who ruled in the 17th and 18th centuries built much of the present old city. Delhi was the capital of British India from 1912 to 1931, when a new city called New Delhi was established at a distance of three miles. The central part of Delhi is within the ruins of walls built in the mid-17th century and 3 of the city's original 14 gates have survived. Delhi is an old and crowded city but its limits have been extended several miles in all directions during the last three decades. The city has seen an unparalleled growth industrially, commercially, and administratively after India gained independence from Great Britain in 1947 when its total population was less than a million. In addition to natural increase, almost one million refugees from Pakistan are believed to have been settled here. Inmigration from neighboring states as well as from other parts of the country continues. Despite attempts by various levels of government to improve living conditions, and direct future urban growth, the housing shortage and lack of common amenities of life have led to the emergence of slums and construction of unauthorized structures on an unprecedented scale. Perhaps no other city has received so much attention in terms of urban planning and programs to improve living conditions and influence the course of future urban growth and development. Evaluations show only partial success. Problems of law and order have become bewildering. Recent statistics show that crime rates have grown, accidental death has become more frequent, and even suicides are on the increase.

As contrasted to the old city, New Delhi is attractive and spacious. The city is basically a planned one where the huge economic base is government offices and employees working in main government buildings which dominate the city's center. The southern part has been extended for several miles, and new neighborhoods and communities have been created. New Delhi has several beautiful gardens, shopping centers, and many tourist attractions. But even in this city, the problem of unauthorized structures and squatter settlements exists on a smaller scale (Delhi Administration, 1977; India, Ministry of Home Affairs, 1983; Mitra, 1970; Rao and Desai, 1965).

Madras, another metropolis, is the heart of the Dravidian culture and claims to have a history and background of five thousand years. Like Calcutta and Bombay, it is a state capital. Madras is a shipping center and has a number of industrial plants; but it is more provincial—a typical regional city and therefore less cosmopolitan. The British are given credit for constructing a fort in 1640 as an administrative center which became the nucleus of the present city after some years. By 1921 its population had grown to more than half a million. Unlike Calcutta and Bombay, Madras has adequate land free from swamps for expansion but its population growth has been rather slower than other giant cities. The original city growth and development show little planning. The city has numerous landmarks including old Hindu temples and Christian churches. Modern buildings, small shops, thatched huts, orchards, open fields, double decker buses, cars, and ox carts exist side by side. Overall living conditions for the vast majority of population are still poor. Among the giant cities, disparity in sex ratio is far less in Madras, attributed to the tradition that most South Indian women prefer to move with their menfolk when they migrate to the city (Balakrishna, 1961; Mudaliar, 1973).

Characteristics of Metropolitan Cities

The metropolitan cities occupy a unique position in Indian society and her economy. The cities are centers of manufacturing, industries, commerce, banking, political life, administration, education, and social services; they also act as agencies for the diffusion of all kinds of social and cultural change; and they are the major loci of Western-educated manpower and political power as well. In addition, these centers have become magnets for displaced rural residents seeking employment opportunities. They are also promoting new expectations and aspirations among longer-term residents.

The metropolitan centers of Calcutta, Bombay, Delhi, and Madras have become so cosmopolitan in the context of Indian culture that people from almost all regions have come to them and established their own little communities based on their caste, religion, and language. All the metropolitan centers are, however, confronted with massive problems of providing adequate employment, housing, transport, water, and drainage systems. More than one-fourth of their population is reported to be surviving at the subsistence level where even basic necessities of life are often lacking. Some major characteristics of metropolitan cities can be easily identified and are discussed below.

Indigenous and European Patterns

The colonial origin and development of almost all major existing cities are responsible for several unique features which deserve special mention. The urban morphology of cities is usually a mixture of indigenous and European patterns (R. L. Singh, 1973; Yadav, 1979). The typical Indian city contains a congested old section having the main market place, with irregular, narrow, and crowded streets, and little open space; often surrounding the commercial streets are residential neighborhoods. The European pattern exhibits a remarkable contrast with wide streets, large homes, parks, and spacious grounds, though this pattern is undergoing considerable change now. Some cities, of course, have exclusively European patterns, while others show varying degrees of blending as well as conflicts with indigenous patterns. This ecological pattern has generally promoted population segregation according to function and social status; it is still being followed and has been even extended in some cases when new urban development has taken place.

Concentration of Migrants

Through rural-urban migration facilitated by push and pull factors, a large number of people from rural areas are uprooted and transplanted in the alien environment of urban areas, especially big cities. Over one-third of all people living in big Indian cities were born outside and their proportion tends to increase in direct relation to the size of cities. Thus, most Indians in cities are in transition—living in cities but still maintaining their village contacts and rural ways of life. This feature is often perpetuated by the existence of rural enclaves of population established by earlier streams of migrants in cities. Moreover, the majority of migrants to cities are usually young males in the early working age group who leave behind in their villages extended kinship ties. This fact explains the disproportionate sex ratio in most Indian cities and also the persistence of attachment to the native place. The noncommitment of labor to urban ways of living has also been attributed to the migratory character of city population to some extent, though other explanations have also been put forward from time to time.

Monster of Unemployment

With population growth and the constant flow of migrants, unemployment is widespread in most Indian cities, especially among unskilled and other marginal workers. Unemployment among the educated classes in urban areas is a peculiar feature of Indian society also. Thousands of high school and college graduates experience varying degrees of unemployment as educational institutions seem to have multiplied faster than industries which can absorb formally educated people. According to one estimate, in the larger cities, of all the employed, 52 percent were literate, although only 3 percent had college educations. Some 46 percent of all the educated unemployed are reported to be concentrated in the four major cities in India. No reliable data on urban unemployment are available. The number of persons registered with employment exchanges gives some idea of urban unemployment. One government publication reports that the number of job seekers registered with employment exchanges increased from 3.4 million in December 1969 to 16 million in December 1980 (India, Ministry of Information and Broadcasting, 1981; Planning Commission, 1981).

From data gathered as a result of city socioeconomic surveys conducted under auspices of the Research Program Committee of the Planning Commission, unemployment was greatest in Calcutta where roughly one-tenth of the labor force of several hundred thousand able-bodied persons was unemployed. In Bombay and Madras, the situation regarding unemployment appears no better (India, Planning Commission, 1981).

There is also a considerable disguised unemployment or underemployment in metropolitan cities as reflected in the rapid growth of the low-productivity service sector and in seasonal occupations in which unskilled, uneducated workers, and especially recent migrants seek means to subsist. All this points to the lack of increase in the creation and availability of jobs despite industrialization at the same rate as the growth of urban population.

Vicious Cycle of Poverty

The problem of poverty is closely related to unemployment and underemployment. According to World Bank studies, annual per capita GNP in India was estimated to be about $260 in the early 1980s, identical with national per capita income (Population Reference Bureau, 1985). Notwithstanding the various limitations of these concepts, it is beyond questioning that incomes are generally very low and vast disparities exist between rich and poor. Moreover, apart from low income, poverty is manifested in illiteracy, ill health, malnutrition, and low expectancy of life. Perhaps half of all Indians are living in poverty conditions (India, Planning Commission, 1981).

Urban residents may have an edge because of the general availability of fundamental necessities in cities, even if limited. But overall economic conditions are somewhat unsatisfactory for the vast majority of urban dwellers and far worse for scheduled castes and tribes.[2]

What was written in 1940 that the average Indian worker was ill-fed, ill-clad, ill-lodged, and led a dull and dreary existence is certainly true even today. No amount of statistics and research studies are needed to establish that income levels are low and living conditions poor. Differences between living standards of the poor and the middle class have become more pronounced in recent years. The structure of poverty tends to generate a vicious cycle which locks millions of people into permanent poverty especially those who are illiterate, unskilled, sick, old, or handicapped. The problem of widespread beggary is yet another dimension of poverty in metropolitan cities.[3] Problems of poverty and beggary are extremely complex and solutions are not easy to find. The report of the seventh Five Year Plan (1985) and all earlier plans have reiterated that the central objective of planning is to eliminate poverty and improve living standards through social and economic programs. Unfortunately the twin problems of unemployment and poverty are still widespread and policies have been unsuccessful in reducing poverty and its major consequences (De Souza, 1978; Singh and De Souza, 1980).

Urban Sprawl

The rapid process of urbanization taking place for the last several decades has been almost uncontrolled, unregulated, and unchecked. In the past, most industries grew without any systematic planning. Existing towns and new cities kept growing haphazardly and failed to have a corresponding growth in housing facilities and other amenities of life. This situation has led to general overcrowding in existing neighborhoods and housing structures and to the emergence of hutment colonies, often called urban villages, within the cities and in the hinterland surrounding them. Municipal authorities often have taken no effective steps to cope with the growing problems of inadequate and poor housing either because of lack of financial resources or for want of suitable regulatory laws. In many cases, the sheer presence of vested interests has been responsible for widespread construction of unauthorized structures and perpetuation of unsatisfactory living conditions. The constant flow of migrants into cities with little capacity to absorb them in their overcrowded and dilapidated housing structures continues to intensify the problem of existing slums.

A number of studies published in the 1950s and 1960s surveying conditions of industrial labor revealed the magnitude of subhuman living conditions prevailing in most large Indian cities. Socioeconomic surveys of major cities also clearly brought out that approximately one-third to one-fourth of the population in Calcutta, Bombay, Delhi, and Madras lived in poor dwellings, considered unfit for human habitation even by applying a criterion of subsistence living. The position of other large cities—Lucknow, Kanpur, Poona, Hyderabad, Allahabad, Jaipur—was found no better. The All India Economic Association and All India Conference on Social Welfare discussed and debated the problems of urbanization and slum clearance in the 1950s. Even the Unesco Center for the Study of the Social Implications of Industrialization and Urbanization, located in India, warned about the tragic consequences in 1952. Findings of these studies and conferences completed several decades ago are still valid. Scholars writing in the 1980s continue to remind us that living conditions have not changed substantially for the poor and in many cases they may have deteriorated. There is no denying that with slums and squatter settlements practically in every city—small or large—problems of congestion, inadequacy of health services, absence of suitable transit systems, and lack of essential municipal amenities are multiplying day by day and further aggravating existing problems.[4]

Centers of Education, Science, and Technology

Education, science, and technology centers are mainly concentrated in metropolitan cities. Specialized centers for advanced study and research have been established in the giant cities of Bombay, Calcutta, Delhi, and Madras. During the past 30 years, several hundred engineering colleges and institutes have been started; medical, business, and law colleges have expanded; and the country has seen a phenomenal growth of arts and sciences colleges. By now more than 5 thousand colleges and institutes with a total enrollment of about 3 million students are functioning under supervision of 120 universities. In addition several dozen independent institutions carry out research in such diverse fields as agriculture, technology transfer, electronics, medicine, public health, telecommunications, ocean development, the environment, atomic energy, public administration, and astrophysics. At least 150 research laboratories undertake research in physical, chemical, and biological sciences.

The philosophy behind this expansion is that India requires a large body of trained manpower to manage and run its programs of economic development, which no one would dispute. But, as noted, the Indian economy has failed to absorb the output of colleges and institutes. The problem of the outflow of Indian-trained scientists, doctors, and engineers to foreign countries and its implications has received little attention, presumably because most migrants are children of the middle and upper classes who hold positions of power to shape the country's educational policies but have persistently refused to reorganize educational priorities. Even after three decades of planning, literacy is still reported to be less than 40 percent. In the age group 14-17 years, only 20 percent of children were in school, and in the age group 17-23, only 4 percent were in colleges by 1980. Moreover, despite high ideals postulated in various five-year plans, adult literacy programs have not received much-needed attention and hence not been given necessary resources.[5]

Monuments and Tourism

In 1981, almost one million tourists visited India, bringing in foreign exchange earnings of about $600 million. To attract more tourists, the national government has established regional offices, mainly in metropolitan cities, and owns hotels and lodges to provide accommodation. The government has approved about two hundred travel agencies and four hundred hotels, primarily in metropolitan cities. India is a land of veritable tourist treasures—temples, mosques, palaces, forts, gardens, ruins of ancient civilizations, magnificent scenery, sculptures, paintings, exquisitely worked handicrafts, handmade silks, carved metal pieces, and so on. The giant cities of Bombay, Calcutta, Delhi, and Madras not only provide these features and dozens of other monuments of interest to tourists but are centers for most visitors. From these cities tourists travel to other places by air-conditioned buses, railways, or air. Within the last 20 years, cultural institutions have been established in these cities which promote music and dance performances and folk arts. In music, Calcutta and Bombay continue to lead; in classical dance, Madras has long held first place; and the Delhi theater has achieved considerable prominence in recent years. Some of the best museums have been established in these cities and are usually full every day. Finally, most programs organized and broadcast by the All India Radio are initiated from the four giant cities of the country.[6]

Concentration of Commerce, Banking, and Finance

The four giant cities perform the bulk of commerce, banking, and finance activities. India has the largest shipping fleet among the developing countries and ranks 15th in world shipping tonnage. The country has the world's fourth largest railroad network. Bombay, Calcutta, and Madras are the leading ports handling almost two-thirds of the total foreign trade covering more than six thousand commodities. Major industrial complexes designated as free trade zones are established in these cities where export-oriented goods are manufactured and shipped to foreign countries. Most imports also enter through these ports. All metropolitan cities, especially the giants, play a significant role in India's internal trade as well, which is many times larger than its external trade.

The giant cities are centers of banking and insurance. The Reserve Bank of India which acts as banker to government, commercial banks, and some other financial institutions has its headquarters in Bombay, with regional offices in Calcutta, Delhi, and Madras. About two hundred commercial banks have their central offices in metropolitan cities with about 36 thousand branches. Some 130 banks are in the public sector as a result of nationalizing banking in 1969; 102 are regional rural banks while the remaining transact all types of commercial banking business. In the public sector, the State Bank of India is the largest bank with six thousand branches, handling one-fourth of the total banking activities. Until recently Calcutta State Bank was the only clearing center for all foreign transactions with India. The banking industry was nationalized on the philosophy that these institutions should function as an instrument for promoting economic and social development in a more purposive manner and that credit requirements of weaker sections of population are given special consideration.

The general insurance industry was also nationalized in January 1973. The industry consists of the General Insurance Corporation of India and its four subsidiaries, all with their main offices in the four giant cities. In these cities India's policies of trade, banking, and insurance are formulated and steps are devised for their implementation throughout the country (India, Ministry of Information and Broadcasting, 1981).

Centers of Power Elites

India's giant cities, we have seen, are focal points of power, wealth, education, political action, and social progress; centers of trade, industry, finance, transport, and communication; places which patronize art, set fashions, and diffuse new ideas. While the per capita income of 95 percent of the people is small and quite inadequate to make ends meet, India has a wide range of variation in the income of different classes of people, and a microscopic section of about 5 percent of the population has considerable income commanding a large portion of the total wealth.

The giant cities are also endowed with scientific, technical, and research bodies enjoying national and often international status as well. Newspapers—the carriers of mass communications in different regional languages including English—are issued from metropolitan centers. The elite usually settle in metropolitan cities with their activities radiating in all directions. In contemporary India, metropolitan cities are centers of innovations, ideas, and inventions in all fields of activity, while in the rural India, it is primarily the feudal society which still thrives and dominates living patterns. The distinct impact of westernization is discernible in almost all walks of life—political, economic, social, and intellectual. The need for indigenous thinking is keenly felt however (Lal, 1974; Saberwal, 1976; Shils, 1961).

Rising Urban Land and Housing Values

Rapid urbanization has stimulated speculative activities in land, raising land and housing prices to a very high level especially in metropolitan cities. In Delhi, the price of land in some suburban areas has increased 20-fold in two decades; in other cities outside metropolitan areas, the increase has been as high as five times during the same period (A. B. Bose, 1969; India, Town Planning Organization, 1970; Indian Institute of Public Administration, 1977b). This holds true in varying degrees for almost all towns and cities. High land prices have made house ownership something of an impossibility not only for the lower middle class but even for the middle class. Housing deficits are, therefore, going up. Because of high land prices, provision of community facilities has also suffered substantially in a period when demand for them has mounted with the ever-increasing influx of population. The question of whether land values are entirely the product of speculation or the result of declining supply and increasing demand continues to be debated in scholarly writings. Nevertheless, there is no denying that land values have increased substantially and have promoted a vicious cycle of higher and higher values in almost every part of the country, especially in cities and metropolitan centers. During the last ten years, rent and prices of apartments and houses have risen to such an extent that widespread corruption is reported in the sale of properties in giant cities. In Bombay the price of an apartment of one thousand square feet is believed to be about $200,000; in Delhi prices of suburban houses have skyrocketed; in Calcutta, housing is not even available; and Madras is also witnessing an increase in prices. The need to take effective steps is fully recognized but little has been done so far.

Heterogeneity and Contrasting Life Styles

In Indian cities we find not only contrasts in technology which are ages apart, but also in social organizations and subcultures. Human-powered rickshas and bullock-carts vie with modern transport vehicles and electric trains; air-conditioned hotels and restaurants rise near dilapidated dormitories and cheap eating places; Western-style shopping centers dazzling with electric lights contrast with petty shops lit by kerosene lamps; and around palatial buildings are scattered low-roofed mud huts for domestic servants. While some parts of India are, economically speaking, medieval if not primitive, others have definitely entered the modern era and display fully urban ways of life. Besides contrasts in physical environment, sharply contrasting life-styles are noticeable. Migrants in a country with vast regional variations in racial groups, religions, languages, castes, living patterns, dress, and food habits bring variety, diversity, and heterogeneity into the cities. This fact is further strengthened by glaring differences between rich and poor. Presence of a Westernized, educated group among the mass of illiterate village-like urbanites presents yet another sharp contrast.

A recent study of Delhi attempted to identify the various processes and strategies developed by low-income migrants to cope with diverse cultural settings in the new physical, social, economic, and political environment of the city and to highlight ways in which traditional family structures, social networks, social values, and life-styles undergo changes. The study found that urbanizing rural migrants tend to form their own little communities and develop complex patterns of social interaction with multiple reference groups to cope with the uncertainties of economic and social life. They emulate customs, rituals, food habits, dress codes, vocabulary, material culture, and other ways of urban life. The study also revealed that urbanizing migrants are not merely puppets of forces beyond their control but have considerable influence in shaping the course of events happening around them. They are increasingly participating in the political process and putting forward their demands for better treatment on the basis of political reciprocity (Bulsara, 1970; Majumdar, 1983; Singh and De Souza, 1980).

Slums and Squatter Settlements

Slums and squatter settlements present a striking feature in the ecological structure of Indian cities, especially of metropolitan centers. There are no reliable estimates of the number of people who live in such places, mainly because there is no general agreement on the definition of a *slum* or even a *squatter settlement,* and partly because of the temporary nature of many squatter settlements. Tentative estimates place almost 20 to 25 percent of the total Indian urban population as residents of slums and squatter settlements. Estimates for the four largest metropolitan centers of Calcutta, Bombay,

Delhi, and Madras are believed to be around one-third of their populations; it is reported that more than six million persons reside in squatter settlements alone in these four giants. No clear-cut distinction can be drawn between slums and squatter settlements in practice except that slums are relatively more stable and located in older, inner parts of cities compared to squatter settlements which are relatively temporary and often scattered in all parts of the city, especially outer zones where urban areas merge with their rural hinterland. The shifting character of squatter settlements comes about by the constant flow of rural-urban migration and movement within the city itself and because of the introduction of periodical programs for their removal and resettlement.[7]

The emergence of slums and squatter settlements in cities, especially large cities and metropolitan centers, is closely linked with growth of cities in every part of this world. Old neighborhoods deteriorate and old housing structures become dilapidated. Rural migrants continue to flow toward cities in search of economic gains and to share fruits of urbanization and modernization. They overcrowd existing low-income residential settlements and tend to establish new structures in small or large numbers near their work places. Some manage to find employment opportunities in the organized sector of labor while most work as laborers in the informal sector including construction of roads, factories, public buildings, and houses. Others engage in semiskilled and unskilled service occupations. Some join the long lines of unemployment and continue to seek manual and menial work on a daily basis. Although they are contributing a great deal in nation-building activities and services essential to the economic and functional organization of cities, their incomes are usually low and they are forced to live at the subsistence level. On the basis of the Delhi Study, we find that squatter settlements are formed as a response pattern of low-income and low-status rural immigrants to the urbanization process. The study points out that these settlements provide an alternative choice of shelter for existing needs of the poor in the absence of adequate public housing programs for the underprivileged. They serve not only as entry points but also express the process by which population growth and social change are affected in the city under circumstances of highly inadequate allocation of resources for housing and social services (India, Town and Country Planning Organization, 1975; Majumdar, 1983).

Available studies provide some data about dwellers of slums and squatter settlements primarily with regard to their employment and income levels, housing patterns, health and educational facilities, availability of civic amenities, and social organization. Despite various limitations of these data, it is useful to present a few highlights concerning the plight of millions of persons who constitute almost one-third of the population of India's metropolitan giants.

Employment and Income

Socioeconomic surveys of metropolitan cities undertaken in the 1950s and 1960s highlighted pockets of unemployment among rural migrants living in slums and squatter settlements. Underemployment has been found to be more common in some cities than unemployment, especially in the south. A recent study of Delhi, however, has pointed out that these migrants do not necessarily swell the ranks of the unemployed and underemployed because most are engaged in nation-building activities of construction, transportation, skilled service occupations, unskilled manual work, and domestic service. Perhaps the same trend prevails in other metropolitan centers. New entrants usually have ties of kinship, caste, religion, language, and region of origin in the cities to which they move. Through this network they find employment opportunities soon after their arrival and also get some type of accommodation to live in. Studies have also revealed that some element of continuity exists between rural and urban occupations of a majority of migrants. Traditional social networks are functional because they provide a ready-made setting to help new migrants find employment and living places. But they also tend to be dysfunctional in limiting their initiative and perhaps new opportunities of mobility. Nevertheless, adjustment patterns to urban life are certainly facilitated by traditional social networks which protect against hazards of ill-health, unemployment, and accident and promote positive mental health.

Average income of slum and squatter settlement households has been reported to be considerably lower than the average for metropolitan centers as a whole. In 1981 average household income in Calcutta and Delhi was estimated at about 30 American dollars per month whereas in Madras it was probably about 20 dollars. Household size in slums is usually higher, four or five members, compared to those living in squatter settlements with three to four members. The same is true for household incomes because some differences exist in favor of the average slum household where incomes are slightly higher. As the ratio of females to males is generally low in most metropolitan centers, a significant number of slum and squatter settlement dwellers do not live in family units. The tradition of migrants' leaving women and children in the village is reported to be more predominant in the north as compared to the south, but continues to be widespread. Earnings of female workers are substantially lower and they have fewer occupational choices. Contrary to prevailing beliefs about slums and squatter settlements, the Kanpur study brought out that family or household income structure was dependent on their size, which meant that low-income families or households had small families with an average of four to five members, and high-income families, large families, with an average of five or more members.[8]

The problem of indebtedness also prevails but has not received much attention as yet. Data show that only a small proportion of households or families use established credit institutions whose interest rates are relatively low. Most are reported to borrow from friends and relatives at a considerably higher rate of interest; others borrow from traditional moneylenders. The Kanpur study showed that almost 40 percent of borrowers had obtained loans from moneylenders to meet shortages in current expenses followed by expenditure on social ceremonies, illness, travel, and business. This study also indicated that only 50 percent of households in slums and 40 percent in squatter settlements had saved a part of their income. Overall findings of studies strike a common note which emphasizes that more than half the families in slums and squatter settlements in metropolitan cities have poor incomes, of which 90-95 percent has to be spent on food alone.

Housing

Streams of migrants are flowing not only into India's giant cities, but also into hundreds of medium-size and smaller cities throughout the nation. This pattern has led to high demand for urban residential accommodation which places greater pressures on slums and squatter settlements where new migrants usually settle assisted by social networks with kin, friends, or fellow villages.

The rapidly accumulating shortage of housing for low-income groups combined with the vicious inflationary spiral has also been pushing people from older slums to squatter settlements because low-income unskilled and illiterate workers cannot afford to live even in slums.

According to socioeconomic surveys based on 1950s data, half the households in Bombay occupied less than 40 square feet per person. In Bombay and Calcutta, households often live jointly in one-room tenements, and a considerable portion of the population is reported to live on porches and beneath the stairs of stores. Recent studies have shown that housing conditions have further deteriorated during the past 30 years in all metropolitan centers. In Delhi, structures of 85 percent of the households in squatter settlements are single rooms and 52 percent have mud walls and thatched roofs. Stories of the deplorable conditions in Calcutta are well known and need no elaboration here. A recent survey of housing conditions in Kanpur found them to be far from satisfactory and worsening.[9]

Migrants usually try to settle near their place of work. This situation has led to the emergence of industrial slums to accommodate lower level industrial workers, residential slums to provide shelter to domestic workers, and construction slums which house workers who follow building activities within the city.

The pattern of ownership of property and land varies from city to city. In Delhi, most buildings in older slums are privately owned, while most squatter settlements exist on public land. In Calcutta, almost all slums are located on privately owned land; in Madras more than 50 percent of land where slum dwellers live belongs to public or semipublic organizations; and in Bombay, squatter settlements are located more outside the central city mainly on public land, whereas privately owned older slums continue to flourish within the city boundaries. The existence of slums and squatter settlements clearly emphasizes that urban planners have usually neglected the needs of housing the informal and unorganized sector of labor in the past. This sector is an essential component in the growth of urbanization, industrialization, and modernization, and should receive proper attention so that housing needs in any future planning are taken into consideration. Even the newly established city of Chandigarh is reported to have remained deficient in providing suitable housing to informal sector laborers (De Souza, 1978; Sarin, 1979).

Education

Despite laws for compulsory primary schooling, studies have shown that a considerable number of children in metropolitan centers do not attend any school. There is also a high rate of dropouts among slum and squatter children at all education levels.

A large proportion of urban adults continues to have higher rates of illiteracy and poorer degrees of skills. Overall literacy for the total Indian population for 1981 is reported to be 40 percent for males and 19 percent for females. Literacy rates in slums and squatter settlements are definitely lower than for the urban population as a whole. In the socioeconomic survey of Calcutta, Bombay, and Delhi, almost 58 percent of the population, excluding children, was reported to be illiterate. The need to promote literacy programs and to introduce vocational training has been stressed time and again, but only marginal success has been achieved on this front thus far. However, primary schooling for children under age ten has shown remarkable progress in recent years.[10]

Health and Nutrition

Very few data are available on health and nutrition of slum and squatter settlement dwellers. It is believed that the degree of malnutrition is fairly high, frequently making the inhabitants susceptible to disease. Health and nutrition are adversely affected by the unhygienic environment of slums and squatter settlements. Respiratory diseases, gastrointestinal disorders, skin diseases, fever, worms; ear, nose and throat problems; and tuberculosis are common. Health conditions in Calcutta are reported to be worse than in Bombay and Delhi. Mortality rates continue to be high, especially infant mortality, which exceeds 100 deaths per 1,000 live births, and the maternal death rate. Among children, nutritional diseases are quite common. Availability and accessibility of health services continue to present problems for the vast majority of the urban poor.

High infant mortality and the desire to have male children considerably influence traditional attitudes reflected in low acceptance of family-planning programs at present.

However, both the Delhi and Kanpur studies found some form of awareness of family planning has begun to take root among slum and squatter settlement dwellers. The Kanpur study struck a more optimistic note by revealing that 22 percent of households had used family planning methods and two-thirds recognized the idea of a small family as a fountain of happiness.

Environmental Quality and Amenities

A separate kitchen, bathroom, and lavatory remains a dream unfulfilled for millions of families in Indian metropolitan centers. Socioeconomic surveys have abundantly shown the inadequacy of basic amenities of life. In Calcutta fewer than 10 percent of households surveyed are reported to have the usual amenities of life; in Bombay and Delhi, the figure is 30 percent. No statistical presentation can give the picture of human suffering brought about by the absence of proper drainage, lavatories, or latrines and inadequate water supply. Some 20 to 30 percent of people in metropolitan centers had no electricity until the late 1950s and probably the same proportion does not enjoy electricity even today. The physical environment of slums and squatter settlements is worse today with regard to water, drainage, latrines, and lighting. Even where community facilities exist, they are shared by dozens of families and their maintenance often remains poor.

Lack of adequate space for recreational and leisure activities is yet another dimension of the poor living conditions persisting on a wider scale. Not only has congestion become chronic and acute, but the very absence of life's basic necessities has created an unhealthy environment responsible for high morbidity and mortality, impairing the efficiency and productivity of the Indian worker (Bhooshan and Misra, 1979; Bijlani, 1977; India, Environmental Hygiene Committee, 1950; Indian Institute of Public Administration, 1977a; United Nations, 1983).

Social and Community Organization

Older residents of cities and metropolitan centers act as agents in bringing new migrants from rural areas or small towns to the slums and squatter settlements. Through this continuous and constant flow of migration we witness the ever-growing phenomena of human settlements inhabited by homogeneous social groups characterized by kin, caste, religion, regional, and linguistic affinities. Some such settlements which grow into relatively large units of over one hundred households do manifest heterogeneity and diversity, though generally consisting of several homogenous dominant castes.

The Delhi Study has identified several basic patterns of social networks reported to function in squatter communities. At the core of these networks lie the kinship relations of six to eight families which are extended toward other such kinship groups of the same caste and village. Religious, regional, and linguistic affinities also considerably influence patterns of relationships. The caste panchayats of rural India exist in squatter settlements though their activities are rather limited. But local groups often have affiliations with city-level associations representing different regions, castes, or religions.

There are also associations having some form of structure, membership procedures, and officialism which generally cut across caste, village, or regional affiliations. Such associations are further linked to city-level regional associations. The Delhi Study further points out that the dominance of primary social networks does not constitute a simple transplantation of the village community in an urban context, but is essentially a modified supportive structure of socialization and social control. Inhabitants establish their base in social networks and forms with which they have been familiar and these provide the mechanisms of spontaneous adjustment to the new urban environment (India, Town and Country Planning Organization, 1975; Majumdar, 1983).

These networks of relationships perform varied functions apart from being adaptive mechanisms. Such patterns of relationships provide rural migrants in squatter settlements with identity, security, cohesion, mutual aid, and support; the large groups and associations function as agencies for political mobilization and loci of power at the settlement level mainly to obtain better amenities for living. Religious sermons and festivals, marriage and death ritual ceremonies, and folk music continue to play an important role in these settlements.

The Delhi Study recognizes that social networks perform some secular functions such as securing economic benefits, jobs, educational and recreational facilities, water, and other urban conveniences, and they also act as interest group organizations cross-cutting particularistic loyalties of kin, caste, village, or region. This function provides the social base for the leadership structure among squatter settlements, found to be both ascribed and achieved.

Emergence of leaders takes place at various levels of grouping, from kinship-cum-caste groups to city-level leadership of settlements. Elements of consensus are found to dominate the Delhi squatter settlements as internal tensions and conflicts are resolved through caste panchayats and other mediators who often occupy roles of leadership. Use of formal and informal elections to identify leaders is reported to be widespread. The study concluded that there are three major functions of leaders at the community level: (1) they act as agents of social stability and change by facilitating adjustment of inhabitants to urban life and incorporating them in the main stream of urban society; (2) they articulate interests and problems of the squatters by lobbying with local government authorities, political power groups, and voluntary agencies to bring basic amenities and services to their settlements; (3) they attempt to prevent removal and eviction of squatter settlements from existing sites (India, Town and Country Planning Organization, 1975; Majumdar, 1983).

Studies of established slums in Delhi and other metropolitan centers have shown somewhat similar patterns of social networks to those described above. Almost all studies have found the vast majority of slum dwellers live in some form of family units despite single-person or male-only migration having been more prevalent for many years in some regions. After kinship, caste and religion have been found to be the most important variables in providing social identity and social base for networks of relationship. Language and regional affinities constitute yet another dimension, though they are often intermixed with caste and religion. Even in slums where there is a considerable amount of heterogeneity, residential and social segregation of different castes and religions have been reported to be fairly common. Most studies have found that the mutual aid system of social and economic interdependence of the village community does not exist in urban slums.

The Calcutta Study has reported greater degrees of social heterogeneity and noted that slum leadership comes mainly from hut owners and well-to-do businessmen in the slums along with a few educated persons. At the same time, emergence of caste and religious leadership in Calcutta slums has been on a much smaller scale than in Delhi and Bombay slums. It has been, however, increasingly recognized that demographic factors such as size, age structure, and sex composition, and social factors such as kinship, caste, religion, and language affect the formation and functions of local associations which exist in the slums of metropolitan centers. Although caste-based associations and informal groupings tend to dominate, the formation of caste-free associations representing different castes has been significant in recent years (Chandra, 1977; Doshi, 1974; A. M. Singh, 1979). Many studies have reported the increasing role of education in local leadership.

A study of women in Delhi squatter settlements highlights the lack of their participation in local leadership at almost all levels of social organization. This issue is further linked to prevailing lower rates of literacy among women. Moreover, in view of most associations having all-male membership, women in general have hesitated to participate in them. Their lack of participation has also been affected by the persistence of traditional roles allocated to women in society. Although the traditional occupational structure in metropolitan centers has changed considerably, and more and more women have been drawn into skilled and professional activities, especially from the middle class, the general participation of women in local groups and associations has remained low. Among poor urban women, participation has been conspicuous by its absence and perhaps will continue to be so until their literacy rates become higher, and the social structure of society undergoes a radical change to accept new roles of women.[11]

The Urban Informal Sector as a Safety Value

The urban informal sector, also called the secondary or unorganized sector, provides a large amount of employment and income to millions of urban dwellers in India and other less industrialized societies. Economic development programs in general and industrial development in particular have made the urban economy more heterogeneous and diversified. The growth of urban population as a result of natural increase and inmigration from rural areas has generated more and more demand for consumer goods and other basic services. The organized sector of the urban economy has in part failed to grow rapidly with the ever-increasing supply of urban labor for want of capital and other resources. Partly the organized urban economy has itself created secondary sources of production and distribution to avoid observance of laws regulating urban wages and working conditions, and partly the informal sector has asserted itself and gained momentum in the context of searching for more employment opportunities. Whatever may have been the source of its growth, the existence of dualism in the urban economy has received considerable attention by scholars, planners, and international organizations such as the International Labor Organization (ILO) and the World Bank. Studies have raised scholarly debates and controversies about its exact nature, functions, and the problems of its measurements.[12] Whether the informal sector gives employment to household heads or secondary members, or to males or females, whether it acts as a buffer between employment and unemployment are questions not answered with any degree of certainty as of yet. All we can say is that the sector provides employment to millions of people and goods and services in an urban economy without which the quality of urban life would become still more inferior for both the middle and lower classes.

The urban informal sector is believed to include small-scale and cottage industries manufacturing machine parts, implements, furniture, and handicrafts, and employing semiskilled and unskilled workers with little or no formal education. This sector is also engaged in trade, transport, and service activities involving such workers as vendors, petty retailers, barbers, tailors, carpenters, drivers, washermen, loaders, cooks, domestic servants, porters, and sweepers. The sector further includes a large number of self-employed entrepreneurs who produce consumer goods and services and make them available to urban dwellers at competitive prices.

Studies of Bombay, Calcutta, Delhi, and Madras all have shown that the informal sector is an integral part of their economy and will continue to be so. Perhaps it can never be eliminated; at best it might be better regulated since the general level of wages in this sector is low and working conditions less than satisfactory. Most of the workers live in the poorer neighborhoods and perhaps several million are pavement dwellers.

The positive functions of the informal sector are beyond question. But dysfunctions in terms of congestion, deterioration, pollution, and unplanned growth, and the resultant forms of social disorganization have not been fully investigated and highlighted. The need to develop programs for guiding and directing the informal sector and for launching massive programs to improve working and living conditions must be recognized and given full attention.

Approaches and Strategies for Better Urban Environment

Before independence, programs for improving cities emphasized largely public health and sanitation. Some programs were also developed for improving roads, water supply, and drainage, partly by local municipalities and partly by specially created town improvement trusts. Redevelopment of old and blighted city areas and general housing shortage received little attention.

Several official committees and commissions made recommendations for improving housing conditions occasionally, but only in the postindependence era was the critical state of the urban environment recognized. The first five-year plan admitted that Indian cities have a large proportion of substandard houses and slums filled with unsanitary mud huts of flimsy construction, poorly ventilated, overcongested, and often lacking such basics as electricity, water, and drainage.

Programs directed toward housing construction and urban conditions first started in connection with rehabilitation and settlement of refugees from Pakistan. As that problem took on staggering proportions, the problem of their housing came to be seen as part of overall national and regional urban planning and development. Slum clearance by demolition was recommended in this context. Establishment of the National Building Organization and state housing boards, creation of suitable legal framework for town and country planning, and enactment of housing standards were other measures advocated for adoption. Some programs for subsidized industrial housing and low-income group housing were included in the national plan.[13]

During the second and third five-year plans, preparation of master plans for urban areas, especially large cities and metropolitan centers, was advocated. The demolition strategy was somewhat de-emphasized for clearing slums; instead some measures were recommended to improve existing conditions and prevent growth of new slums by enforcing municipal laws and building codes. Difficulties reported for implementation of slum clearance and improvement programs were the high cost of acquiring slum areas, nonavailability and high costs of alternative sites near work places, inability of slum dwellers to pay even subsidized rent and their reluctance to move from the areas selected for clearance. Both plans made a series of recommendations for balanced regional planning, expansion of housing facilities, and development of trained civic administration. The third plan also recognized a need for drastic measures to freeze urban land values, undertake large-scale land acquisition, and devise suitable taxation for urban land and property. The fourth and fifth plans went further and emphasized the formulation of urban land policy to prevent concentration of land ownership, check undue rise and speculation in urban land values, and introduce fiscal measures to end unearned increases in land values. These plans also strengthened the programs for improving the urban environment, especially in slum areas.

Earlier activities to offer subsidized housing to the poor were further extended to provide developed plots and building material for house construction by the poor. The priority in the sixth plan (1980-85) was also to focus attention on housing needs of the poor. The plan admitted, however, that it would be impossible to provide a house to every family at public cost in view of competing claims of other priority problems facing the country.

Broadly speaking, within the framework of national and regional economic planning, it is fully recognized that the problems of slums and squatter settlements must be tackled as part of overall urban develpment and to minimize the differentials in living conditions between the rural and urban areas. Although the rate of urbanization is now moderating, the problem of sheer numbers is overwhelming.

Even if the urban growth rate slows, it is projected that 30 percent of India's population will be urban by 2001, numbering almost 400 million. Urban problems will become more serious unless tackled through suitable programs and strategies. Among the most widely enacted programs and strategies used to cope with the ever-expanding problems of the urban poor are those of (1) formulating master plans for selected cities and metropolitan centers; (2) urban redevelopment and relocation strategies for slums and squatter settlements; (3) introducing minimum environment improvement plans; (4) providing subsidized housing; (5) legislating for land acquisition and to control urban land values and speculation; and (6) research, development, and training strategies.

Preparing Master Plans

Even before initiation of the first five-year plan in the 1950s, several cities had prepared and partially implemented master plans from time to time. However, during the third five-year plan, the national government provided 100 percent assistance to the states for preparation of master plans of selected cities. These plans essentially provided necessary guidelines for regulating land-use, demolishing slums and squatter settlements and relocating their inhabitants, and preventing unrestricted growth of cities, especially metropolitan ones. In 1974, the draft fifth five-year plan stated there had generally been a growing awareness in the states of the problems of urban development but not much headway had been made in taking comprehensive action.

The Delhi Study examined the effectiveness of its master plan. We are told that the plan's physical framework failed to recognize the interdependence of the dual socioeconomic systems in the metropolis and that planned intervention has been ineffective in large areas of the city, developed since the plan was made. In Delhi, the master plan seems to have been replaced by the National Capital Region Plan which envisages an integral development of the present city of Delhi together with selected adjoining areas from the states of Haryana, Rajasthan, and Uttar Pradesh.[14]

The fate of master plans in other metropolitan cities has been no better. In Bombay, several master plans have come and gone, and eventually a new plan to cover the entire metropolitan region has been formulated which seeks to develop a new Bombay to house an additional two million residents. We were told in 1981 that high-rise modern buildings jostle with rickety structures, while huts proliferate and squeeze in where there is space; behind the new buildings along Bombay's main roads, the old Bombay lies crumbling and dilapidated.[15] For Calcutta we have an excess of published and unpublished material concerning its master plans but living conditions continue to be unhealthful for a majority of its residents.[16]

Apart from the inadequacy of financial resources, it has been increasingly realized at least in some quarters that needs of the urban poor who live in slums and squatter settlements have not been fully understood and taken into consideration in finalization of the master plans. Both the plans and housing designs have generally been conceived by Western-educated elites and considerably influenced by foreign thinking, scholarship, and literature to serve the needs of the middle and upper-middle classes in metropolitan centers. The bottom of the social and economic hierarchy, including rural migrants who live in these centers, suffers from the urban environment characterized by the absence of basic amenities. Almost every committee and commission, official and unofficial, has acknowledged the depressing picture as it exists in the metropolitan centers during the past three decades. Scholarly as well as popular writings warn that conditions are likely to deteriorate further in the future. India seems to have reached a dead end, and its leaders are failing to cope with the mounting problems of urban development.

Urban Conservation, Redevelopment, and Relocation Strategy

The first five-year plan advocated removal and demolition of all slums and squatter settlements from metropolitan centers. However, the impossibility of the task was soon realized because relocation presented a number of difficulties. A shift in the policy of demolition therefore took place and some steps were taken to emphasize conservation of existing housing structures and improvement of living conditions. Along with this plan, programs of urban community development were also introduced to develop self-help projects and enlist cooperation of residents of the slums and some selected squatter settlements for their implementation. Undoubtedly, some marginal changes came about in the urban environment, but these activities did not bring any substantial improvement; nor did they stop the growth of additional squatter settlements within metropolitan center boundaries or in outskirt jurisdictions.

In Delhi, a special program of urban community development was launched with active financial and technical assistance from several foreign sources. Though the program had some initial success, it failed eventually because it had many weaknesses such as lack of cooperation from local politicians, absence of integration within existing welfare activities, organizational inadequacies, involvement of overqualified postgraduates from the field of professional social work instead of high school trained workers, and dominance of foreign technical guidance.[17] In almost all metropolitan centers, the shortage of housing structures continues primarily because of a constant flow of new migrants, slow housing construction, emphasis on middle-class housing, ever-rising land values, and failure to acquire suitable land for relocation. Under the master plans, Delhi and Calcutta vigorously followed the conservation and redevelopment strategy, and thousands of existing structures were improved. Steps were also taken to provide some of the most basic amenities of life in slums and squatter settlements.

Strategy of Minimum Needs Program for Environmental Improvement

Several steps were taken to introduce environmental improvements in slums and squatter settlements during the second, third, and fourth plans. During 1972-73 an intensified program was launched to provide a minimum level of services like water supply, sewerage, drainage, paving, and street lighting in the slum areas of 20 large cities.

The fifth five-year plan went a step further to recognize the need for a new strategy—the Minimum Needs Program—under which improvement activities were extended to all cities with a population of 300 thousand and above, and in addition to one city in every state where no such large city existed. Operation of this activity was moved from the national to the state sector, though financial resources were provided by the central government. Under the fifth plan, the central government initiated a new project to supplement efforts of state governments in providing urban infrastructure like roads, water supply and sewerage, land acquisition, and development. The states were also required to spend matching funds in selected cities toward the approved program of integrated development. During 1979-80, this project was replaced by another which placed considerable emphasis on integrated development of small and medium cities. All these activities constituted a bold step and were long overdue. But neither the strategy of demolition and conservation nor any plan for providing subsidized housing for low-income groups can ever meet the challenge presented by the formidable, complex, and growing phenomena of slums and squatter settlements in metropolitan centers. Under the circumstances, the strategy to introduce activities of the Minimum Needs Program seems to be the most feasible alternative to the existing conditions which have been declared substandard for the past three decades.[18]

Strategy for Public and Subsidized Housing

As part of the five-year plans, a number of programs were introduced which were generally classified as subsidized housing for industrial workers, low-income group housing, middle-income group housing, plantation labor housing, dock labor housing, and public-housing schemes. Under these programs, provisions were made to advance loans and grants at a reasonable interest rate; promote cooperative societies which were given urban land to develop new colonies; and undertake direct construction of housing activities for public employees.

During the first and second plans, several new towns were also established for rehabilitation and resettlement of refugees who had migrated from West Pakistan. It was also recognized that the government could not provide directly all the financial resources needed for construction of subsidized housing. Several types of new institutional arrangements were made under which state housing boards and the Life Insurance Corporation assumed a major responsibility for providing loans for new housing construction. Creation of the Central Housing and Urban Development Corporation under the fourth five-year plan in 1970 was another milestone to provide financial support to state and local governments.

Despite some programs in creating new public and private housing, the basic shortage has continued and the problems of slums and squatter settlements have kept on multiplying. In 1974 the five-year plan estimated the current housing shortage in urban areas to be six million houses without making any allowance for increasing deterioration in existing structures. The plan was forced to admit the massive constraints and impediments which existed in the way of achieving any modest progress and improvement in living conditions, especially of the urban poor. These conditions continue today after the seventh five-year plan has been finalized.

Strategy of Legislative and Administrative Improvements

The first five-year plan had clearly recognized the need for suitable legislation in different states as the most essential step in developing the legal structure to introduce town and country planning. It was emphasized that proper legal powers were required to control haphazard growth and ribbon development, regulate location of industries, eliminate slums and squatter settlements, and complete socioeconomic surveys. In subsequent years, the legislative base was further widened to include land acquisition for relocation of slum and squatter settlement dwellers, demolition of dilapidated private housing structures, control of speculation in urban land values, placement of ceilings on urban land prices, and dispersal of selected industries away from the metropolitan centers. In accordance with democratic principles, every attempt was made to follow due process of law in the formulation and implementation of urban development policies. Special legislative steps were also taken to ensure the provision of industrial housing by large-scale industrial establishments. Legislation to maintain and enforce tolerable minimum standards for housing and other sevices was also developed. In some metropolitan cities, rent control laws were enacted which created special problems, particularly in Bombay. Finally, new forms of legislation were proposed, formulated, and in some cases experimented with to devise fiscal measures for eliminating unearned increments in land value and for the enhanced stamp duty on the sale or transfer of urban land and property. Success of these legislative measures appears to be marginal, though no comprehensive study has been taken to examine their effectiveness.

Through the strategy of the five-year plans, India's central government has played a vital role in meeting existing problems of urbanization and in guiding the future course of urban development. India is federally organized and the central government has provided basic directives of policy and large financial resources, but implementation of most of the planned programs has remained the responsibility of state and local governments. At this level the need was felt for strengthening the existing administration and creating new administrative organizations on the other. Several expert committees offered recommendations to augment financial resources of local bodies so that steps were taken to introduce the service-orientation needed for improving living conditions in large cities.

In addition to establishing housing boards in every state, a new type of nonelective organization known as the urban development authority was created in large cities, especially in metropolitan centers. Activities of the urban authorities functioning in Calcutta and Delhi have received widespread attention, though not without criticism. The Calcutta Metropolitan Planning Organization set up in 1961 completed the master plan for Calcutta (1961-86), under which several types of activities were introduced to improve living conditions. This organization has been enlisting guidance and cooperation from a number of foreign experts, and substantial financial resources have also come from several foreign sources. The Delhi Development Authority is another organization which has completed a master plan and has produced a considerable amount of study and research material on urban development in general and guided urbanization in particular. The success and effectiveness of these organizations have remained a matter of debate and controversy. Notwithstanding their inadequacies, one can only say that it is through the intervention of these organizations that attention has increasingly been focused on the problems of slums and squatter settlements and some steps have certainly been taken to bring marginal improvements in living conditions there (Misra et al., 1974; Sundaram, 1977).

Research and Training Strategy

As part of the preparation of master plans, the role of systematic social research to complete socioeconomic surveys of large cities was fully recognized. In addition, within metropolitan centers, surveys of housing and living conditions and other social aspects of urban planning were also carried out. The extent to which urban planners took these surveys into consideration in formulating final master plans has remained unclear. These plans have been labelled blueprints for land use patterns and physical planning but without blending in the socioeconomic aspects.

Apart from social research, a considerable volume of research on building and construction materials was also undertaken, mainly at the Central Building Research Institute. The National Building Organization established in 1954 has been responsible for initiating several projects regarding techniques and materials for waterproofing mud wells, using waste products, producing cheap construction material, and low-cost housing designs.

Accordingly the fifth-year plan stressed the need for cheaper building material, new building technology, and use of local material, such as bricks and lime mortar, on a more organized basis by housing boards in different states. Training of social researchers, community workers, and urban planners has also received considerable attention. Training of different types of construction workers received some attention, though the third five-year plan admitted that it was not done so extensively and systematically. Several special training programs were organized for municipal administrators and other personnel working both at the state and local levels whose activities were connected with development and provision of urban services. Although the need for research and training exists in many spheres of activities relating to urban development, its actual organization has often neglected some of the indigenous approaches and basic knowledge of working with the urban poor in metropolitan centers. Some specialized centers for the study of demography and urbanization functioning in the giant cities have published a considerable amount of study material on basic problems of population growth and increasing levels of urbanization, highlighting their social, economic, and political implications for the general Indian economy and the urban economy in particular.

Revolution of Rising Expectations Generates Frustrated Hopes

There has been a revolution of rising expectations all over the world for the past three decades partly generated by various proclamations of liberty, equality, and justice made by the United Nations and partly because of member nations attaining independence from Western countries. Through the diffusion of these new ideals via the mass media, a new climate of thinking, planning, and action has been established to promote better levels of living. In India, long before independence and as early as the 1930s, the national political leadership had pledged to bring about general social and economic improvements and had spoken with one voice to eradicate poverty.[19] The Indian constitution which came into force in January 1950, while embodying the new values, expressed lofty ideals to create a welfare state with a bias toward the socialistic pattern of society.[20] The first five-year plan (1951-56) ushered in an era of planned economy, and the noblest ideals of liberty, equality, and justice for all were repeated. With India now launching its seventh five-year plan, poverty and unemployment remain widespread not only in the metropolitan centers but throughout the country. Educated and uneducated persons are restless everywhere. People find neither work nor shelter; and if they do find some employment in cities, people are forced to live under the most unhealthful and degrading conditions as are found in most parts of Bombay, Calcutta, Delhi, Madras, and other metropolitan cities. The goals and ideals proclaimed in the constitution and the various five-year plans have remained unfulfilled for a vast majority of India's population. New thinking and new orientation must encompass the stark realities of beggary, poverty, and unemployment; inadequate housing; deficient amenities of life; and unhealthful environment.

Reducing Overurbanization Through Rural Reconstruction

The agriculture sector continues to contribute nearly half of India's national income. Time and again, various committees and commissions have produced voluminous documents advocating the total reconstruction of the country's 600 thousand agriculturally oriented villages. Developing India's rural areas has been one of the paramount objectives of the five-year plans and in the early 1950s special programs of community development were introduced not only to increase agricultural productivity, employment opportunities, and cheaper credit facilities, but to improve general living conditions by providing more elementary education, better health services, adequate drinking water, proper drainage, extensive electrification, improved roads, and additional house sites.[21]

Recognizing that rural development programs have been partially successful, India launched the Integrated Rural Development Program in 1978-79. This program aims at reducing unemployment in rural areas and providing assets and inputs to the rural poor to enable them to rise above the poverty line on a lasting basis. Of course, unless the goals of rural reconstruction and development are achieved, more employment opportunities are created and living conditions are improved, the continuous and phenomenal outmigration to urban areas in general and metropolitan cities in particular cannot be slowed (see India, Planning Commission, 1981; Misra and Sundaram, 1980; Panchandikar, 1978). When the infant mortality rate in rural India is still more than one hundred per thousand live births, literacy continues to be less than 40 percent even in cities, malnutrition is widespread, the drinking water supply is inadequate, electricity is lacking, and more than half the population live in mud huts without proper drainage, one wonders whether the ideal goals as enumerated in the constitution and the five-year plans will ever be reached (see India, Planning Commission, 1981).

Population Control Through Family Planning

Perhaps no other subject has received so much attention as population control in India during the past 30 years.[22] The Indian government introduced a series of activities as part of the first five-year plan in 1952 to reduce population growth in order to stabilize the population within the next few decades. By the end of 1984, India's population was an estimated 760 million with an annual birthrate of 34 and a death rate of 13 per thousand—yielding a growth rate of 2.1 percent a year. At this rate the population is likely to double every 35 years and is projected to be about one billion by the year 2000 (India, Registrar General, 1981; Mitra, 1978; Population Reference Bureau, 1985; United Nations Department of International Economic and Social Affairs, 1983). No program or strategy to promote better urban environments can be successful unless the annual growth rate is reduced to about 1 or 1.5 percent with which the population doubles in about 70 years. The fifth five-year plan objective of reducing the birthrate from 35 to 30 per thousand by 1979 was not achieved. The sixth plan contained a policy to reduce the birthrate to 21 and the death rate to 9 per thousand by the turn of the century with a projected 1.2 billion population by 2050 (see the sixth five-year plan). In metropolitan cities, the Minimum Needs Program is implementing the necessary infrastructure for promoting family planning among slum dwellers and residents of poor neighborhoods, where birthrates continue to be fairly high.

Need for Dewesternization and Indigenous Foundations

Not all strategies incorporated in the various five-year plans were equally successful or totally implemented to achieve the goals set by the Indian leadership. Today after 30 years of the planned era, conditions in slums and squatter settlements are far from satisfactory. It is estimated that no less than one-third of the total estimated Indian population of one billion by 2001 will become urban; the number of cities of a million plus will increase from 12 to 20 with a total population of 70 million. All this means that Indian cities will witness more concentration of people and centralization of activities. The need to take concrete and effective steps can hardly be overemphasized. All the strategies used to cope with the increasing flow of rural migrants and improve general living conditions in metropolitan cities have contributed in one way or another toward creating a better environment. But at the same time, one cannot escape the harsh realities of life when one finds the utterly inadequate housing and deplorable, subhuman living conditions in slums and squatter settlements. The only workable alternative with some prospect of success relates to the provision of minimum services and extensive introduction of the general environmental improvement strategy to make existing housing structures more livable and neighborhoods more healthful places.

The five-year plans have often expressed ideals and aspirations to promote and create better living conditions in rural and urban areas. The third five-year plan stated that "luxury housing and waste of urban land should be prevented so that a larger number of modest dwelling units can be constructed for the same investment." The fifth five-year plan emphasized optimum land use, preventing concentration of land ownership, and checking speculation in urban land values. It also recommended intensifying research and development of suitable housing designs and production of cheap building materials. In subsequent years the sixth plan and now the seventh plan have also reiterated the same goals of providing shelter to low-income groups on a priority basis. However, no serious efforts have been made to translate these ideas and ideals into practice.

Elitist ideologies based on Westernized styles of urban life continue to dominate Indian plans and programs while needs of three-fourths of the urban population are subordinated. One cannot look forward to substantial improvements unless radical changes in thinking come about soon. There is a desperate need for indigenous thinking about urban planning, housing designs, and construction material and about urban administrative structures and social organizations. Unfortunately Western-educated elites have often shown little concern for the vast informal urban sector of the economy which has been contributing to India's economic and industrial growth but has not been able to derive even bare amenities of life for itself so far.

In recent years problems of urban slums and squatter settlements have received considerable attention at the international level.[23] The United Nations Conference on the Human Environment (Stockholm, 1972), the World Population Conferences (Bucharest, 1974, and Mexico City, 1984) and the United Nations Conference on Human Settlements (Vancouver, 1976) have been important developments in the evolution of global thinking for improvement of urban living conditions in giant cities. Reports of these conferences together with many specially prepared documents have highlighted the need for a unified approach for the betterment of living conditions in all human settlements, urban and rural.

The United Nations and its allied agencies have not only brought into focus the problems of human settlements but have also provided technical assistance in terms of financial resources and consultants.[24] It is difficult to evaluate the impact and effectiveness of international activities. Even if one admits the usefulness of these activities, it can hardly be overemphasized that the national governments concerned have to make serious attempts themselves to resolve the problems of the urban poor. No foreign aid and technical know-how can bring permanent solutions unless indigenous foundations are laid for the provision of basic amenities within the broader framework of socioeconomic planning. Bureaucratic organizations dominated by Western-educated elites must recognize the importance of the informal sector of the economy and devise indigenous solutions to cope with the ever-increasing problems of urban development, housing, transportation, education, and social welfare, especially in metropolitan cities.

International agencies might initiate and set some climate for thinking and discussion, but suitable plans for action have to be formulated by national governments after considering cultural requisites and availability of local building materials. There is an urgent need for a second look at existing policies and programs so that workable solutions are found to reduce the abject poverty of the masses, growing inequalities between rich and poor, and the high rate of population growth. In this context, dependence on foreign capital and technical assistance will have to be kept to the minimum.

Advocating indigenous planning models, the Vancouver Declaration on Human Settlements and Vancouver Action Plan clearly noted that a nation's character is made visible in its settlements. Foreign models must not be allowed to dominate planning decisions which should be guided by national goals and implemented by local people making the best possible use of indigenous resources, in the context of local environment and culture. Reorganization of the higher education system must be undertaken so that Indian experts on economics, geography, political science, public administration, public policy, social welfare, sociology, and urban planning can better understand the underlying causes of massive poverty, unemployment, and substandard living and then formulate more realistic solutions.

In summary, I have tried to identify the poor living conditions prevailing in all metropolitan cities and to survey the major problems and strategies launched for improving them as a part of India's various five-year plans. A case is made for the dewesternization of Indian thinking and the framework concerning urban development and redevelopment, housing the poor, and urban planning through reorganization of higher education on an indigenous basis. The priorities envisaged in the sixth five-year plan and now in the seventh plan covering rural reconstruction and development, population control and family planning, urban housing and shelter, and providing life's amenities to all residents in metropolitan cities are indeed laudable. But effective steps need to be taken to implement these programs more vigorously through a public administration system which is more responsive to local requirements of shelter in a healthful environment with clean water, drainage, and electricity. All of this presupposes successful economic and social development.

One of the weakest links in regulating urban development pertains to urban land policy. Existing legislation and enforcement strategies have definitely failed. Urban land and housing have already become very expensive, and corruption is reported to be rampant. An effective plan must be formulated and implemented to control speculation in urban land and housing. Perhaps, no less a surgical operation is called for than the one undertaken in the case of the banking and insurance industries several years ago. As admitted in the sixth five-year plan, suitable measures would be needed "to promote a more effective and equitable functioning of the urban land market so that these enormous and unwarranted increases in land values are checked." The need is for action, not for additional new models and glorified social research on urban development.

Notes

1. See Nagpaul (1979). From time to time, several Indian periodicals have brought out special numbers on such themes as urban development, urban policies, and metropolitan cities. Among the most important are *Civic Affairs, Indian Journal of Public Administration, Indian Journal of Social Work, Journal of the Institute of Town Planners, Nagarlok, National Geographical Review of India,* and *Social Action.* In addition, a considerable amount of literature has become available in the last 15 years. The most widely quoted books are Bhardwaj (1974); Bhooshan and Misra (1979); A. B. Bose (1981); Ferreira and Jha (1976); India, Ministry of Health and Family Planning (1966); M.S.A. Rao (1970, 1974); and Trivedi (1975, 1976).

2. Under the Indian constitution, some population groups are classed as scheduled castes and tribes; in 1981 about 25 percent of the population was reported to be in these categories.

3. In the 1950s and 1960s, several surveys of the problem of begging in metropolitan cities were completed; see Gore (1956).

4. Several official and nonofficial committees and commissions have reported about slums and squatter settlements in the past 50 years before and after independence. Many books and monographs are available on the subject; see Arangannal (1971); Bharat Sewak Samaj (1958); Desai and Pillai (1970, 1972); India, Town and Country Planning Organization (1975); Iyer and Verma (1964); Majumdar (1983); Nambiar (1961); Venkatarayappa (1972); and Wiebe (1975).

5. See selected publications on education by the Indian Ministry of Education.

6. See selected publications by the Indian Ministry of Tourism.

7. In addition to the citations in note 1, the following are useful: Majumdar (1983); Mukherjee (1975); Ramachandram (1972); Siddiqui (1968); Singh and De Souza (1980).

8. Socioeconomic surveys of major Indian cities sponsored by the Research Program Committee of the Planning Commission give considerable attention to income and poverty. The various five-year plans also contain very valuable data, especially the sixth plan. See also Bhatt and Chawda (1979); Dantwala (1970); Fonseca (1972); and Indian Institute of Public Administration (1977a).

9. As part of the various five-year plans, many study groups produced reports on housing problems during the last two decades. The Indian Ministry of Works, Housing and Supply, and Indian National Building Organization have brought out a considerable amount of material on the subject. Earlier books on living conditions of industrial labor reported poor housing in all major cities. Socioeconomic surveys of metropolitan cities have given special attention to housing problems. See also Chakrabartty (1960); Indian Institute of Constitutional and Parliamentary Studies (1971); Indian Institute of Public Administration (1960, 1977a); and B. B. Rao (1979).

10. The best statistical data are available in government reports of the Ministry of Education, Ministry of Health and Family Planning, the National Institute of Health and Family Welfare, the National Institute of Nutrition, and the Planning Commission. See also A. B. Bose (1981) and Pathak (1979).

11. More material is available on rural than urban leadership. See India, Ministry of Education and Social Welfare, 1974; and Lal (1974).

12. See Sethuraman (1981). Both the ILO and World Bank have published studies on the role of the informal sector in selected major cities including Calcutta; these studies are widely quoted and much debated. See also De Souza (1978).

13. See the various five-year plans for the best available material.

14. Various reports of the Delhi Development Authority and the Town and Country Planning Organization are most useful. See also Sundaram (1978) and Yadav (1979).

15. A number of publications of the Bombay Metropolitan Regional Development Authority are relevant here. See also Arunachalam (1978).

16. Recent publications of the Calcutta Metropolitan Planning Organization provide basic data. See also Banerjee and Roy (1967); A. Ghosh (1966); M. Ghosh et al. (1972); and Lahiri (1978).

17. A number of publications are available on the urban community experiment in Delhi, published by the Municipal Corporation of Delhi. For a critical review, see Nagpaul (1962).

18. See the various five-year plans for the best sources.

19. At least two decades before independence, national leaders pledged to reduce poverty by raising general levels of living through a planned economy. The National Committee on Planning, established in the 1930s under the chairmanship of Professor K. T. Shah, published a number of volumes on Indian economic and social planning. After independence, the First Five-Year Plan (1952) was one of the most important documents providing the basic framework for subsequent social and economic planning.

20. Many commentaries are available on the Indian Constitution. See Basu (1978) and Singhvi (1971).

21. An extensive program of rural community development was launched in the early 1950s. Hundreds of monographs and dozens of books are available on the subject. See S. N. Bhattacharyya (1970) and V. R. Bhattacharya (1982).

22. A voluminous amount of literature is available on population control and family planning. See Banerji (1971) and K. G. Rao (1974).

23. During the last two decades hundreds of articles and many books have been published surveying the poor living conditions in almost all countries of the world. Among the most widely quoted are Gilbert and Gugler (1983); International Christian University (1976); Jakobson and Prakash (1974); Linn (1983); Lloyd (1979); McGee (1971); Misra (1979); Pacione (1981); Taylor and Williams (1982); and Weitz (1973).

24. The United Nations and its allied agencies have issued a large number of publications on urbanization; see all those listed in the bibliography for this chapter. In addition, *Habitat,* an official newsletter of the United Nations Center for Human Settlements, Nairobi, Kenya, publishes periodically national and international developments in the field of human settlements.

Bibliography

Arangannal, Rama. 1971. *Socio-economic Survey of Madras Slums.* Madras: Tamil Nadu Slum Clearance Board.

Arunachalam, A. B. 1978. "Bombay: An Exploding Metropolis." In *Million Cities of India,* edited by Rameshwar Prasad Misra. New Delhi: Vikas.

Balakrishna, Ramachandra. 1961. *Report of the Economic Survey of Madras City.* Delhi: Manager of Publications, Government of India.

Banerjee, Bireswar, and Debika Roy. 1967. *Industrial Profile of the Calcutta Metropolitan District.* Calcutta: Indian Publications.

Banerji, Debabar. 1971. *Family Planning in India: A Critique and a Perspective.* New Delhi: Peoples Publishing House.

Basu, Durga Das. 1978. *Introduction to the Constitution of India.* New Delhi: Prentice-Hall of India.

Bharat Sewak Samaj. 1958. *Slums of Delhi.* Delhi: Atma Ram.

Bhardwaj. 1974. *Urban Development in India.* Delhi: National.

Bhatt, Mahesth, and V. K. Chawda. 1979. *The Anatomy of Urban Poverty: A Study of Slums in Ahmedabad City.* Ahmedabad, India: Gujarat University.

Bhattacharya, Vivek Ranjan. 1982. *New Face of Rural India: March of New 20-Point Programme.* New Delhi: Metropolitan Book Co.

Bhattacharyya, S. N. 1970. *Community Development. An Analysis of the Programme in India.* Calcutta: Academic Publishers.

Bhooshan, B. S., and Rameshwar Prasad Misra. 1979. *Habitat Asia: India,* vol. 1. *See* Misra, 1979.

Bijlani, H. U. 1977. *Urban Problems.* New Delhi: Indian Institute of Public Administration.

Bose, Ashish B. 1969. *Land Speculation in Urban Delhi.* Delhi.

———. 1973. *Studies in India's Urbanization, 1901-1970*. New Delhi: Tata McGraw Hill.
———. 1981. *Social Statistics in Indian Health and Education*. New Delhi: Vikas.
Bose, Nirmal Kumar. 1968. *Calcutta, 1964: A Social Survey*. Bombay: Lalvani.
Buchanan, Daniel Houston. 1934. *The Development of Capitalistic Enterprise in India*. Bureau of International Research, Harvard University and Radcliffe College Publications. New York: Macmillan.
Bulsara, Jal Feerose. 1964. *Problems of Rapid Urbanisation in India*. Bombay: Popular Prakashan.
———. 1970. *Patterns of Social Life in Metropolitan Cities (with Particular Reference to Greater Bombay)*. Bombay: Research Programme Committee of the Planning Commission.
Chakrabartty, Syamal. 1958? *Housing Conditions in Calcutta*. Calcutta University Socio-economic Survey Section Monograph 1. Calcutta: J. Basu for Bookland Private Ltd.
Chandra, Subhash. 1977. *Social Participation in Urban Neighbourhoods*. New Delhi: National Publishing House.
Dantwala, Mohanlal Lalloobhai. 1970. *Poverty in India, Then and Now, 1870-1970*. Delhi: Macmillan India.
Delhi Administration. 1977. *The Gazetteer of Delhi*. Delhi.
Desai, Akshyakumar Ramanlal, and S. D. Pillai. 1970. *Slums and Urbanization*. Bombay: Popular Prakashan.
———. 1972. *A Profile of an Indian Slum*. Bombay: University of Bombay.
De Souza, Alfred, ed. 1978. *The Indian City. Poverty, Ecology and Urban Development*. New Delhi: Monohar.
Doshi, Harish. 1974. *Traditional Neighbourhood in a Modern City*. New Delhi: Abhinav.
D'Souza, Victor S. 1968. *The Social Structure of a Planned City, Chandigarh*, Bombay: Orient-Longman.
Dutt, Ashok K. 1977. "Planning Constraints for the Calcutta Metropolis." In *Indian Urbanization and Planning: Vehicles of Modernization*, edited by Allen G. Noble and Ashok K. Dutt. New Delhi: Tata McGraw Hill.
Ferreira, J. V., and S. S. Jha, eds. 1976. *The Outlook Tower. Essays on Urbanization in Memory of Patrick Geddes*. Bombay: Popular Prakashan.
Fonseca, A. J., ed. 1972. *Challenge of Poverty in India*. 2d ed. New Delhi: Vikas.
Gadgil, D. R. 1971. *The Industrial Evolution of India in Recent Times*. 4th ed. Calcutta: Oxford University Press.
Ghosh, A. 1966. *Calcutta, The Primate City*. Delhi: Office of the Registrar General, India.
Ghosh, Murai, Atok K. Dutta, and Biswanath Ray. 1972. *Calcutta: A Study in Urban Growth Dynamics*. Calcutta: Mukhopadhyay.
Ghurye, Goving Sadashiv. 1963. *Cities and Civilization*. Bombay: Popular Prakashan.
Gilbert, Alan, and Josef Gugler. 1982. *Cities, Poverty and Development. Urbanization in the Third World*. New York: Oxford University Press.
Gore, Madhav Sadashiv. 1956. *The Beggar Problem in Metropolitan Delhi*. Delhi: Delhi School of Social Work.
———. 1970. *Immigrants and Neighbourhoods. Two Aspects of Life in Metropolitan Society*. Tata Institute of Social Sciences no. 21. Bombay: Tata Institute of Social Sciences.
India. Environmental Hygiene Committee. 1950. *Report of the Environmental Hygiene Committee, October 1949*. Delhi: Manager of Publications.
India. Ministry of Agriculture. 1976. *Report of the National Commission on Agriculture*. New Delhi.
India. Ministry of Education. 1967. *The Indian Gazetters*. Vol. 1. New Delhi.
India. Ministry of Education and Social Welfare. 1974. *Towards Equality. Report on the Status of Indian Women*. New Delhi.
India. Ministry of Health and Family Planning. 1966. *Report of the Rural-Urban Relationship Committee*. 3 vols. New Delhi.
India. Ministry of Home Affairs. Bureau of Police Research and Development. 1983. *Report on Accidental Deaths and Suicides in India 1980*. New Delhi.
———. 1984. *Crime in India, 1979*. New Delhi.

India. Ministry of Information and Broadcasting. 1981. *India: A Reference Annual, 1981*. New Delhi.
India. Planning Commission. 1952. *The First Five Year Plan*. New Delhi.
———. 1981. *The Sixth Five Year Plan, 1980-85*. New Delhi.
———. 1985. *The Seventh Five Year Plan, 1985-90*. New Delhi.
India. Registrar General. 1961. *Census of India*. New Delhi.
———. 1982. *Census of India. 1981*. New Delhi.
India. Town and Country Planning Organization. 1975. *Jhuggi Jhompire Settlements in Delhi. A Sociological Study of Low Income Migrant Communities*. Parts 1 and 2. New Delhi.
India. Town Planning Organisation. 1970. *A Report of Land Values in Delhi*. Delhi.
Indian Institute of Constitutional and Parliamentary Studies. 1971. *Report of the National Seminar on Housing and Slums*. New Delhi.
Indian Institute of Public Administration. 1960. *Problems of Urban Housing*. Bombay: Popular Book Depot.
———. 1977a. *Slums and Squatter Settlements in Kanpur*. New Delhi.
Indian Institute of Public Administration. Centre for Urban Studies. 1977b. Special Issue on Urban Studies. *Nagarlok* 9(Apr.-June).
International Christian University. Social Science Research Institute. 1976. *Asia Urbanizing. Population Growth and Concentration and the Problems Thereof: A Comparative Symposium by Asian and Western Experts in Search of Wise Approaches*. Tokyo: Simul Press.
Iyer, S. S., and Krishna Kumar Verma. 1964. *A Roof over Their Heads; A Study of Night-Shelters in Delhi*. Delhi: Delhi School of Social Work, University of Delhi.
Jakobson, Leo, and Ved Prasash, eds. 1974. *Metropolitan Growth. Public Policy for South and South East Asia*. New York: Wiley.
Lahiri, T. B. 1978. "Calcutta: A Million City with a Million Problems." In *Million Cities of India. See* Arunachalam, 1978.
Lakdawala, Dansukhlal Tulsidas. 1963. *Work, Wages and Well-Being in an Indian Metropolis; Economic Survey of Bombay City*. Series in Economics no. 1. Bombay: University of Bombay.
Lal, Sheo Kumar. 1974. *The Urban Elite*. New Delhi: Thomson Press.
Linn, Johannes F. 1983. *Cities in the Developing World: Policies for Their Equitable and Efficient Growth*. New York: Oxford University Press for the World Bank.
Lloyd, Peter Cutt. 1979. *Slums of Hope? Shanty Towns of the Third World*. New York: St. Martin's Press.
Lokanathan, Palamadai Samu. 1935. *Industrial Organization in India*. London School of Economics and Political Science Studies in Economics and Commerce no. 4. London: Allen & Unwin.
Majumdar, Tapan K. 1983. *Urbanising Poor in Delhi: A Sociological Study of Low-Income Migrant Communities in the Metropolitan City of Delhi*. New Delhi: Lancers.
McGee, Terence G. 1971. *The Urbanization Process of the Third World. Explorations in Search of a Theory*. London: Bell.
Misra, Rameshwar Prasad., ed. 1978. *Million Cities of India*. New Delhi: Vikas.
———., ed. 1979. *Habitat Asia. Issues and Responses*. 3 vols. New Delhi: Concept.
Misra, Rameshwar Prasad, and Kavasseri Vanchi Sundaram. 1980. *Multi Level Planning and Integrated Rural Development in India*. New Delhi: Heritage.
Misra, Rameshwar Prasad, Kavasseri Vanchi Sundaram, and V. S. Prakasa Rao. 1974. *Regional Development Planning in India. A New Strategy*. New Delhi: Vikas.
Mitra, Asok. 1970. *Delhi—Capital City*. New Delhi: Thomson Press.
———. 1978. *India's Population. Aspects of Quality and Control*. New Delhi: Abhinav.
Mudaliar, M. 1973. "Metropolitan Madras, Tradition and Cultural Changes." *Bulletin of the Institute of Traditional Cultures* (India) (Jan.-June): 42-51.
Mukherjee, S. 1975. *Under the Shadow of the Metropolis. They Are Citizens Too!* Calcutta: Metropolitan Development Authority.
Murphy. Elaine M. 1981. *World Population. Toward the Next Century*. Interchange and Teaching Module Series. Washington, DC: Population Reference Bureau.
Nagpaul, Hans. 1962. "Sviluppo di communita in zone urbane dell' India" (Community

Development for Urban Areas in India). *International Review of Community Development* 9:95-111.
———. 1969. "Urbanization in India. Growth in Metropolitan Cities and Need for Urban Community Development." *Indian Journal of Social Research* 10:103-16.
———. 1979. "Approaches and Strategies for the Improvement of Slums and Squatter Settlements in Metropolitan India: A Bibliographical Essay." *Indian Journal of Social Research* 20:105-18.
Nambiar, P. K. 1961. "Slums of Madras City." In *Census of India, 1961*, vol. 9 (Madras), Part II-C. Delhi: Madras Branch, Registrar General, India.
Pacione, Michael, ed. 1981. *Problems and Planning in Third World Cities*. New York: St. Martin's Press.
Panchandikar, Krishna Chintamani. 1978. *Rural Modernization in India. A Study in Developmental Infrastructure*. Bombay: Popular Press.
Pathak, Shankar. 1979. *Social Welfare, Health, and Family Planning in India*. New Delhi: Marwah.
Piggott, Stuart. 1945. *Some Ancient Cities of India*. London: G. Cumberledge, Oxford University Press.
Population Reference Bureau. 1985. "1985 World Population Data Sheet." Washington, DC.
Ramchandran, P. 1972. *Pavement Dwellers in Bombay City*. Bombay: Tata Institute.
Rao, B. B. 1979. *Housing and Habitat in Developing Countries*. New Delhi: Newmay Group.
Rao, Kamala Gopal. 1974. *Studies in Family Planning: India*. New Delhi: Abhinav.
Rao, M.S.A. 1970. *Urbanization and Social Change. A Study of Rural Community on the Metropolitan Fringe*. New Delhi: Orient-Longmans.
———. 1974. *Urban Sociology in India: A Reader and a Source Book*. New Delhi: Orient-Longmans.
Rao, Vijendra Kasturi Ranga Varadaraja, and M. B. Desai. 1965. *Greater Delhi: A Study in Urbanization, 1940-1957*. Bombay: Asia Publishing House.
Saberwal, Satish. 1976. *Mobile Men. Limits to Social Change in Urban Punjab*. New Delhi: Vikas.
Sarin, Madhu. 1979. "Urban Planning, Petty Trading, and Squatter Settlements in Chandigarh, India." In *Casual Work and Poverty in Third World Cities*, edited by Ray J. Bromley and Chris Gerry. Chichester, England: Wiley.
Sen, Saurendra Nath. 1960. *The City of Calcutta, A Socio-Economic Survey, 1954-1955 to 1957-1958*. Calcutta: Bookland Private.
Sethuraman, S. V., ed. 1981. *The Urban Informal Sector in Developing Countries: Employment, Poverty and Environment*. Geneva: International Labour Office.
Shils, Edward A. 1961. *The Intellectual between Tradition and Modernity: The Indian Situation*. Comparative Studies in Society and History, suppl. 1. The Hague: Mouton.
Siddiqui, M.K.A. 1968. "The Slums of Calcutta. A Problem and Its Solution." *Indian Journal of Social Work* 29:173-81.
Singh, Andréa Menafee. 1979. *Neighbourhood and Social Network in Urban India: South Indian Voluntary Association in Delhi*. New Delhi: Marwah.
Singh, Andréa Menafee, and Alfred De Souza. 1980. *The Urban Poor. Slums and Pavement Dwellers in the Major Cities of India*. New Delhi: Manohar.
Singh, R. L., ed. 1973. *Urban Geography in Developing Countries: Proceedings of IGU Symposium no. 15, Varanasi, November 22-29, 1968*. Varanasi, India: National Geographical Society of India.
Singhvi, L. M., ed. 1971. *Fundamental Rights and Constitutional Amendments*. Delhi: National Publishing House.
Sundaram, Kavasseri Vanchi. 1977. *Urban and Regional Planning in India*. New Delhi: Vikas.
———. 1978. "Delhi, the National Capital." In *Million Cities of India*. *See* Arunachalem, 1978.
Taylor, John L., and David G. Williams. 1982. *Urban Planning Practice in Developing Countries*. Urban and Regional Planning Series, vol. 25. Oxford, England: Pergamon.
Trivedi, Harshad R. 1975. *Urbanization and Macro-Social Change*. Allabahad, India: Chugh Publications.

———. 1976. *Urbanism, A New Outlook.* Delhi: Atma Ram.

United Nations. 1975. *Urban Slums and Squatter Settlements in the Third World.* New York.

———. 1976. *Report of Habitat: United Nations Conference on Human Settlements. Vancouver, 31 May-11 June 1976.* A/CONF.70/15. New York.

———. 1983. *Report on Population in India 1980.* New York.

United Nations. Department of Economic and Social Affairs. 1969. *Growth of the World's Urban and Rural Population 1920-2000.* Population Studies no. 44. ST/SOA/SER.A/44. New York.

United Nations. Department of Economic and Social Affairs. Centre for Housing, Building, and Planning. 1976. *Global Review of Human Settlements: A Support Paper for Habitat: United Nations Conference on Human Settlements.* A/CONF.70/A/1/1 ADD 1. New York: Pergamon.

United Nations, DESA. *See* United Nations. Department of Economic and Social Affairs.

United Nations. Department of International Economic and Social Affairs. Statistical Office. 1983. *Demographic Yearbook, 1983.* 35th ed. ST/STAT/SER.R/13. New York.

United Nations Secretariat. Office of Technical Cooperation. 1971. *Improvement of Slums and Uncontrolled Settlements.* ST/TAO/SER.C/124. New York.

United Nations Secretariat. Public Administration Division. 1970. *Administration Aspects of Urbanization.* ST/TAO/M/51. New York.

Venkatarayappa, K. N. 1972. *Slums. A Study in Urban Problems.* New Delhi: Sterling.

Weitz, Rananan, ed. 1973. *Urbanization and the Developing Countries; Report on the Sixth Rehovat Conference.* Praeger Special Studies in International Economics and Development. New York: Praeger.

Wiebe, Paul D. 1975. *Social Life in an Indian Slum.* Durham, NC: Carolina Academic Press.

Yadav, D. S. 1979. *Land Use in Big Cities: A Study of Delhi.* New Delhi: Inter-India Publications.

10

Giant and Secondary City Growth in Africa

Dennis A. Rondinelli

Few of the world's giant cities are found in Africa. Its relatively slow pace of urbanization during the first half of the 20th century and the large percentage of population remaining in rural areas make the African continent one of the world's less urbanized regions.

The largest cities in Africa, with the exception of Cairo, Alexandria, Kinshasa, and Jos, are less than two million in population. The average size of the largest cities in all African countries is a little more than 950 thousand, relatively small by world standards. In 1980, less than half the African countries had any cities with more than one million residents, and 14 countries had no cities with a population of more than 500,000.

Although Africa is relatively late in urbanizing—Nairobi, Harare, Addis Ababa, Kinshasa, Lome, Abidjan, and Bamako, for example, all emerged as significant urban agglomerations only in the late 19th and early 20th centuries—cities have a long tradition in African history. O'Conner (1983) points out that Cairo and Alexandria are ancient cities; Mombasa, Zaria, Kano, and Mogadishu were all in existence by the 16th century; Accra, Kumasi, and Ouagadougou began to emerge by the 17th century, and by the early 19th century Banjul, Freetown, Monrovia, Ibadan, Sokoto, and Khartoum all had significant population concentrations. Some of these cities were indigenous urban centers that emerged from commercial and trading functions or as defensive or administrative centers. Others were Islamic cities providing a combination of religious and trading functions. Many were colonial cities created by foreigners for their own economic and political purposes and shaped by European aesthetic principles. Still others have emerged as dualistic or hybrid cities that accommodate or integrate foreign and indigenous activities with varying degrees of harmony.

The level of urbanization in Africa remains lower than in most other parts of the world, but its pace has increased markedly over the past two decades. The growth of urban population is expected to continue at a high rate well into the early years of the 21st century. The United Nations estimates that in 1980 nearly 35 percent of the population in central Africa were living in urban places, but that by the end of the 1990s about half will be urban residents. Nearly 44 percent of the population in North Africa now dwells in cities, and by the end of the century the percentage will increase to more than 58. The number of urban dwellers in western, northern, and southern Africa is expected more than to double over the next 20 years. In eastern Africa urban dwellers are likely to triple. In all, more than 345 million Africans are expected to be living in urban places by the end of the 1990s. (See table 10.1.) Moreover, although there are now relatively few "giant cities" in Africa, the United Nations (1980) projects an increase in the number of cities with more than a million in population from the two that existed in 1950 to 37 by 1990. These "million cities" will have about 38 percent of Africa's urban population.

Whether they are urbanizing rapidly or staying predominantly rural, most African countries are now facing important decisions about how to influence their patterns of urban development over the next two decades. Unlike governments in most other countries in the developing world, those in Africa have a unique opportunity to intervene through public policy to influence the pace and pattern of urbanization. Whether giant cities emerge in a polarized urban settlement structure as in many other parts of the world, or urbanization is more diffuse and balanced, is yet to be determined on much of the African continent. But the importance of formulating policies before urbanization has advanced too far is underlined by a recent World Bank study (1979:77) that points out that "excessive urban concentration is difficult to correct once it has occurred." The study concludes that if African governments want to avoid the familiar adversities associated with highly concentrated patterns of urbanization in developing countries, they will have to take actions soon to promote or encourage a more diffuse pattern of urban development.

This chapter examines the pace, pattern, and structure of urbanization in Africa. I focus on the distribution of population among urban places, and the demographic characteristics and economic and social functions of primary and secondary cities in African countries. Special attention is given to the relationships among population distribution, urbanization, and economic development, and to the implications for urban development policy.

Urban Structure in Africa

African countries show two distinct patterns of urban development. In northern and southern African countries, the concentration of urban popula-

TABLE 10.1
Number and Percentage of Urban Population—Africa, 1950-2000
(in thousands)

	1950		*1960*		*1970*		*1980*		*1990*		*2000*	
Region	*N*	%	*N*	%	*N*	%	*N*	%	*N*	%	*N*	%
Eastern Africa	3.4	5.5	5.8	7.5	10.7	10.7	21.3	16.1	40.3	22.7	70.5	29.4
Middle Africa	3.8	14.6	5.8	18.1	10.2	25.2	17.6	34.4	29.1	43.7	45.2	51.6
Northern Africa	12.7	24.6	19.8	29.8	31.3	36.6	49.6	43.8	76.9	51.4	111.9	58.3
Southern Africa	5.3	37.3	7.6	41.7	10.6	43.8	14.9	46.5	21.9	51.4	32.6	57.9
Western Africa	6.6	10.2	10.8	13.5	17.5	17.3	29.5	22.3	50.8	28.7	85.5	35.9
Total	31.8	14.5	49.5	18.2	80.4	22.9	132.9	28.9	219.2	35.7	345.8	42.5

tion in large metropolitan areas is creating a highly polarized settlement structure that is similar to urbanization patterns found in much of Southeast Asia and South America. Often such a structure is the result of, and maintains, a dualistic economy that creates large disparities in income and wealth between people living in the major metropolitan area and those living in other regions. In eastern and western Africa, and especially in sub-Saharan countries, population is distributed so widely in settlements of very small size that an articulated urban system capable of supporting those modern services, facilities, infrastructure, and productive activities requiring relatively large population concentrations, has not yet emerged. Experience suggests that in developing countries with weak, unarticulated, and poorly integrated urban systems it is difficult to generate and support agricultural and rural development or to provide people living in rural regions with those services and facilities requiring an urban base (See Rondinelli and Ruddle, 1978).

Pace and Pattern of African Urbanization

Although a majority of all Africans still live in rural areas, the pace of urbanization on the continent has been rapid since the 1950s. Urban development is likely to continue at a rapid pace over the next two decades. As table 10.1 indicates, eastern and western African countries are the least urbanized, with fewer than one-quarter of their populations living in urban places in 1980. About one-third of the population in middle African countries live in urban places; while more than 40 percent of the population of northern and southern African countries are now urbanized.

Of the 45 less-developed African countries, more than half now have fewer than 25 percent of their populations living in urban places (table 10.2). Sixteen countries have from 25 to 49 percent, and only Mauritius, Tunisia, and Libya have more than half their populations urbanized.

Yet, urban population growth has been, and continues to be, strong. The World Bank calculates that in the 34 less-developed African countries shown in table 10.3, the average annual urban population growth rate between 1970 and 1982 was 5.6 percent. In 16 of those countries the average annual growth rate was above the mean, ranging to over 8 percent a year in Mozambique, Tanzania, Cameroon, the Ivory Coast, and Mauritania. In all 34 countries, with the exception of Burkina Faso (Upper Volta), the average annual increase in urban population was higher than that of the total population.

All population projections for Africa show continued rapid urbanization for the rest of this century. The urban population in Africa is expected to increase by over 200 million people between 1980 and the end of the 1990s (table 10.1). By the beginning of the 21st century, the number of countries with less than 25 percent of their population in urban places is likely to drop from 28 to 10, and the number of countries with more than half their population urbanized is expected to increase from 3 to 17. (See table 10.2.)

If migration continues as it has in the past, many who move from rural areas to urban places will likely go to the largest cities. Of the 36 developing

TABLE 10.2

Urban Population in Selected Developing Countries—Africa, 1980 and 2000

Percentage of Population Living in Urban Places	*Country and (Percentage Urban)*			
	1980		*2000*	
Less than 25%	Burundi (2)	Botswana (16)	Burundi (4)	
(1980: 26 countries;	Rwanda (4)	Gambia (18)	Rwanda (9)	
2000: 10 countries)	Swaziland (8)	Madagascar (19)	Lesotho (11)	
	Uganda (9)	Mali (19)	Swaziland (16)	
	Mozambique (9)	Chad (19)	Burkina Faso (16)	
	Malawi (10)	Guinea (20)	Benin (18)	
	Burkina Faso (11)	Nigeria (21)	Mozambique (18)	
	Lesotho (12)	Togo (21)	Botswana (24)	
	Tanzania (12)	Angola (22)	Uganda (24)	
	Niger (13)	Sierra Leone (22)	Niger (24)	
	Ethiopia (14)	Zimbabwe (23)		
	Kenya (15)	Guinea-Bissau (23)		
	Benin (15)	Mauritania 924)		
25% - 49%	Sudan (26)	Zaire (36)	Tanzania (25)	Angola (36)
(1980: 16 countries;	Central African Republic (29)	Ghana (37)	Kenya (26)	Senegal (37)
2000: 18 countries)	Somalia (31)	Ivory Coast (41)	Ethiopia (28)	Zimbabwe (38)
	Liberia (34)	Morocco (41)	Togo (30)	Guinea-Bissau (39)
	Senegal (34)	Zambia (44)	Gambia (31)	Sierra Leone (40)
	Gabon (36)	Algeria (44)	Madagascar (32)	Sudan (42)
	Cameroon (36)	Egypt (44)	Nigeria (33)	Somalia (46)
		Namibia (45)	Guinea (33)	Liberia (49)
		Congo (46)	Chad (33)	
			Mali (34)	
50% - 74%	Mauritius (52)		Congo (50)	Cameroon (46)
(1980: 3 countries;	Tunisia (53)		Mauritania (50)	Zaire (56)
2000: 16 countries)	Libya (54)		Ghana (51)	Egypt (57)
			Gabon (54)	Central African Republic (58)
			Zambia (54)	Namibia (63)
			Malawi (54)	Tunisia (66)
			Ivory Coast (55)	Mauritius (67)
			Morocco (55)	Libya (72)
75% or more			Algeria (76)	

SOURCE: United Nations, Department of International Economic and Social Affairs (1980); World Bank (1982, 1983).

TABLE 10.3
Population Characteristics for Selected Developing Countries—Africa, 1970-2000

African Region	Population (millions) 1982	Population (millions) 2000[a]	Percentage of Average Annual Population Growth 1970-82	Percentage of Average Annual Population Growth 1980-2000[a]	Percentage of Average Annual Population Growth Urban 1970-82	Percentage of Urban Population in Largest City 1980
Eastern						
Burundi	4	7	2.2	3.0	2.5	–
Ethiopia	33	57	2.0	3.1	5.6	36.6
Kenya	18	40	4.0	4.4	7.3	57.4
Madagascar	9	16	2.6	3.2	5.2	36.6
Malawi	7	12	3.0	3.4	6.4	18.8
Mozambique	13	24	4.3	3.4	8.1	83.2
Somalia	5	7	2.8	2.4	5.4	34.2
Uganda	14	25	2.7	3.4	3.4	51.5
Tanzania	20	36	3.4	3.5	8.5	50.4
Zambia	6	11	3.1	3.6	6.5	35.4
Middle						
Angola	8	13	2.5	2.8	5.8	63.6
Central African Republic	2	4	2.1	2.8	3.5	36.6
Cameroon	5	7	2.0	2.5	6.4	39.3
Chad	9	17	3.0	3.5	8.0	21.5
Zaire	31	55	3.0	3.3	7.6	27.9
Northern						
Algeria	20	39	3.1	3.7	5.4	11.5
Egypt	44	63	2.5	2.0	2.9	39.0
Libya	3	7	4.1	4.3	8.0	63.7
Morocco	20	31	2.6	2.5	4.1	26.6
Sudan	20	34	3.2	2.9	5.8	30.6
Tunisia	7	10	2.3	2.3	4.0	30.8
Western and Southern						
Benin	4	7	2.7	3.3	4.4	62.9
Burkina Faso	7	10	2.0	2.4	1.1	–
Ghana	12	24	3.0	3.9	5.0	34.5
Guinea	6	9	2.0	2.4	5.2	79.8
Ivory Coast	9	17	4.9	3.7	8.2	32.6
Liberia	2	4	3.5	3.5	5.7	–
Mali	7	12	2.7	2.8	4.7	34.2
Mauritania	2	3	2.3	2.6	8.1	38.9
Niger	6	11	3.3	3.3	7.2	31.2
Nigeria	91	169	2.6	3.5	4.9	16.9
Senegal	6	10	2.7	3.1	3.7	64.9
Sierra Leone	3	5	2.0	2.4	3.9	46.5
Togo	3	5	2.6	3.3	6.6	60.4

SOURCE: World Bank (1984: Tables 1, 21, 22).
a. Projections.

countries shown in table 10.4, the population of the largest city more than doubled between 1960 and 1970 in 11 countries. It increased by more than 100 percent in 15 countries between 1970 and 1980. If African countries develop in ways similar to many poor countries in other parts of the world, they are likely to spawn large "primate cities"—those with substantially larger portions of the urban population than other cities within the country—that continue to grow to enormous size. Strong patterns of urban primacy are already beginning to emerge in many African countries. In ten countries—Kenya, Mozambique, Uganda, Tanzania, Angola, Libya, Benin, Guinea, Senegal, and Togo—more than half and up to 83 percent of the urban population is concentrated in the largest city (table 10.4). In 13 other countries at least one-third of the urban population lives in a single city. In many countries, the bulk of the nation's modern economic activities, capital facilities, physical infrastructure, industry, and social services is also concentrated in the largest city.

The intermediate level of the urban system—secondary cities with more than 100,000 population, but smaller than the largest city—is quite weak in most African countries. In 1980, only Zambia, Zaire, Algeria, Egypt, Morocco, and Nigeria had more than five cities, other than their largest urban center, with a population greater than 100,000. Four of these countries accounted for nearly 70 percent of Africa's secondary cities. Eleven other countries had only one or two secondary cities. Only in Zambia, Nigeria, and Morocco do all of the secondary cities combined have a larger percentage of the urban population than the largest city. (See table 10.4.)

The weaknesses of intermediate-sized cities may explain the continued attraction of the largest city to rural migrants. Cairo's population in 1980 was estimated to be nearly three million larger than the combined populations of Egypt's 14 secondary cities. There were no cities in the 500,000 to one million range in 1975, and those with from 100,000 to 500,000 accounted for a little more than 18 percent of Egypt's urban population. Few cities or towns in Egypt can match the services, facilities, productive activities, and employment opportunities provided in a city of Cairo's size.

In African countries that are still predominantly rural, the problem is not so much the overconcentration of urban population in one city as the wide dispersal of people in settlements too small to support even basic social services and nonagricultural economic activities. In Tanzania, Mauritania, Niger, Burkina Faso (Upper Volta), Mali, and Yemen, for example, the few urban centers are of such small size that population thresholds are too low to support the variety of services needed to stimulate the commercialization of agriculture, meet basic needs of the rural poor, or increase productivity and income of the rural population. In Mauritania, for example, the population of the largest city was less than 150,000 during the 1970s, and major towns ranged in size from a little more than 2,000 to less than 23,000 people.

Thus, although the pace of urbanization is rapid in nearly all African countries, many remain largely rural. Two dominant patterns of urbanization appear. In some countries one or two cities have grown to relatively large size and dominate the urban settlement system in a "primate city" structure; in

TABLE 10.4
Number, Population, and Growth Rate of Secondary and Largest Cities in Developing Countries—Africa, 1970 and 1980

Region, Country	Secondary Cities								Largest Cities					
	Number		Population (thousands)		Population (thousands)		Percentage Population Growth		Population (thousands)		Percentage of Urban Population		Percentage Population Growth	
	1970	1980	1970	1980	1970	1980	1960-70	1970-80	1970	1980	1970	1980	1960-70	1979-80
Eastern Africa														
Burundi	0	0	0	0	–	–	–	–	74	98	–	–	15.6	32.4
Ethiopia	1	1	226	439	9.7	9.6	21.0	94.2	784	1668	33.8	36.6	103.6	112.7
Kenya	1	1	256	396	22.3	17.8	58.0	54.6	550	1275	48.0	57.4	131.1	131.8
Madagascar	0	0	0	0	–	–	–	–	373	625	38.1	36.4	49.8	67.6
Malawi	0	0	0	0	–	–	–	–	148	352	36.4	18.8	–	137.8
Mauritius	0	0	0	0	–	–	–	–	136	153	39.3	30.2	22.5	12.5
Mozambique	0	0	0	0	–	–	–	–	375	750	80.1	83.2	106.0	100.0
Somalia	0	0	0	0	–	–	–	–	190	377	29.5	34.2	–	98.4
Uganda	0	0	0	0	–	–	–	–	357	813	45.6	51.5	134.9	155.7
Tanzania	0	0	0	0	–	–	–	–	375	1075	40.8	50.4	128.7	186.7
Zambia	5	5	732	1323	56.7	59.2	–	59.2	299	791	23.2	35.4	–	164.5
Middle Africa														
Angola	0	0	0	0	–	–	–	–	465	959	54.8	63.6	115.3	106.2
Central African Republic	0	0	0	0	–	–	–	–	187	297	37.2	36.3	55.8	58.8
Cameroon	1	1	178	352	15.0	14.4	–	97.8	250	526	21.1	21.5	44.5	110.4
Chad	0	0	0	0	–	–	–	–	155	313	37.4	39.3	–	101.9
Zaire	9	9	1916	3379	29.0	30.5	130.6	76.4	1367	3089	20.9	27.9	168.0	125.9

Northern Africa																
Algeria	6	6	1194	1682	18.3	13.9	53.6	40.8	1075	1391	16.5	11.5	23.1	29.4
Egypt	14	14	4577	5911	32.5	30.9	44.7	29.1	5480	7464	38.9	39.0	47.1	36.2
Libya	1	1	213	396	32.1	28.7	104.8	85.9	388	880	58.4	63.7	122.9	126.8
Morocco	9	9	2861	4292	54.6	51.9	42.6	50.0	1525	2194	29.1	26.6	38.5	43.9
Sudan	2	2	246	452	9.5	8.5	–	83.7	771	1621	29.9	30.6	110.1	110.2
Tunisia	1	1	243	305	10.8	8.9	58.8	25.5	760	1046	34.0	30.8	26.7	37.6
Western and Southern Africa														
Benin	0	1	0	114	–	10.4	–	–	204	685	47.4	62.9	–	235.8
Botswana	0	0	0	0	–	–	–	–	–	110	–	47.0	–	–
Burkina Faso	0	0	0	0	–	–	–	–	–	123	–	–	–	–
Ghana	2	2	512	775	20.4	18.8	46.2	51.4	754	1416	30.0	34.5	90.4	87.8
Guinea	0	1	0	174	–	8.2	–	–	330	736	60.8	79.8	182.1	131.2
Ivory Coast	0	0	0	0	–	–	–	–	356	685	29.8	32.6	97.8	92.4
Mali	0	0	0	0	–	–	–	–	249	440	33.2	34.2	70.8	76.7
Mauritania	0	0	0	0	–	–	–	–	–	198	–	38.9	–	–
Namibia	0	0	0	0	–	–	–	–	–	135	–	29.0	–	–
Niger	0	0	0	0	–	–	–	–	–	206	–	31.2	–	–
Nigeria	22	24	5049	9671	56.0	65.3	133.1	91.5	1389	2517	12.6	16.9	9.4	8.1
Senegal	0	2	0	218	–	17.2	–	–	559	821	60.1	64.9	50.3	46.9
Sierra Leone	0	1	0	145	–	17.4	–	–	202	388	42.5	46.5	96.1	83.0
Togo	0	0	0	0	–	–	–	–	150	273	58.4	60.4	–	82.0

SOURCE: United Nations, Department of International Economic and Social Affairs (1980: Table 48).
NOTE: All 1980 estimates based on number and population of cities in 1970.

others, no cities have grown to very large size and the vast majority of the population is scattered in very small settlements.

Large Cities in Africa

By 1980, 19 African cities had reached a population of more than one million. Based on 1970 population counts, Lagos, Ado-Ekiti, East Rand, Durban, Tunis, Kananga, Dar es Salaam, and Nairobi were projected to reach one million population by 1980. Accra, Johannesburg, Cape Town, and Addis Ababa were expected to surpass 1.4 million in population, and Jos, Casablanca, and Alexandria were projected to reach more than two million residents. Kinshasa grew to more than three million, and the Cairo-Giza metropolitan area increased its population to more than seven million. United Nations projections indicate that 12 of these 19 cities are expected to grow to more than three million by the end of the 1990s.

Although in 1950 only 9 cities in Africa had a population of more than one million, the United Nations estimates there will be 38 cities reaching or exceding that level by 1990 (table 10.5). By the end of the 1990s, 29 cities will have a population of between one and two million, another 17 cities will have between two and four million residents, and 11 more will exceed four million.

Most of the cities projected to reach populations of more than one million in northern, western, and southern Africa are coastal cities and ports. Nearly all of the largest cities inland are political capitals of their countries.

Economic and Social Functions of Large Cities

As in other parts of the world, large cities in Africa perform important social and economic functions. Huge cities provide economies of scale and proximity that have been conducive to industrialization, allowing the cities to absorb significant numbers of workers in manufacturing jobs, and governments to construct modern infrastructure and provide those advanced health, educational, commercial, and other facilities that require large population concentrations in order to operate efficiently. Most large African cities now play crucial roles as communications and transport hubs for their surrounding regions, providing modern international ports, harbors, and air facilities. Most have become national financial centers. Nearly all serve as important commercial, service, and administrative centers, providing large numbers of managerial, clerical, and professional jobs. The informal or small-scale service and manufacturing sectors of these cities absorb thousands of low- or nonskilled workers who cannot find employment in large industries or in the formal commercial sector.

In most of the poorer African countries, the largest cities are the centers of whatever national economic growth occurs. Hasselman (1981:139) notes, for example, that in Liberia "the Monrovia region contains nearly 50 percent of

TABLE 10.5
Estimated Population Size of Cities of One Million or More—Africa, 1990 (in millions)

Region	*City*	*Country*	*1990 Projected Population*
Eastern Africa	Addis Ababa	Ethiopia	3.243
	Nairobi	Kenya	2.628
	Tananarive	Madagascar	1.076
	Lourenco Marques	Mozambique	1.438
	Dar es Salaam	Tanzania	2.480
	Kampala	Uganda	1.663
	Lusaka	Zambia	1.524
	Harare	Zimbabwe	1.718
Western Africa	Luanda	Angola	1.755
	Cotonou	Benin	1.512
	Accra	Ghana	2.470
	Conakry	Guinea	1.397
	Abidjan	Ivory Coast	1.189
	Ado-Ekiti	Nigeria	1.955
	Ibadan		1.296
	Ilorin		1.084
	Jos		4.156
	Kaduna		1.096
	Lagos		2.542
	Mushlin		1.447
	Ogbomosho		1.028
	Dakar	Senegal	1.879
	Kinshasa	Zaire	5.556
Northern Africa	Algiers	Algeria	1.954
	Alexandria	Egypt	3.633
	Cairo-Giza		9.991
	Tripoli	Libya	1.556
	Casablanca	Morocco	3.236
	Rabat-Sale		1.586
	Khartoum	Sudan	3.144
	Tunis	Tunisia	1.479
Southern Africa	Cape Town	South Africa	2.019
	Durban		1.444
	East Rand		1.563
	Johannesburg		2.310
	Port Elizabeth		1.116
	Pretoria		1.010

SOURCE: United Nations, Department of International Economic and Social Affairs (1980).

the country's total urban population (and nearly 15 percent of the entire population); this is Liberia's only non-exploitive, non-subsistence region. This is also the location of the 'motor' of the country's economy." He points out that the government collects more than 90 percent of the country's taxes from profits and income in the Monrovia area, the location of nearly 99 percent of the international trade generated within the country. More than 93 percent of property taxes are collected in Monrovia, and the city has more than

80 percent of the registered vehicles and issues more than 81 percent of the nation's general licenses.

In Tunisia, according to Lawless and Findlay (1981:94), the restructuring of the national economy since the colonial period has

> established Tunis more strongly than ever before as the economic, social, and political core of the nation. The relationships between the urban "core" and the regional "periphery" have become increasingly strongly manifested in the flow of raw materials, people, and investment toward Tunis from the less fortunate areas of the country.

Concentration of the banking system in Tunis pulls capital and savings from all over the country, and the "flow of human resources toward the capital has been equally dramatic." During the 1970s, more than one-third of all of Tunisia's food, construction, and textile industries were concentrated in Tunis, as were 43 percent of the chemical and rubber and 61 percent of the mechanical and electrical engineering firms, as well as 47 percent of all other industries. Although manufacturing in Tunis has grown slowly, more than 55 percent of those employed in the Tunisian capital find work in the administrative, commercial, and service sectors.

Reviewing the structure of urbanization in Kenya, Richardson (1980) concluded that despite the government's fear that Nairobi is growing too large, its attractions for firms and households are difficult to challenge:

> As in industrial location, Nairobi's advantages include the following: access to the country's largest market; the second best site (inferior only to Mombasa) for export-oriented industries (and the best site for those using air freight); the depth of the metropolitan labor pool, especially for professional, technical, and skilled workers; easier linkages with other industries already concentrated there; proximity to government to facilitate lobbying for protection and subsidies; the social and cultural amenities of the national capital, a major locational pull on managers; and communication economies, especially with other countries.
>
> As for households, the quality of public services and infrastructure endowment are far superior to those at other locations. Incomes are higher, at least for those jobs, even after discounting for the much higher living costs. There are some job prospects, even though the potential supply of labor far exceeds the demand. The large informal sector offers consolation prizes to the migrant unemployed (p. 103).

Nairobi provides about 28 percent of all of the wage employment and nearly 53 percent of the wage employment among major towns in Kenya. The manufacturing sector provided less than one-fifth of the wage employment in Nairobi during the early 1980s, but the city's wholesale and retail trade, and social, personal and business services sectors employed over 60 percent of its wage earners (Kenya, 1983). In 1980, wage earnings in Nairobi were nearly four times those of Mombasa and more than six times those of the next four largest cities combined.

Also, studies of urbanization in Egypt point out that Cairo continues to account for the largest amount of growth of all the cities and towns in the country; and it is the national hub through which all important transport lines and facilities connect other major urban settlements with the capital. Together with Alexandria, Cairo shares 65 percent of the country's economically active labor force and 50 percent of the nation's industrial enterprises (Jenssen et al., 1981).

These important, and sometimes dominant, economic functions performed by large African cities continue to make them the most attractive locations for rural migrants. In western Africa, the pattern of internal migration has been and continues to be out of the poorest rural regions into the largest cities. Zachariah and Conde (1981:59) report that "in general, the principal areas of in-migration in each country is also the location of the capital city of the country and near the seacoast. This is true in Ghana, Liberia, the Ivory Coast, Sierra Leone, the Gambia and Senegal . . .," the exceptions being Burkina Faso (Upper Volta) and Togo.

The attraction of the largest cities to rural migrants remains strong because of their perceived advantages in providing employment, and because of the higher incomes city residents receive compared to earnings of people living in rural regions or smaller towns. Studies of Sudanese migration to Greater Khartoum report findings similar to those of studies of rural migration to other large cities in Africa: the pull factors are "higher average annual earnings, job availability, better educational opportunities and the low cost of migration due to the presence of friends and relatives" in the metropolitan area (Oberai, 1977:214). These pull factors are reinforced by the low wages derived from subsistence or near-subsistence agricultural production, lack of adequate nonfarm jobs, the paucity of services and facilities, and lack of educational opportunities in rural areas.

Average annual earnings in Greater Khartoum during the 1970s were more than nine times higher than in rural villages, while the cost of living was only a little more than twice as high. Although migrants to Khartoum were mostly poorly educated and unskilled workers, less than one-quarter of whom could find jobs in manufacturing, more than half could find employment in services, trade, and transport jobs within six months of arriving. In a pattern similar to that in many other large African cities, newly arrived migrants in Khartoum stayed with friends and relatives until they could find jobs, entered low-paid occupations or small-sector activities, and improved their income and occupational status over time as they gained experience, improved their skills, or raised their level of education.

Similarly, the per capita income of Lusaka's residents is estimated to be over six times that of rural people in Zambia, explaining in large part why the city's population grew by more than 10 percent a year during much of the 1960s and by about 7 percent a year since the early 1970s (Bryant, 1980).

The largest cities in Africa also support very large small-scale commercial and trade sectors, absorbing a substantial proportion of the urban labor force. About 60 to 70 percent of the urban labor force in Kumasi is employed in the city's informal sector. About half of Nairobi's and one-third of Abidjan's workers are also engaged in small-scale commercial activities (Setheruraman, 1981).

These economic and social functions are crucial to the national economies of nearly all African countries and will assure the continued growth of the largest cities well into the next century.

Adversities of Large-Scale Urban Concentration

But the rapidity of growth of large cities in relatively poor African countries has also created serious social problems, with which many African governments find it increasingly difficult to cope. Most large African cities cannot provide enough jobs for even their current work forces let alone the large numbers of migrants who continue to flow into them each year in search of employment. Many of the migrants who come to large cities are unskilled and uneducated, and can only find informal sector occupations providing subsistence incomes. The litany of problems associated with large cities in most other parts of the developing world can also be heard clearly in Africa. A description of Liberia's largest city, Monrovia, notes that

> The strong influx of people has caused multiple pressures on the city. Traffic conditions have become critical; slum areas have proliferated; there are periodic shortages of water and electricity; and the number of unemployed and underemployed persons has reached alarming proportions. Some 23-27 percent of the labor force is estimated to be without jobs in Monrovia. The average migrant to Monrovia stays there for 2 to 4 years before obtaining work. Understandably, the social services have been stretched beyond their capacities, and people living in Monrovia find it extremely difficult to obtain adequate housing and opportunities to improve their education (Hasselman, 1981:160).

Studies of Tunis report that demand for labor in the city has lagged far behind the supply of workers for many years, and that "few parts of the city offer more jobs than there are residents of working age.... Major deficits occur in the majority of the city's suburbs. Large areas of western Tunis are almost entirely devoid of any local job opportunities and are far from the main zones of employment" (Lawless and Findlay, 1981:114).

Rapidly growing urban populations are placing increasing demands on housing and such public facilities and services as health care, education, transportation, sanitation, and drinking water systems that are already overstrained. Continued migration of the rural poor to large cities in Africa results in greater concentration of people in slums and squatter settlements, to which basic services and facilities are rarely extended. An estimated 79 percent of the population of Addis Ababa, for example, lives in slum and squatter settlements, as do about 70 percent of Casablanca's residents, and 60 percent of Kinshasa's. More than one-third of the populations of Nairobi and Dakar are slum dwellers (Donohue, 1982). In Nairobi, the population of the lowest income squatter settlements has been growing more than twice as fast as that of the city as a whole (Mermon, 1982).

Few African cities can provide adequate urban services for their growing populations. The World Bank (1981:115) points out in its report on sub-Saharan Africa, that

> In Freetown . . . 95 percent of the population use shared pit latrines, while in Abidjan, 65 percent use open pits or unlined water courses. Only 20 percent of Abidjan's residents are served by a sewage system—and this is much higher than is common throughout the region.

Studies of urbanization in Ghana conclude that because of steady high levels of migration into Accra, household formation has for many years outstripped the provision of urban facilities and generated a large and constant housing shortage (Asiama, 1984:172).

Cairo, the largest city in Africa, with between seven and eight million inhabitants in 1981, suffers all of the problems of a metropolis in which urban infrastructures and public facilities were constructed for two million people. One study of the city concludes that "as a consequence of the immense growth in the population of Cairo, the agglomeration diseconomies and disadvantages have become more and more a challenging problem for the national as well as for the city government." It points out that

> regularly, public and private transport collapses in the city; energy supply and water provision services have become inefficient; the capacity of the telecommunication system has long since been outstripped by demand; land values and housing costs have increased far beyond any reasonable level (Jenssen et al., 1981:11).

All of these problems are exacerbated by the rapid increases in land costs as growth continues in the largest African cities. Urban land costs in Accra in 1979 had increased by ten times the costs at the beginning of the decade, making housing extremely expensive for the vast majority of the city's poorer residents (Asiama, 1984:173-74). A similar explosive increase in land costs in Tunis has not only made it more difficult for lower income families to obtain decent housing in the city's center, where many have jobs or are engaged in informal sector activities, but has driven many long-term lower income residents from the city's core to the periphery, where services are scarce and transportation costs to their sources of livelihood are higher (Lawless and Findlay, 1981).

These economic and social problems in large African cities also cause serious environmental damage. Examining the situation in Nigeria, Ayeni (1981:137) points out that

> environmental deterioration is one of the most alarming and uncomfortable problems of metropolitan Lagos, because it can be seen almost everywhere, in the form of heaps of garbage and other waste material and also in the structure of houses in the central slums of Lagos Island and the peripheral bidonvilles of Ojota, Maroko, Ajegunle and elsewhere.

But weak revenue bases or inefficient revenue collection practices and the limitations on revenue-raising placed on them by central governments make it impossible for most municipal governments in the largest cities to keep pace with expanding urban service needs (Rondinelli, 1986).

Thus, although the largest cities in Africa play crucial roles in the national economy and are important seats of political power, serious social, economic, and physical problems exist in most of them and are exacerbated by their continued growth.

Urbanization and Economic Development in Africa

Many countries in Africa will face serious problems of economic development during the next two decades as the result of their high population growth rates and low levels of economic expansion. The large majority of the world's poorest countries, by the World Bank's classification (1983:148-99), are now in Africa. More than 204 million Africans live in countries with per capita gross national products (GNP) of less than $270 a year and 210 million more live in countries with per capita GNP of less than $880 a year.

The adverse consequences of high rates of population growth in developing countries have been assessed by the World Bank in its 1984 *World Development Report.* The report points out that for Africa, often considered an underpopulated region, these consequences could be dire during the next two decades. The Bank estimates that population growth rates will remain at or near 3 percent a year in much of eastern, western, and southern Africa, and well above 2 percent a year in North Africa for the rest of this century. The report notes that sub-Saharan Africa's population is now growing faster than that of any other region in the world, and it has the highest fertility rates. The sub-Saharan population is expected to continue increasing at a rapid rate at a time when economic growth remains sluggish and agricultural production is low. During the 1970s, per capita income in sub-Saharan countries—excluding Nigeria—declined. In 1981 and 1982 gross domestic product grew hardly at all, while population increased by 2.7 percent a year. The labor force in most of Africa will grow by at least 3 percent a year over the next two decades, but economic growth is not expected to be high enough to absorb the growing population in productive jobs.

Economic conditions in many African countries have deteriorated in the past few years and social progress has been slow. While gross domestic product (GDP) grew at about 3.3 percent a year in sub-Saharan Africa and at about 3.6 percent in northern Africa and middle eastern countries from 1973 to 1980, GDP grew by only 1.7 percent in sub-Saharan Africa and fell by –0.5 in north African and middle eastern countries in 1981. Life expectancy in low-income African countries has risen only from 39 to 46 years and in middle-income African countries from 41 to 50 years since 1960. Child mortality rates in these countries are the world's highest and literacy rates are among the lowest. The number of people living in what the World Bank (1982) calls "absolute poverty" has been increasing in much of Africa.

Population is growing and poverty is increasing in Africa at the same time that resources for development are becoming more scarce. Exports from African countries to world markets have been decreasing, debt servicing problems have become more serious, and many governments depend heavily on declining sources of foreign aid.

Thus, a major problem national economic planners in many African countries now face is how to allocate scarce investment resources—sectorally and spatially—in ways that will promote economic growth, balance the distribution of population more evenly, ensure an equitable distribution of benefits and alleviate high levels of poverty within their countries.

Some development theorists argue that the heavy and continued concentration of people and economic activities in the largest cities often drains human, financial, and natural resources from already poor rural hinterlands, perpetuating or enhancing regional income differences, retarding economic development of those areas, and preventing development and growth of other cities and towns in rural regions.

Emergence of a primate city structure in many African countries is the result of political and economic policies that initially established dominant urban centers and have reinforced their growth. Often such a pattern of mutually reinforcing economic and spatial development has been the result of colonial policies aimed at exploiting natural resources and postcolonial investment strategies aimed at maximizing economic growth through export industrialization and import substitution. This pattern has resulted from political preferences for investment in the national capital rather than in interior regions or secondary cities and towns (Mabogunje, 1981). Even in highly rural countries such as those in the Sahel, the heavy concentration of government investments in the capital cities has accentuated differences in income and wealth between urban and rural areas generally and has made it difficult for smaller cities and towns to grow and diversify their economies to promote development in regions outside the capital.

But little solid evidence supports the contention that large primate cities are beneficial to the development of poor countries in Africa. Growing evidence suggests that if alleviation of rural poverty, increased agricultural productivity, and employment and income generation are important goals of development policies, then growth of a few large cities into massive metropolitan centers may in fact be detrimental. If governments in developing countries seek to promote socially and geographically equitable development, then investment in intermediate cities and regional market centers may be a more effective way of achieving that goal (see Rondinelli, 1983).

The rationale for allocating strategically greater shares of investments for infrastructure, social services, and productive activities to secondary cities and market towns in Africa is based on two major arguments. The first is that excessive concentration of population and investment in a single metropolitan area does not necessarily lead to higher levels of national economic growth and, indeed, over the long run can produce adverse consequences for national economies and for regional and social equity. The second is that African secondary cities and market towns can, and do, perform important functions in promoting rural and regional development that should be strengthened.

The validity of most of the arguments made by macroeconomists for concentrating investments in a single large metropolitan area in developing countries have come to be seriously questioned over the past decade.

Indeed, the increasing interest in strengthening the economies of secondary cities and market towns has arisen in large part from dissatisfaction with conventional economic growth theory. Arguments for concentrating investments in primate cities are based primarily on the assumption that maximizing growth of the GNP is the main objective of development, and that therefore investments should be made in those sectors and places within the country that provide the highest returns, regardless of the distributional effects.

Many economists argue that in capital-scarce developing countries the highest rates of return can be achieved by investing heavily in the industrial sector and in the largest city, where economies of scale and proximity are higher and marginal costs of investment are lower than in other parts of the country (Mera, 1973). Theoretically, higher returns to investment in the primate cities would stimulate growth of national economic output and the benefits would spread and trickle town from the primate city to stimulate agricultural development and incorporate the rural poor into the national economic system.

The implication of the argument is that diversion of investments to other cities and towns within the country would reduce the overall rate of national economic growth and incur social opportunity costs.

But nearly all of the assumptions underlying this argument are now questioned. First, in much of Africa, economic development is not likely to occur through massive industrialization. Since nearly three-quarters of the labor force in eastern, western, and southern Africa are employed in agriculture, economic growth in those countries will depend more on agricultural development as a base for expanding internal demand for domestic consumption goods than on export-oriented manufacturing. If agriculture is the key to development in Africa, then most of the arguments for concentrating investment in primate cities seem irrelevant. Economic growth is more likely to be stimulated by investments that support and facilitate agricultural development and strengthen internal markets for agricultural goods. Such investments must be located within easy reach of farm households and rural communities if they are to be effective.

Second, even if industrialization is a major factor in development strategies, the assumption that the concentration of investment in primate cities will generate benefits that automatically "trickle down," and thereby stimulate economic growth throughout the country, is contradicted by experience. Many developing countries have seen just the opposite: growth of primate cities often produces "backwash effects" that drain rural regions of their capital, labor, raw materials, and entrepreneurs to feed the growth of metropolitan centers. Gugler and Flanagan (1978:40) observe in their study that "anyone who ventures beyond [the national capitals] to the vast expanse of West Africa is stunned by the disparity between the concentration of resources in the capital cities and the neglect that is the fate of much of their hinterlands." The stimulative effects of concentrated investment have usually been restricted to the metropolitan areas surrounding the primate cities, thereby creating or exacerbating severe regional disparities between the metropolitan center and the rest of the country. In his assessment of "growth pole" policies Richardson (1978:135) argues that "experience has shown that this type of strategy is doomed to fail, at least as an instrument of *spatial* development," because the urban enclaves it usually creates are not beneficial to a large majority of people living within the growth poles or in rural areas surrounding them.

Third, while conventional arguments for concentrating investments in giant cities—they provide greater economies of scale—may be valid for high-technology, capital-intensive industries and massive infrastructure projects, these are only a small part of the investments needed to promote sustained and widespread economic development in most African countries. Once cities reach populations of more than about 500 thousand, it is not clear that they are beneficial for a whole range of small- and medium-scale processing, commercial, service, and manufacturing activities needed for development in Africa. By promoting the growth of a few metropolitan areas to very large sizes to achieve greater economies of scale for large, capital-intensive industries, economic planners may be creating diseconomies of scale and adversities for a wide range of small and medium enterprises that are more important for economic development in most African countries. Allocating investments to other areas of the country may be more beneficial for generating economic growth in rural regions, for as Anderson and Leiserson (1978:8) indicate, "improvements to social and economic infrastructure in rural areas and towns are a particularly important form of support to nonfarm activities."

Fourth, if employment generation is an important goal of development, recent International Labour Organization studies suggest that marginal increases in employment absorption capacity decline rapidly with the growth of Third World cities beyond a million in population. Bairoch (1982) argues that the marginal benefits for employment generation of cities larger than a million in population are so slight that they are substantially outweighed by the magnitude of economic and social problems and diseconomies of scale that come with continued urban population growth and concentration.

Fifth, the implication that smaller cities and towns are inefficient and uneconomical locations for investment in essential urban services, infrastructure, and productive activities has never been proven. Indeed, scattered evidence from developing countries suggest that the assumption is false. Although a great deal more research needs to be done in Africa, studies in other parts of the developing world indicate that economies of scale for most urban services and utilities and for most support services for industry are reached in towns of about 100 thousand population. The marginal per capita costs of providing urban services in many developing countries seem to follow the familiar U-shaped curve—declining with city size down to a population of around 50 to 70 thousand, depending on local circumstances, then rising again for settlements of smaller size (Rondinelli, 1983:chaps. 4 and 5).

All of this suggests that widespread economic growth, more balanced distributions of population, and more diffuse patterns of urbanization are unlikely to be achieved by investing heavily in a single metropolitan area and expecting spread and trickle down effects automatically to raise average income levels and reduce high levels of poverty throughout the country. If African governments want geographically widespread economic development and a more balanced distribution of population, they will have to invest in a more geographically dispersed pattern.

Secondary City Development and the Diffusion of Urbanization

Although experience in developing countries suggests that highly concentrated investment in primate cities is unlikely to produce equitable distributions of the benefits of development and more balanced patterns of population growth, neither will overly dispersed investment. The former creates polarization and the latter spreads investments too widely to have sufficient economic impact on the areas in which they are made. As noted earlier, in many African countries the vast majority of the population lives at very low densities in villages so small that services and facilities cannot be efficiently provided. In the Sudan, Tanzania, and Burkina Faso (Upper Volta), for example, the highly dispersed rural population remains largely unserved by basic commercial, personal, or social services because of the lack of regional market centers and intermediate cities large enough to support them. Studies of the Sudan conclude that "without access to markets, inputs, education, health and other services, development of the traditional market sector cannot occur" (USAID, 1979a:8). Studies of Tanzania note that because of poor access to cities in towns where markets and agricultural supply facilities are located, "the inability of rural farmers to market their goods wipes out any incentives offered by the government to increase production. The same problems exist with inputs supply. As a result, many of the fertile areas are vastly underutilized" (USAID, 1979b:31).

The need in much of Africa is to allocate investments in services, facilities, infrastructure, and productive activities in a pattern of "decentralized concentration;" that is, in settlements with sufficient concentrations of people and a hinterland market large enough to be able to support them economically and efficiently. But secondary cities and regional market centers must be distributed widely enough geographically, and linked strongly enough with their hinterlands, to provide access for a large rural population throughout the country.

Functions of Secondary Cities and Market Towns

Recent studies of secondary cities and market towns in Africa, and in other regions of the developing world, indicate that they can play important roles in supporting agricultural and rural development, distributing services, providing off-farm employment, and serving as focal points for the decentralized concentration of population and investment. In much of Africa, market centers with as few as 10 to 20 thousand residents play important roles in the rural economies of their regions, and towns and cities of 70 thousand or more offer agricultural support and urban service and employment functions essential to regional development.

Among the more important functions of intermediate-sized cities and market towns are the following (Rondinelli, 1983:chap. 4; Rondinelli and Ruddle, 1978).

Public and Social Service Centers

Most African secondary cities are public service centers, providing convenient locations for decentralizing public services through field offices of national ministries or agencies, or regional, provincial, or district offices, thereby offering greater access for urban and rural residents to public services and facilities that must be located in towns. They offer sufficient economies of scale to allow the concentration within them of basic and intermediate-level health, education, social, welfare, and public services, and to act as regional or district centers for a variety of public facilities.

In his study of secondary urbanization in Kenya, for example, Obudho (1976:103) notes that

> provision of public utilities like water supplies, electricity, and rail and road traffic was an index of the importance of towns. From the inception of the colonial domination the hierarchical importance of any central place could be calculated on the presence or absence of these utilities.

The concentration of primary and secondary schools in Tlemcen made that Algerian secondary city an educational center for its region and complemented its commercial activities (Lawless and Blake, 1976). In Nigeria, as states acquired greater political and administrative authority, their capitals grew into secondary cities and became increasingly important public service centers, supporting regional universities, technical colleges, other institutions of higher education, hospitals, and other social services requiring large population thresholds (Salau, 1979).

Small-Scale Manufacturing Centers

Most African secondary cities provide conditions conducive for small- and medium-scale manufacturing and artisan and cottage industries that serve local markets and satisfy internal demand for low-cost consumer goods. Some secondary cities can also support large industries.

What is most striking about the economic structures of secondary cities in most African countries is that they have had, and continue to have, relatively small shares of the nation's manufacturing employment compared to the largest cities. Yet they have a greater proportion of manufacturing, commercial, and service activities per capita than do small towns and rural villages. This situation has been the case historically in colonial countries of the developing world because, as noted earlier, many colonial governments and postindependence regimes concentrated investments in infrastructure and productive activities in the largest cities, especially in the national capital. Smaller cities and towns often received little of the nation's industrial investment.

There are also discernible differences in the structures and characteristics of secondary cities' industrial sectors from those in the largest cities. Studies comparing Lagos with Kano, for example, indicate that the industrial structure of the two Nigerian cities is markedly different. Investment by multinational corporations in Nigeria has been concentrated primarily in Lagos, with far less investment in Kano. Production processes in the secondary city are more labor-intensive, management is more personalized, and labor unions have less influence. Labor is less skilled and educated and the Kano manufacturing sector is more heavily dominated by small-scale producers (Lubeck, 1977).

When import substitution and petroleum industries began to play a larger role in the Nigerian economy, their locational preference for port cities greatly influenced the growth and diversification of Lagos. But in spite of larger secondary cities such as Port Harcourt, Kaduna, and Kano developing strong competitive advantages over smaller middle-sized cities in attracting capital-intensive industries, these firms play a less important role in the economies of secondary cities than they do in the capital city (Mabogunje, 1977).

Although most secondary cities in Africa grew without the benefit of large-scale industrialization, they exhibit a good deal of economic vitality stemming from numerous small-scale enterprises, ranging from traditional operations that provide their owners with only small amounts of income to more modern establishments. In Fayoum City, a secondary Egyptian urban center with a population of nearly 200 thousand, for example, employment in informal manufacturing is extensive. Hoffman's studies (1986) show that the city has developed into a regional center for furniture manufacturing, a sector that absorbs almost one-quarter of all workers in local informal enterprises. Cabinetmakers, shoemakers, and clothing producers also account for much of the city's employment.

Lubeck's studies of Kano, Nigeria, show that most migrants to the city initially find work in the tertiary sector, but many informal sector activities are partially integrated with larger manufacturing establishments, providing a channel of upward mobility, and allowing migrants to obtain industrial jobs (Lubeck, 1977).

Commercial Centers

Secondary cities usually offer a wider variety of basic household and consumer goods, commercial and personal services, and opportunities for off-farm employment in formal and informal commercial activities than do rural villages.

Although small- and medium-scale manufacturing plays an important role in most African secondary cities, their economies are dominated by commercial and service activities. Most secondary cities and market towns in Africa came into being as commercial and trade centers and quickly took on personnel and business service functions. Social, business, and public services secondary cities provide are no less important than commerce and trade in stimulating their growth and economic diversification. In Fayoum City, Egypt, for example, about 65 percent of the labor force is engaged in small-scale trade, service, financing, and transport activities (Hoffman, 1986).

Historically, Algerian secondary cities like Tlemcen grew primarily as agricultural markets, and diversified into commercial centers as farmers turned more of their fields to the production of cash crops. During the French colonial period, when both domestic and foreign demand for agricultural goods was high, larger market towns became regional centers for banking, transport, storage, and wholesale and retail trade and—depending on the number of foreign residents—for luxury goods shops and import trade. After a short period of settlement by colonists, agriculture was transformed almost entirely to cash-crop production, and market town economies became dominated by trade and service activities (Lawless and Blake, 1976).

Regional Marketing Centers

Many secondary cities also act as regional marketing centers offering a wide variety of distribution, transfer, storage, brokerage, credit, and financial services and important outlets for the sale and distribution of agricultural goods grown in surrounding areas.

Studies of the rice-marketing system in Ghana indicate that market towns and secondary cities play a crucial role in organizing and providing access to the system for rice farmers (Okoso-Amaa, 1975). After retaining crops for family use, paddy farmers sell whatever surpluses remain to traditional rice traders in nearby villages, illegal trader-smugglers, and itinerant traders, who resell the rice in market towns or to state sales agents located in larger cities. Surpluses gathered in village markets are assembled by traders in "feeder markets" located in district towns and resold to millers and wholesalers in larger cities. Some of the rice is sold in feeder markets to state sales agents and then resold in private stores in the cities. Private urban wholesalers also distribute the rice through city markets to retailers and institutions.

Traeger observed that in Ilesha, Nigeria, urban middlemen play an important role in bulking agricultural products in villages and on individual farms. They resell some products to other intermediaries for redistribution to more distant cities and towns, buy manufactured goods in larger cities such as Ibadan and Lagos and sell them to traders in Ilesha and to farmers in smaller towns and villages in the city's trading area, as well as in the Ilesha market (Traeger, 1976-77). These marketing activities also create new economic opportunities for women that may not be available or acceptable in smaller towns or rural villages. In many African secondary cities, women participate heavily in market trade and associated activities. In Ilesha, for example, women participate actively not only as market traders, but also as brokers, intermediaries, wholesalers, and lenders. They dominate the operation of food distribution activities within the city's internal marketing system.

Centers of Demand for Agricultural Goods

Population growth and economic diversification in secondary cities create new demand for cash crops and commercial agricultural goods that can be grown in their rural hinterlands, and in so doing stimulate commercialization of agriculture and increase productivity in nearby rural areas.

Mortimore (1970) points out that as Kano, Nigeria, grew in population, farmers in the area immediately surrounding the city shifted from subsistence to cash crop production; land tenure became more individualized; and farming methods changed. Heavy manuring replaced fallowing, farmers began intercropping legumes and grains, and there was a marked increase in agricultural labor hiring, all of which intensified as commercial agriculture spread in the rural towns and villages in Kano's hinterland.

Studies of rural food-processing industries in northern Nigeria report that growing demand in the large markets of the city of Zaria stimulated the production of *fura*—a cooked ball of grain eaten with sour milk—by women in nearly rural villages. The women who made fura began to employ other women in the village to process and prepare it and to hire children and students to carry their goods to market. The expanding market in Zaria also created demand for rural crafts, apparel, mats, and decorative household goods made from agricultural products (Simmons, 1975). Although growing urban markets increased the demand for rural products in early stages of urban growth, the city also spawned competition for rural enterprises as it reached a size that allowed larger establishments to produce the same or substitute goods using more modern technology or mass production.

As secondary cities in Africa have grown larger, their marketing areas have expanded and their marketing linkages with other cities and towns have become more extensive. Jones (1976:313) notes that as Ibadan grew into an intermediate city, its "staple supply hinterland wound round and leaped over the supply hinterland of neighboring cities," encompassing them in much the same way that larger markets encompassed smaller ones in many Western countries. Ibadan's growth created demands for food and other agricultural goods that were beyond the capacity of its rural hinterlands to meet, and urban merchants reached out to other cities and towns to obtain supplies.

Agro-Processing and Supply Centers

Many secondary cities act as agro-processing and agricultural supply centers for inputs, such as fertilizers, seeds, cultivation and harvesting implements, irrigation components, and pesticides for farmers in their regions.

In Senegal, nearly half the industrial activity in secondary towns is related to agro-processing. Agro-processing provides about 46 percent of the commerce and 44 percent of the nonfarm jobs. In Kaolack, Ziguinchor, and Droubel, oil mills offer employment for rural nonfarm workers, as do sugar- and rice-processing activities in Richard Toll, tomato processing in Dagana, fish processing in Saint Louis and Mbour, and shrimp processing in Zinguinchor (Senegal, 1984).

In many African countries there is a symbiotic relationship between secondary cities and their rural hinterlands. Obudho and Waller (1976) note that Kisumu serves as a processing center for one of Kenya's most productive agricultural regions. Kisumu benefits by its proximity to the tea plantations in Kericho, coffee farms in Abagusii, South Nyanza, Kakemega, Busia, and Bungoma districts, and sugar cane-processing and allied industries in Muhoroni and Mwani. The industrial base of the city—composed of grain mills, groundnut crushing factories, hide- and skin-curing plants, timber yards, fish-packing plants, and agricultural equipment assembly plants—is closely related to the agricultural economy of the surrounding areas. During the 1960s and 1970s, food processing dominated Kisumu's manufacturing sector and the city served as a center for agricultural processing in its region.

Off-Farm Employment Centers and Sources of Wage Remittances

Most secondary cities in Africa are sources of off-farm employment and supplementary income for people living in nearby rural areas and, through remittances earned by migrants, provide additional income for household members remaining in rural areas.

Describing the function of medium-sized towns in southeastern Nigeria, Okafor (1985:155) points out that they "provide the first attraction to school leavers and migrants where they engage in several off-farm activities. Such towns facilitate the adaption of migrants from rural to urban life. The majority of the traders, taxi drivers, carpenters, are usually people from within a radius of about ten kilometers from the urban base of the operations." He observes that many of these migrants work in the towns part-time and continue to farm nearby lands to supplement their incomes. Secondary towns also provide employment for residents from surrounding villages who take jobs as traders, or as casual or daily-paid laborers.

Studies in Kenya indicate that about 13 percent of the income men earned in a sample of recent migrants to urban centers was remitted home (Rempel and Lobdell, 1978). In Ghana, 40 percent of rural households with members in cities have received cash or in-kind remittances (Connell et al., 1976).

Centers of Transportation and Communications

African secondary cities often serve as regional centers of transportation and communications, linking their residents and those in nearby rural areas to larger cities and other regions of the country.

Transportation networks have played a crucial role in the growth of secondary cities and towns in much of Africa, and their role as transportation and communications centers is still crucial to rural economies. In Nigeria, Mabogunje (1977:583-84) observes that the extension of railway and road networks not only created new secondary cities, but also transformed the economies of existing ones from administrative and marketing to commercial, service, and manufacturing centers. Roads and railways built at the turn of the century to exploit resources of the interior by-passed some traditional centers such as Ile-Ife, Ilesha, Benin City, Sokoto, Katsina, and Yauri, and turned what were previously small villages into important nodal points in the colonial spatial economy. Kaduna, Jos, Enugu, and Port Harcourt, all located along new rail lines, grew in population and became more economically diversified.

Gugler and Flanagan (1978:27-28) found a similar pattern in other African countries; they point out that "for West African towns, fortune rode the trains. Those that received terminals grew but those that did not stagnated or declined, as did many river ports."

Centers for Social Transformation

Many secondary cities and towns in Africa also function as centers of social transformation: absorbing rural migrants who might otherwise go directly to the largest city or national capital; accommodating and encouraging the integration of diverse social, ethnic, religious, or tribal groups; supporting organizations that help socialize and assimilate rural people into city life; and providing new economic opportunities for people seeking social and economic mobility.

Religious and social organizations in many secondary cities offer migrants from rural areas shelter and contacts through which they can seek jobs. Many of these organizations mediate conflicts and maintain communications among their members and between them and people who have risen to positions of influence in the city and who would be sympathetic to their advancement. Ethnic associations in African secondary cities play a variety of roles, including providing small amounts of credit to establish businesses and loans to help members over difficult economic times. Islamic institutions, for example, serve not only as channels for acculturation into city life, but also provide subsistence for temporary or permanent migrants, find jobs for rural youth studying in the city, and recruit labor for their members who have businesses (Lubeck, 1977). Islamic and Christian organizations in secondary cities substitute for the close family and community ties rural migrants often leave behind. They help the sick, bury the dead, and celebrate weddings or christenings. These associations may also provide a channel for remitting income or mobilizing capital for investments in members' home villages (Mabogunje, 1977).

In sum, although secondary cities in Africa are still relatively few, not widely dispersed geographically, and often poorly linked to market towns and their rural hinterlands, where they exist they seem to perform a wide variety of social, economic, and service functions important for meeting basic needs of their residents and for stimulating agricultural development and off-farm employment.

Policy Implications

Studies of migration in Africa show that people migrate to cities and towns primarily for economic reasons—either adequate job opportunities do not exist in rural areas or they perceive employment opportunities to be better in the cities (Findlay, 1977). The numbers of new secondary cities and the strength of the economies of existing ones can be increased only when they offer employment opportunities sufficient to attract the rural migrants now going directly to the largest cities. Employment expansion requires investment in agro-processing, agribusiness, services, market facilities, and small-scale manufacturing. Such enterprises will locate in market towns and secondary cities, however, only if they have adequate physical infrastructure, good transport access to rural areas and larger cities, and adequate housing and social services. With proper investment, secondary cities and market towns can facilitate development of the regions in which they are located.

Clearly, from historical experience the creation of large primate city growth poles based on investment in capital-intensive, export-oriented industrialization is neither appropriate nor sufficient to generate widespread and sustainable economic growth in most of Africa. Investing in the service, distribution, commercial, marketing, agro-processing, and other functions that secondary cities and market towns can and do perform offers a far better base for stimulating widespread economic growth and balancing population distribution.

But in formulating strategies for allocating a larger share of investments to secondary cities and regional urban centers, a number of important policy implications must be kept in mind.

First, experience in other regions of the developing world suggests that if secondary cities and market towns are not to become "enclaves"—mini-growth poles that drain the resources of the rural areas surrounding them—the economic activities encouraged within them should be related to the agricultural economies of their rural hinterlands and stimulate productive activities in their regions. Spread effects and beneficial interaction do not happen automatically with urbanization and investment in cities. Economies of secondary cities, and the linkages between them and their rural hinterlands, must be structured in ways that will stimulate internal production and demand and raise incomes of people living in the region. This structure requires careful regional analysis and planning.

Urbanization policies in African countries should focus not only on secondary cities, but also on stimulating and diversifying the economies of smaller towns with potential for development in order to increase the number and geographic distribution of urban settlements. Richardson (1977) has suggested four specific ways of building up the intermediate level of the settlement hierarchy in developing countries: (1) by promoting growth of small- and middle-sized cities close enough to major metropolitan centers to benefit from their agglomeration economies, yet not so close that they will be swallowed up in the continued growth of giant cities; (2) by promoting development of regional urban centers or large market towns far away from the major metropolis as countermagnets for rural migrants and high-population-threshold economic activities that would normally locate in the largest city; (3) by developing small cities in underdeveloped and sparsely populated rural regions through investments in growth-generating activities that will allow these towns to expand to a size that will begin generating internal economies of scale; and (4) by developing transportation axes that connect existing and potential secondary cities and create conditions conducive to growth of multiple midpoint or nodal centers at terminal or break points in the transportation network.

Second, although appropriate spatial allocation of investment can provide the physical conditions for more widespread economic development, it alone cannot change the structure of the economy. National policies on international trade, foreign investment, population growth, migration, agricultural prices, and wages must all support and reinforce spatial policies if they are to have an impact in generating equitable economic growth. This phenomenon requires a stronger integration of national and regional planning and better coordination of sectoral and spatial planning.

Third, significant changes can be brought about without massive new investment. Careful locational analysis and planning of current investment to promote a pattern of "decentralized concentration" focused on existing secondary cities and market centers can begin to strengthen their capacity to facilitate regional development. Incremental changes in allocation and location of already-planned investments can be the basis for building a stronger network of secondary cities and market towns from which to provide the services, facilities, and productive activities needed to stimulate rural economies. Methods of analysis and planning for regional allocation and location of investment have been developed that provide a process and framework for regional development that can be adopted to the needs of African governments wishing to pursue an investment strategy of decentralized concentration (Rondinelli, 1985).

In brief, two decades of experience with development indicates the need to allocate and locate investments strategically in a spatial pattern of decentralized concentration to stimulate widespread economic growth and a more balanced distribution of urban population. If African countries are to avoid the adversities of spatial polarization and economic dualism that often accompany the growth of giant cities in poor countries, and if they are to manage their inevitable urban population growth more effectively, they must make strategic decisions about allocation of investments among settlements and regions while those decisions can still influence the pattern of urbanization and economic development.

Bibliography

Anderson, Dennis, and Mark W. Leiserson. 1978. "Rural Enterprise and Nonfarm Employment." A World Bank Paper. Washington, DC: World Bank.

Asiama, Seth Opuni. 1984. "The Land Factor in Housing for Low Income Urban Settlers: The Example of Medina Ghana." *Third World Planning Review* 6(1):171-84.

Ayeni, Bola. 1981. "Lagos." In *Problems and Planning in Third World Cities,* edited by Michael Pacione. New York: St. Martin's Press.

Bairoch, Paul. 1982. "Employment and Large Cities: Problems and Outlook." *International Labour Review* 121:519-33.

Bryant, Coralie. 1980. "Squatters, Collective Action and Participation: Learning from Lusaka." *World Development* 8:73-85.

Connel, John, Biplab Dasgupta, Roy Laishley, and Michael Lipton. 1976. *Migration from Rural Areas. The Evidence from Village Studies.* Institute of Development Studies Discussion Paper. Delhi: Oxford University Press.

Donohue, John J. 1982. "Some Facts and Figures on Urbanization in the Developing World." *Assignment Children* no. 57-58:21-41.

Findlay, Sally Evans. 1977. *Planning for Internal Migration. A Review of Issues and Policies in Developing Countries.* Publication no. ISP-RD-4. Washington, DC: International Statistical Programs Center, U.S. Bureau of the Census.

Gugler, Josef, and William G. Flanagan. 1978. *Urbanization and Social Change in West Africa.* Urbanization in Developing Countries. Cambridge, England: Cambridge University Press.

Hasselman, Karl-Heinz. 1981. "Liberia." In *African Perspectives,* edited by Harm J. de Blij and Esmond B. Martin. New York: Methuen.

Hoffman, Michael. 1986. "The Informal Sector in an Intermediate City: A Case in Egypt." *Economic Development and Cultural Change* 34:263-77.

Jenssen, B., K. R. Kunzmann, and S. Saad-El Din. 1981. "Taming the Growth of Cairo." Working Paper no. 3. Dortmund, West Germany: Institute of Urban and Regional Planning, University of Dortmund.

Jones, William O. 1976. "Some Economic Dimensions of Agricultural Marketing Research." In *Regional Analysis,* edited by Carol A. Smith. Vol. 1, *Economic Systems.* Studies in Anthropology. New York: Academic Press.

Kenya. Central Bureau of Statistics. 1983. *Statistical Abstract, 1983.* Nairobi.

Lawless, Richard I., and Gerald H. Blake. 1976. *Tlemcen: Continuity and Change in an Algerian Islamic Town.* London: Bowker for the Centre for Middle Eastern and Islamic Studies of the University of Durham.

Lawless, Richard I., and Allan M. Findlay. 1981. "Tunis." In *Problems and Planning in Third World Cities,* edited by Michael Pacione. New York: St. Martin's Press.

Lubeck, Paul M. 1977. "Contrasts and Continuity in a Dependent City: Kano, Nigeria." In *Third World Urbanization,* edited by Janet Abu-Lughod and Richard Hay, Jr. Chicago: Maaroufa Press.

Mabogunje, Akin L. 1977. "The Urban Situation in Nigeria." In *Patterns of Urbanization. Comparative Country Studies,* edited by Sidney Goldstein and David F. Sly. Vol. 2. Dolhain, Belgium: Ordina Editions.

———. 1981. *The Development Process: A Spatial Perspective.* New York: Holmes & Meier.

Memon, P. A. 1982. "The Growth of Low Income Settlements." *Third World Planning Review* 4(2):145-58.

Mera, Koichi. 1973. "On the Urban Agglomeration and Economic Efficiency." *Economic Development and Cultural Change* 21:309-24.

Mortimore, Michael J. 1970. "Population Densities and Rural Economies in the Kano Close-Settled Zone, Nigeria." In *Geography and the Crowding World,* edited by Wilbur Zelinsky, Leszek A. Kosiński, and R. Mansell Prothero. New York: Oxford University Press.

Oberai, A. S. 1977. "Migration, Unemployment and the Urban Labour Market: A Case Study of the Sudan." *International Labour Review* 115:211-32.

Obudho, Robert A. 1976. "Spatial Dimension and Demographic Dynamics of Kenya's Urban System." *Pan African Journal* 9:103-24.

Obudho, Robert A., and Peter Waller. 1976. *Periodic Markets, Urbanization, and Regional Planning: A Case Study from Western Kenya.* Contributions in Afro-American and African Studies no. 22. Westport, CT: Greenwood Press.

O'Connor, Anthony Michael. 1983. *The African City.* London: Hutchinson.

Okafor, Francis C. 1985. "The Functional Role of Medium-Sized Towns in Regional Development." *Third World Planning Review* 7(2):143-59.

Okoso-Amaa, Kweku. 1975. *Rice Marketing in Ghana.* Uppsala: Scandinavian Institute of African Studies.

Rempel, Henry, and Richard Lobdell. 1978. "The Role of Remittances in Rural Development." *Journal of Development Studies* 14:324-41.

Richardson, Harry W. 1977. "City Size and National Spatial Strategies in Developing Countries." Working Paper no. 242. Washington, DC: World Bank.

———. 1978. "Growth Centers, Rural Development and National Urban Policy: A Defense." *International Regional Science Review* 3:133-52.

———. 1980. "An Urban Development Strategy for Kenya." *Journal of Developing Areas* 15:97-118.

Rondinelli, Dennis A. 1983. *Secondary Cities in Developing Countries: Policies for Diffusing Urbanization.* Sage Library of Social Research, vol. 145. Newbury Park, CA: Sage.

———. 1985. *Applied Methods of Regional Analysis: The Spatial Dimensions of Development Policy.* A Westview Special Study. Boulder, CO: Westview Press.

———. 1986. "Extending Urban Services in Developing Countries: Policy Options and Organizational Choices." *Public Administration and Development* 6:1-21.

Rondinelli, Dennis A., and Kenneth Ruddle. 1978. *Urbanization and Rural Development: A Spatial Policy for Equitable Growth.* New York: Praeger.

Salau, A. T. 1979. "Urbanization, Planning and Public Policies in Nigeria." In *Development of Urban Systems in Africa,* edited by Robert A. Obudho and Salah El-Shakhs. New York: Praeger.

Senegal. Ministry of Urbanism and Housing. 1984. "The Senegalese Experience in Urbanization: The Role of Secondary Towns in Economic and Social Development." Dakar.

Sethuraman, S. V., ed. 1981. *The Urban Informal Sector in Developing Countries: Employment, Poverty and Environment.* A WEP Study. Geneva: International Labour Office.

Simmons, Emmy B. 1975. "The Small-Scale Rural Food-Processing Industry in Northern Nigeria." *Food Research Institute Studies* (Stanford University) 14:147-61.

Traeger, Lillian. 1976-77. "Market Women in the Urban Economy." *African Urban Notes* 2(3):1-9.

United Nations. Department of International Economic and Social Affairs. 1980. *Patterns of Urban and Rural Population Growth.* Population studies no. 68. ST/ESA/SER.A/68. New York: United Nations.

United States Agency for International Development. 1979a. *Country Development Strategy Statement FY 1981: Sudan.* Washington, DC.

———. 1979b. *Country Development Strategy Statement FY 1981: Tanzania.* Washington, DC.
USAID. *See* United States Agency for International Development.
World Bank. 1979. *World Development Report, 1979.* Washington, DC.
———. 1981. *Accelerated Development in Sub-Saharan Africa: An Agenda for Action.* Washington, DC.
———. 1982. *World Development Report 1982.* New York: Oxford University Press for the World Bank.
———. 1983. *World Development Report 1983.* New York: Oxford University Press for the World Bank.
———. 1984. *World Development Report 1984.* New York: Oxford University Press for the World Bank.
Zachariah, Kunniparampil Curien, and Julien Conde. 1981. *Migration in West Africa: Demographic Aspects.* Oxford, England: Oxford University Press.

11

Latin American Cities and Their Poor

Guillermo Geisse and Francisco Sabatini

The large cities of Latin America have been a long-standing concern of urban planners, social scientists, and technocrats. All have considered the problems affecting these cities, primarily in terms of their population size. A common argument is that excessive urban concentration causes slow economic growth, social marginality, marked regional differences, and so on. The anti-big city argument is applied across the size spectrum from Tegucigalpa, Honduras, a "large city" of 300 thousand inhabitants, to Mexico City, with a population of 16 million.

In previous studies we have advocated an entirely different approach, which may be summarized as follows: the concentration of the Latin American population in great cities has contributed to the expansion of internal markets, increased general productivity of national economies, and creation of better opportunities for social and political integration. The problem of giant cities is not their large size but the pace of their growth and the social mechanisms through which the benefits of concentration in space are accessible only to a minority, whereas costs must be borne by the whole of the society and, particularly by the poor.

The rapid pace of urban growth gives time for only spontaneous and partial solutions which, in developing economies, typically become permanent. Moreover, urban sprawl has proved to be an incentive for land speculation, which then becomes the main mechanism for the private appropriation of urban benefits in the form of land rents. It might seem as if there were no room for planning between the spontaneity of the informal sector, expressed by irregular settlements, and that of the formal urban sector where speculative markets prevail. The play of these opposing forces leads to markedly unequal and extremely inefficient urban forms and structures. Planning might not have prevented excessive growth, but it might have oriented growth to avoid such forms and structures that are the underlying causes of the problems.

Conditions in giant cities will become even more serious in the next 20 years when widespread economic crisis reduces the public expenditure allocation for infrastructure and urban services and democratization trends open up participation channels for poorer sectors to stake their urban claims. The shortage of development funds and social pressures on the democratic system and the state offer grounds to expect a hard and unstable future. Our thesis is that it will be necessary to resort to yet untapped resources and energies in the marginal areas of large cities where the germ of alternative development exists.

The ideas and general background that follow are expected to promote a discussion about prospects of a development alternative for Latin America's giant cities.

The Germ of Alternative Development in Marginal Areas of Giant Cities

Survival Mechanisms

Unemployment, defenselessness, and deplorable living conditions have led the poor living in peripheral areas of cities to develop a large variety of survival mechanisms. Of these, the most widespread and important are an informal economy, reciprocal assistance networks, and progressive development of housing. Other less frequent mechanisms are informal credit systems, soup kitchens, and family orchards.

These mechanisms have a series of typical characteristics that, on the whole, permit them to reach a high level of efficiency for attaining their main goal, survival. These characteristics are, in our opinion, one of the bases to promote policies tending to convert survival mechanisms into development mechanisms.

The first characteristic is *massiveness*. In general, survival mechanisms allow for free and unlimited incorporation. No skills are required—or, if so, they can be easily acquired—and no initial capital is needed. Neither do the mechanisms impose restrictions on numbers. On the contrary, any new member gained generally represents a new source of income for the family group: labor for self-help construction, or a reduction in the margin of social insecurity since the newcomer is likely to join local networks for the exchange of favors or gifts. The purpose of the latter is to provide solutions for the recurring emergencies the poor must face. This phenomenon is not easily grasped by outsiders who view poverty simply as a lack of material things: one more person is nothing but one more mouth to feed and one more body to clothe and shelter.

Another characteristic of survival mechanisms is that they are structured in terms of *territory*. This is the reason why access to land is not only the starting point for a house but, what is more important, the base that allows families to become incorporated into survival mechanisms. There appears a network of social and economic relations on the basis of subjective spatial facts: belonging to a neighborhood, identifying with a certain place, and objective spatial facts—physical proximity and accessibility to activity centers. When there are no alternatives to gaining access to land through the market or public programs, the consequences are land invasions or the doubling up of existing precarious settlements. The relation between home owners and families with whom they share their houses may be considered as reciprocal assistance in the face of emergency situations. In general, both search for the territorial basis to build survival mechanisms of cooperation and reciprocity within a given territory.

A third characteristic of survival mechanisms is their *flexibility*. Because of the freedom of access they offer, survival mechanisms are highly adaptable and flexible. Each family has mobility between territorial groups and social structures in their quest for ensuring survival and progress possibilities. This fact accounts for the high rate of residential mobility observed in marginal urban areas and the enormous capacity to adapt to fluctuations in the urban economy, labor markets, housing policies, and so on.

Finally, a fourth characteristic is the autonomous nature of survival mechanisms, as reflected in the families' capacity to make the most of their own resources such as labor and organization and in the way in which external resources are managed and adapted to local needs. Examples of the latter are local redistribution of the income of those who operate in the "formal" economy and innovations introduced to basic housing units provided by state assistance programs.

The marginal residential areas of the large cities possess some environmental, spatial, socioeconomic, and political characteristics which may be of help in the search for a development alternative for large Latin American cities. In spatial terms, marginal settlements represent an urban development of a human scale that may not be found anywhere else in cities. One can observe in them the presence of a significant social interaction or "neighborhood life," of decentralized nonresidential activities such as services and workshops enjoying a high degree of autonomy in decisions for physical and social development.

In environmental terms, there is a large capacity to recycle materials with consequent low production of waste, an open contrast with the environmental inefficiency of the "formal city." Moreover, because of the challenge posed by survival, it is possible to observe in marginal neighborhoods a more deeply rooted environmental conscience than that of the elites of the formal city which in most cases does not go beyond more ecological rhetoric.

In socioeconomic terms, it is possible to observe the development of labor-intensive activities with a low marginal cost. The most important of these are self-help construction; retail and informal services through which income brought in by some from the formal sector of the economy is distributed to benefit the community; and reciprocal assistance networks which, in practice, constitute highly flexible, low cost "informal" social security systems.

In political terms, the kind of autonomous cooperative development in marginal areas of cities is likely to reach local consensus and pluralistic organizations which, although changeable in their concrete expressions may come to be stable. This development contrasts with overall Latin American political systems which are extremely unstable since the social consensus needed to back sustainable growth is seldom reached.

The positive performance and results of the peripheral neighborhood remind us of the general principles that have inspired the recommendations of Latin American forums for global policies to tackle the economic crisis in our countries. However, this also reminds us that, as in the case of global development, there are obstacles that "alternative development" of giant cities must overcome.

Obstacles to Alternative Urban Development

Some obstacles are inherent to peripheral settlements. The most serious is insufficient use of the marginal areas' own work force and creative resources. Horizontal cooperation, which might contribute to overcome this obstacle with local effort, spreads only in cases of extreme emergency. This situation is the case of reciprocal assistance networks. However, in such important spheres as economic subsistence and housing, individual actions and initiatives prevail (individual insertion in the "informal" economy circuits; individual self-help construction). The rest of the obstacles are primarily external: urban land and housing markets, access of local products to the market, political systems, and access to public investment and programs.

Urban Land and Housing Markets

Latin America's urban poor regard the land as a condition for survival and the vehicle for their integration to the economy of the city, whereas land agents of the formal urban sector consider it a source of wealth. We have thus two opposing social strategies vis-à-vis the land, whose point of conflict is price: a rise in prices is the necessary condition for profits and, at the same time, the main obstacle for the poor to have access to this basic commodity.

In the giant cities of Latin America, the land money-making strategies are controlled by the financial capital of the real estate promoter. The justification of financial investment in real estate is not the "productive" profit generated by construction itself or the normal interest rate that may be obtained from money loaned. What attracts financial capital is the chance to get a share of revenues from the land. These revenues become larger with each urban development operation because of the changes in land use that each operation implies. By integrating in one single project different land-related operations from land development to sale of finished houses, the real estate promoter gets the entire land rent increase.

Profits made by the promoter depend on two factors. The first is the sale price of finished houses. The higher the price, the higher the remainder after deducting construction costs, subcontractor fees, financial costs, publicity expenses, and other costs. The promoter will try to influence this factor, orientating projects to such effective demand groups as may permit an increased sale price per "improved" square meter.

The second factor affecting real estate profits is that part of the remainder that the promoter must cede in advance to the owner, namely, the price paid for the land to be used in the project. The promoter will attain control over this factor by buying land as much ahead of time as possible to be reserved for future projects.

In sum, it was in such a highly speculative land market that cities became more exclusive of the poor and the polarity of their structure became more marked. The poor were indirectly excluded from the land because of the high prices of houses sold by the promoter and directly excluded by the rise in prices of land held in reserve.[1]

When access to land is blocked and overcrowding becomes the only possible alternative, the territorial base for social integration of families of marginal urban areas is seriously undermined. Under these circumstances a proposal for "alternative development" to be supported by the urban poor would be unrealistic, to put it mildly. A different political-economic context is needed.

The Formal Sector of the Urban Economy

The relation between the marginal sectors and the formal (or modern) city is contradictory: the poor are close at hand but set apart. On one hand, the poor are spatially excluded as a sine qua non for maximizing land revenues; residential segregation is a necessary condition for increased land value—the fuel that keeps the real-estate business going. On the other hand, the "excluded poor" are integrated to urban markets as suppliers of cheap labor and services and consumers of manufactured goods. They are also integrated to the power structures from which they are regarded political clientele.

In general, the labor supply, products, and services from marginal urban areas show an extreme dependence on market fluctuations. Ups and downs in the activity of the formal sector of the economy have serious repercussions on the economic possibilities of marginal urban areas. This situation is aggravated by most of the urban poor having little knowledge of the characteristics of the formal sector and its markets, which limits their possibilities to take advantage of whatever slight and occasional opportunities may arise.

Vulnerability of the informal economy of marginal urban areas vis-à-vis these two obstacles (dependence on fluctuations of the formal sector and lack of knowledge of its characteristics) becomes greater in the case of collectively managed local economic activities (workshops for goods production, labor pools for supplying services, etc.). The individual supplier of labor, goods, and services stands a better chance and shows a greater flexibility to adapt to the small and changeable gaps in the wake of the formal economy. This fact explains why individualism prevails in marginal urban areas, and why its chances to change survival dynamics into development are practically nil.

To sum up, the relations between the formal and informal sectors of the urban economy may be either competitive or complementary and are, in either case, dominated by the formal sector. The challenge is, precisely, to redefine links, mechanisms and dependencies that may reduce vulnerability of the informal sector and may provide a better bargaining position and development prospects. We shall argue that weakening the formal sector as a result of the emergent political-economic scenario may open opportunities for progress in this direction.

Traditional Forms of Political Integration of the Urban Poor

The political action of the urban poor of Latin America has historically fluctuated between clientelism and struggle. In both cases this activity has been mediated and organized by political parties. However, in the case of authoritarian regimes, mediation is also the responsibility of public agencies, either the regular state bureaucracy or state-sponsored ad hoc entities. In these cases clientelism takes the form of a passive "ceremonial participation": small rewards in exchange for applause.

Political struggle in democratic and authoritarian regimes has been channelled through political parties. However, in many authoritarian regimes, activities of these parties are repressed and fear of any form of association spreads to the whole of the population, particularly to the very poor. In these conditions and despite repression, original forms of organization appear in the peripheral sectors out of necessity. Such is the case of local organizations in which the neighborhood becomes the meeting point for collective participation and generation of initiatives aimed at common objectives: survival and avoiding extreme deterioration of the quality of life. With this purpose, the neighborhood's own resources are pooled and there is collective negotiation of external assistance and resources.

Although these new forms of organization and political participation are isolated and loose in many large cities, they are becoming highly significant in influencing future democratic forms. Perhaps the most striking characteristic of democracy in Latin America has been the exaggerated importance assigned to the confrontation of political projects and ideologies to the detriment of the other component of democracy: the capacity to generate an ample social and political consensus.

Thus, political dynamism—nurtured by confronting without consensus—has resulted in chronic political instability. The merit of the new forms of local organization and political activity resides precisely in their being based on the neighbors' consensual interests vis-à-vis integral improvement of the neighborhood. These interests range from housing to increasing income through economic activities—development by cooperation and collective local management. The main obstacles preventing their expansion in the future are the deeply rooted habits which have been the legacy of traditional Latin American democracies.

We should add that it is safe to assume that the initial consensus reached by different political forces, which has permitted the redemocratization of countries such as Brazil, Uruguay, and Argentina will become untenable if grass roots politics remain unchanged. The challenge is, therefore, to manage to spread the new forms of political organization originating in the marginal urban areas and, at the same time, find new forms of combining social participation and political parties.

Urban Policies

Urban policies orientated to the marginal sectors have been dominated, at the diagnostic stage, by the idea of deficit, chaos, and irregularity, and, at the action stage, by the principle of professional autonomy and restrictions imposed by economic and social interest. Let us develop these points.

The idea of deficit prevails when poverty is understood as the mere lack of possessions, disregarding the resources, capabilities, and distinctive features also present in poverty situations. The ideas of chaos and irregularity generally used to characterize popular settlements in the marginal areas of cities derive from ignorance of the laws, processes, and peculiar rationales that shape the behaviors, physical structures, and social relations typical of poverty.

The principle of professional autonomy, according to which the expert knows what is best for his client and has the right to decide, acquires an exaggerated importance when there is a lack of confidence in the participation of the urban poor in finding solutions to their own problems.

Finally, funds for urban policies are always limited by the competition of demands originating in other more powerful social sectors. Also, when decisions are taken to finance a given urban program or policy oriented to the urban poor, it is generally because concrete entrepreneurial interests are being favored, especially building and land development enterprises. This action is presented as a contribution of policies to the general objectives of economic recovery and increasing the number of jobs. This reason explains, for example, the insistence on development of conventional low-cost housing policies despite their acknowledged lack of significance in the context of housing problems of large cities.

Thus, discouragement and the impression that little or nothing may be done to improve matters set in: whatever resources are available seem to be extremely limited to offset growing deficits. Chaos and irregularity prove resistent to attempts to introduce order through official projects and the effect of technical assistance does not transcend the life span of the projects.

In short, urban policies are based on serious misunderstanding of the reality of marginal urban areas. They also suffer from structural shortcomings because of competition for financial resources arising from other more influential demands and to the prevalence of economic interests that distort the social objectives stated in social interest public programs and investments.

The challenge of making the state recover its role of public service requires use of a new approach so that its activities may be directed to promoting self-supporting development. This challenge however, is not a merely technical issue.

The New Historical Scenario Affecting Development of Large Cities

From the beginning of the present decade a new historico-political scenario has been taking shape in Latin America. This scenario will condition efforts made to influence the form and structure of urban growth. Its characteristics are the widespread economic crisis and the tendency to political democratization in countries which until recently were governed by military regimes. Experts on international economics agree that effects of the crisis will last for more than a decade.

The international crisis reduced considerably the flow of imported capital while, at the same time, it produced a rise in interest rates in the new loans negotiated to repay interests on original loans that financed consumerism waste and land speculation. Paying interests on its debt only, meant that Latin America had to spend nearly 50 billion dollars in 1985. Moreover, to guarantee that the debt is wholly repaid and maturities are met, the creditors, through the International Monetary Fund (IMF), have imposed a series of internal measures among which is reduction of public spending. This reduction mainly affects investments in infrastructure and urban services.

We must not forget that this restrictive context will be characterized by prevalence of basic urbanization tendencies in Latin America. Large cities will continue to grow. Population projections for the 1980s prepared by the United Nations Centro Latinoamericana de Demografia (CELADE) point out that the 20 larger cities (all with over a million inhabitants in 1980) will grow at an average rate of 4.36 percent. This growth is equivalent to an average of 4.6 million new inhabitants per year and an average growth of 230 thousand new residents per city. This growth rate means that, on average, those 20 cities will have doubled their populations in 16 years. To a large extent, their growth will continue to be produced by migration from the country and smaller cities.

Growth will be even more evident in the case of the urban population in the lower income brackets.[2] The number of urban poor in large Latin American cities will grow at a rate that will double the city's growth rate taken as a whole and projections for the next 15 years show that this tendency is likely to increase.

It is expected that by the year 2000 the four large cities—Santiago, Lima, Bogota, and Caracas—that in 1980 had populations ranging between three and five million, will have become giant cities like São Paulo, Mexico, Buenos Aires, and Rio de Janeiro. All will be populated by a vast majority of urban poor living in irregular settlements. When irregular settlement becomes the rule rather than the exception, the time may have come to modify our concept of city development.

Not even authoritarian regimes will attempt to tackle the crisis in the giant cities with merely quantitive measures such as reduction of public spending and other similar measures. Fundamental qualitative changes will be needed so that these changes may bring about a new development modality. In all fairness, this new development is just beyond the reach of authoritarian regimes such as traditional democracies. Perhaps the proof of this situation is that since the onset of the economic crisis, military government after military government has been replaced by democracies which, in turn, foresee serious political difficulties handling the inherited situation.

Together with generating new, acute problems, the economic crisis has opened up the field for fundamental changes in public policies and social conducts. Three changes are relevant to our proposal for alternative development.

Opportunities for State Intervention in Land Development

The crisis reduced and, in some cases, wholly arrested the growth of aggregate demand for land and, consequently, land prices and prospects for promotional gains. For this reason in some giant cities there has been a general fall in land prices comparable to situations of financial panic in prewar periods.

The drop in prices may produce important changes in land money-making strategies. Before the crisis that has been affecting Latin America since 1982, it was always more convenient for land owners to engage in land speculation, and nothing prevented them from internalizing the unearned increment derived from public investments. In situations of contraction of demand and public spending, it may be more attractive for the land owner to negotiate with the state rather than with the real estate promoter. This situation is even more valid if public investment decisions related to infrastructure are used as incentives to land development projects of corporations representing public as well as private interests.

Decentralization in Urban Planning

Public officials should get used to the idea that budgets for infrastructure and services will become more and more reduced. Given this reduction of public spending on infrastructure and urban services in giant cities that will double their population in a brief period of time, governments have two tasks:

(1) Orienting growth of giant cities toward urban structures capable of minimizing indirect costs by making use of the state's comparative advantages in coordinating and promoting investment decisions of different social agents. Coordination is facilitated by the concentration of urban growth decisions in a reduced number of agents whose behavior is easily foreseeable, since many are state entities, and because of the weakening of real estate promoters resulting from the impact of the economic crisis on the aggregate demand for land.

(2) Using whatever scanty funds are available to support local initiatives capable of mobilizing local resources and organizing capabilities in a complete and steadily flowing way. Experiences at a microeconomic level hint at the possibility of great achievements in the economic and political efficiency of families in the marginal areas of the cities. This situation is especially so if the individual and reactive nature of self-help local development is susceptible to being replaced by collective programmed action, of which community organizations may become the nuclei.

Similar development possibilities may emerge from cooperative management of thousands of individual public transport enterprises that compete for passengers in city streets. The expense involved in constructing several underground train transport systems, the preferred solution of the precrisis governments, might never have been avoided if a different alternative, based on coordination of numerous small enterprises engaged in overground public transport, had been opted for.

There are also prospects that could prove fruitful vis-à-vis programming the time of use of cities' material facilities. These facilities are submitted to successive over- or underuse every day, week, and month. Such a plan might make it necessary to build many new roads—or add new lanes to existing roads—and buildings.

Strengthening Community Organizations

Survival mechanisms contain the seeds of the qualitative changes for the whole of the society. These mechanisms appear to have become stronger with the crisis. However, there is the risk of their becoming weaker (with nothing to take their place) if the new democracies, as in the past, fall into the trap of generating expectations of action by the state instead of opening up the field to a variety of social change agents, including grass roots organizations.

In this connection, several questions come to mind: What will become of the survival mechanisms based on local effort in the face of new opportunities for democratic participation? Are the mechanisms used merely for temporary adaptation, to be dropped as soon as possible? Or are they permanent cultural expressions of poverty, transmitted from one generation to the next, as stated by Oscar Lewis? If this is so, these mechanisms will exist for as long as there is poverty, but for survival and not for development.

Our views differ from those of Lewis: the cultural patterns of the poor show no substantial differences from those of the rest of society; their survival mechanisms are of a merely adaptive nature and respond to the difficult living conditions they have to contend with. When real life conditions are incompatible with realization of the overall cultural pattern, survival mechanisms emerge.

In conclusion, what distinguishes the Latin American urban poor is not their scheme of values and ends but *the means* to reach the latter. In their fight for survival their identity is related to means, not ends. The ends they aim toward are, on the whole, to become integrated to the urban markets and power structures from which they have been excluded or placed in a position of disadvantage. Our thesis is that this integration will never be the result of individual effort, as posited by ideologies based on the compromise between political authority and a free market.

On the contrary, we believe that integration can only materialize if it is undertaken collectively by means of grass roots organizations. Only organizations may guarantee the poor a certain degree of efficiency in the use of their scanty resources for development and the necessary political efficiency to negotiate resources with authorities and the market. Until this moment, collective action has been a mere means of survival. Our purpose is to give it the necessary organization to change it into a means for development.

Possible Lines of Action

The combination of economic crisis and return to political democracy opens up possibilities for proposing another type of urban development to become workable in Latin America. This scheme depends on the use made of the historical opportunity that the conjunction of crisis and redemocratization represent for recovery of the role of the state as promoter of development. It goes without saying that state participation this time would have new orientations because of changed circumstances.

We propose three lines of action for the public sector, whose viability increases with the new context and which would contribute in a very important way to transforming the survival mechanisms of the urban "peripheral" poor into alternative development mechanisms.

The first line of action aims at interrupting the existing connection between the process of urban concentration and rises in land prices by means of planning measures. These measures should be part of a plan that considers the whole of the metropolitan region. The main objective of the plan would be to remove the pressure exerted by demand for urban land in the capital city by diverting growth toward medium or small-sized cities of the metropolitan periphery. This plan has been called decentralization within the core region.

In almost all core regions of Latin America there are, besides large metropolitan areas, medium-sized cities with 20 to 100 thousand or more inhabitants at a distance of less than 200 kilometers from the center. Some of these cities could be declared priority urbanization zones and could attract some of the economic activities and population that would otherwise continue to crowd the capital city. The public cost measured in terms of subsidies for a policy of decentralization from within the center would be much lower than the cost of interregional decentralization policies, which, in any case, have proved to have little success so far.[3]

The main problem of a policy of decentralization within the core region is how to avoid land speculation in priority urbanization zones. If land speculation occurs, it would impose additional costs that would render the policy of spatial decentralization unmanageable. There are good reasons to think that it is politically more feasible to apply instruments of state control on speculation in land in the periphery of the core region than in the periphery of the metropolitan area. In the latter, the eminence of urban growth stimulates expectations vis-à-vis increased value of the land in anticipation to its urban use. In the regional periphery this kind of expectation does not exist until a policy for spatial decentralization is announced. Socialization of the unearned increments resulting exclusively from public action depends, therefore, on the thorough and integral design of the state policy.

For the poor, whose accessibility to land has been steadily deteriorating and has reached limits that pose a serious threat to their quality of life, decentralization from within the center represents an advantage for those who remain in their present locations in the urban periphery and for those who settle in the priority urbanization zones of the regional periphery. In the former case, there would be advantages because the interurban transport linking the urban and regional periphery could be less expensive than intraurban transport in terms of money, time, and trouble. In the latter case advantages lie in the reduced size of cities in the regional periphery.

The second line of action aims at facilitating security of tenure of accessible and served land. Location of a site that meets the requirements of safety, services, and accessibility is the necessary condition for survival mechanisms to appear and develop. The aim should be, then, to substitute the principle of the right to housing, so often used as a political platform but never put into practice by the governments, for a more modest one, within the reach of public resources: the right to the land. If the more progressive governments in times prior to the crisis were unable to build enough houses for everybody and if the houses those governments do build do not satisfy the needs of the users, it stands to reason to reorient public spending toward the generation of necessary minimum conditions for families themselves to take an active part in building their own cheaper homes. The reorientation of public policy from house-product to land-resource is no trivial matter: it implies substantial changes in the very concept for urban development.

The concentration of attention on house-product led public-housing institutions to look for the cheapest possible land, which was the most distant and poorly served. All this was done without a global plan incorporating indirect costs (public and private), accessibility, and maintenance costs of the dwelling units. The urban structure was the residual result of public and private housing projects that lacked coordination in time and space.

With plans aiming at providing land for the majority rather than houses for a few and with the land speculator withdrawing from the scene, the state is in a privileged position with respect to negotiating with landowners and structuring growth of the city. Moreover, planning offers a wide range of alternatives for association with private individuals or entities for provision of services, thus avoiding excessive public action.

Once the state has the power to affect land prices, public-housing development decisions would not be based on acquiring the cheapest land but on acquiring land that meets the requirement of accessibility and low-cost services. At the same time, it will be the state and not the land speculators who reap the benefit of the external and complementary economies resulting from planning.

The third line of action is to support local actions and collective organization of urban development. Local urban development includes housing, facilities, production of goods and services, training, cultural activities, child care, and so on.

Only some survival mechanisms deal with these crucial aspects on the basis of reciprocity and cooperation and very few have incorporated programming methods for their integral development. Moderate progress in this respect might enormously increase efficiency in the use of the scanty resources of the poor and the political efficiency to obtain complementary resources from authorities.

This is not to say that self-help construction is a panacea without problems. For example, families often pay high prices for materials purchased individually in the nearest place and they frequently lack the range of skills or abilities required by house building. Also, given that this is an isolated effort, the degree of self-determination of the settlement is restricted by the difficulty of access to external resources and by family action being limited to one single allotment: the conformation of the neighborhood occurs either in a haphazard way or is determined by external forces. Finally, families move from one stage into the next one without anticipating what may happen in the stages following construction or the indirect effects that their housing decisions may have on the neighborhood or area. Thus, as the different stages are covered, development options become more restricted because of previous action taken by the families themselves with respect to house building and local development.

To overcome these problems it is necessary to incorporate planning (until now virtually reserved for the formal sector) in self-help housing developments. However, this planning cannot follow conventional patterns. Unlike government institutions or agencies, community organizations lack budgets, cadres of experts, statutes, and bureaucracy. Their only resources are labor and the will to become organized in a collective way. The limitation resides in overcoming the barriers that prevent poor families from forming politically mature representative organizations.

It is precisely this third line of public action that must start by stimulating the local capacity for the formation of organizations that meet these conditions. This line of reasoning is valid not only for self-help construction but for the whole of the survival mechanisms and, particularly, for those where individual effort prevails.

The consolidation of local autonomy in urban development is incompatible with conventional technical assistance. Conventional help is based on centrally planned homogeneous solutions, which typically do not coincide with local needs, aspirations, and resources which differ from one community to the next. Technical solutions should be sought *with* the organized community to permit the emergence of a large diversity of local developments and of integration modalities between the community and central policies. The new relationship will be based on the reciprocity generated when all realize that local government is a process of shared learning in search of common objectives of qualitative changes, in which the decentralization of decisions will work to the benefit of both local development and the general consolidation of democracy.

`This new type of technical assistance appears to be indispensable to overcome the isolated nature of other efforts originating in the base—which for this reason are restricted to mere survival. In such different areas as self-help building and local production of goods and services, formation of collective organizations based on cooperation and reciprocity produces better results, from the qualitative and quantitative points of view. The few experiences worth noting show that relative success involving any issue immediately attracts the interest of the population to apply this experience to other problem areas, thus giving impetus to innovative forms of local development.

Notes

1. As a paradox, economic growth—resulting from promoting the land market—brought about the emergency of social exclusion from access to the land of such a generalized nature as unparalleled in the history of the extra large cities. This phenomenon took place because the economic growth corresponded, to a large extent, to expansion of financial capital, which implied a growth in demand for urban land, both indirect (housing loan systems) and direct (speculative land investment).

2. The urban poor are that segment of the city population who are excluded from the land and housing markets and whose integration to the consumption and labor markets and political life is

one of inequality—all of which contribute to the formation of parallel survival economies.

3. In cities like São Paulo, costs of congestion are so high that decentralization toward the periphery of the core region has already been taking place without benefit of explicit policies. However, in smaller cities like Santiago, Bogota, Caracas, and Lima, indirect subsidies for decentralization within the core region in the form of investments in infrastructure would be unavoidable. But in any case, this kind of investment would represent a much lower cost than investment in regions a long way from the capital.

Bibliography

Geisse, Guillermo. 1984. "Conflicting Land Strategies in Large Latin American Cities." *Land Use Policy* 1:309-29.

Hirschman, Albert O. 1984. *Getting Ahead Collectively: Grassroots Experiences in Latin America.* New York: Pergamon.

Iglesias, Enrique V. 1981. "Development and Equity: The Challenge of the 1980s. *CEPAL Review* no. 15(Dec.):7-46.

Lewis, Oscar. 1966. "The Culture of Poverty." *Scientific American* 215(10):19-25.

Lomnitz, Larissa Adler de. 1975. *Cómo sobreviven los Marginados.* Mexico City: Siglo Veintiuno Editores (in Spanish).

———. 1979. "Organización Social y Estrategias de Sobrevivencia en los Estratos Marginales Urbanos de América Latina." In *Proyecto CEPAL/PNUMA, Estilos de Desarollo y Medio Ambiente en América Latina.* Santiago: Comisión Economica para América Latina de las Naciones Unidas (in Spanish).

Mathur, Om Prakash, and Caroline O. N. Moser, eds. "The Urban Informal Sector (Role of the Informal Sector in Centrally Planned, Socialist Economies and Developing Countries and Its Relationship to City-Size). *Regional Development Dialogue* 5(2).

Portes, Alejandro. 1985. "Latin American Class Structures: Their Composition and Change during the Last Decades." *Latin American Research Review* 20(3):7-39.

Rodriguez, Alfredo. 1983. *Por una Ciudad Democrática.* Santiago: Editorial Sur (in Spanish).

Singer, Paul, and Vinicius Caldeira Brandt, eds. 1980. *São Paulo: O Povo em Movimento.* Petropolis, Brazil: Vozes (in Portuguese).

Tokman, Víctor E. 1983. "Wages and Employment in International Recessions: Recent Latin American Experiences." *CEPAL Review* no. 20(Aug.):113-26.

Tokman, Víctor E., and Emilio Klein, eds. 1979. *El Subempleo en América Latina.* Buenos Aires: El Cid Editores (in Spanish).

Wasserstrom, Robert. 1985. *Grassroots Development in Latin America and the Caribbean; Oral Histories of Social Change.* New York: Praeger.

12

Vulnerability of Giant Cities and the Life Lottery

Ignacy Sachs

Like people, cities have personalities. Each represents a unique mix of history and natural setting, cultural patterns, and life-styles. Some are ugly yet attractive, others beautiful but dull. Under such circumstances modeling and theorizing about cities is risky, if even possible, as Italo Calvino (1974) reminds us in the imaginary dialogue of Khublai Khan with Marco Polo:

> "From now on, I'll describe the cities to you," the Khan had said, "in your journeys you will see if they exist."
>
> But the cities visited by Marco Polo were always different from those thought of by the emperor.
>
> "And yet I have constructed in my mind a model city from which all possible cities can be deduced," Kublai said. "It contains everything corresponding to the norm. Since the cities that exist diverge in varying degree from the norm, I need only foresee the exceptions to the norm and calculate the most probable combinations."
>
> "I have also thought of a model city from which I deduce all others," Marco answered." It is a city made only of exceptions, exclusions, incongruities, contradictions. If such a city is the most improbable, by reducing the number of abnormal elements, we increase the probability that the city really exists. So I have only to subtract exceptions from my model, and in whatever direction I proceed, I will arrive at one of the cities which, always as an exception, exist. But I cannot force my operation beyond a certain limit: I would achieve cities too probable to be real."

With this caveat one feature common to practically all large Third World cities can be observed. They act like a gigantic Las Vegas in the sense that the bulk of their populations are gamblers, though the games are different.

Instead of roulette or blackjack, their names are job security, individual social mobility, better access to education for the children, and hospitals for the sick. Wonderful stories circulate about the happy few who made it in a big way.

There are also consolation prizes for many: *panem et circenses* (bread and circuses) play in modern Third World countries a role at least as important as in ancient Rome (see Veyne, 1976). Governments are afraid of urban mob violence, so they subsidize bread and occasionally distribute food to the destitute to keep them quiet. The International Monetary Fund (IMF) may be releasing from the bottle a dangerous djinn through its stubborn insistence on abolishing food subsidies—recall the recent food riots in Egypt, Tunisia, and Morocco, as well as the earlier experiences in Poland (where the IMF had no responsibility). Access to food for the urban poor may be very limited, but at least they do not starve to death as a mass phenomenon once they reach the city. The poor suffer and starve on an individual basis[1] which represents a security valve for governments. As for circuses, large cities offer free of cost the daily spectacle of themselves, the opportunity for people to mix with so many other people even if it is a lonely crowd, not to speak of pageants, carnivals, military parades, public executions, soccer championships, and, increasingly, television.

Nouakchott, capital of Mauritania, is not a giant city—around 300 thousand people as of 1981. But 20 years earlier, it had barely a population of 6 thousand and the latest government projections put it between 650 and 850 thousand at the turn of the century. Unless drastic changes occur, Nouakchott will become one of the 77 African cities with populations projected to exceed a million by 2010. Meanwhile Nouakchott deserves closer scrutiny because the problems it faces now are a preview of how it will look as a metropolis.

At present 37 to 42 percent of Nouakchott's population live in shantytowns and another 31 to 35 percent in organized settlements—for all practical purposes as deprived in services as the shantytowns. An ambitious program of site sales and sanitation proved a failure: the first seven thousand parcels of land were immediately sold out by people who found it more profitable to get some cash and move to new shantytowns. No more than 7 percent of the population can afford to pay for the low-cost and very poor quality housing built through official programs; the cost per square meter equals a worker's pay for four months.

The situation is even worse for water supply. Only one family in ten has access to piped water. Peripheral quarters are served by 41 water points (an average of one for 850 families) which means that many among the poorest people are forced to buy water from water peddlers, paying 7 to 40 times the original price. Under such circumstances, average water consumption in the poorest sections of Nouakchott does not exceed eight liters per person per day. No wonder one of every four infants born in the shantytowns soon dies.

The acute water shortage also prevents people from cultivating vegetable gardens. For food, the poor depend on their earnings and charity. However, of the economically active population—conservatively estimated at less than 40 percent of the total—little more than a third have stable jobs. More than half the families living in shantytowns do not have one single person stably employed; 39 percent of the families eat only once a day.

Why, given this catastrophic situation, do people continue to flow to Nouakchott? Conditions elsewhere may be even worse. Nouakchott acts as a distributing center of public assistance. In the capital one eats precious little, but at least one eats once a day thanks to the food distribution and its sale at highly subsidized prices. Access to health services and education is far better than in the rest of the country: 85 percent of the national health budget is spent in the capital (the central hospital alone absorbs 30 percent of the outlay) and more than half the children go to school. Over half of Mauritania's total public investment is concentrated in the capital. The ratio between public investment per capita in Nouakchott and the whole country is 5:1.[2]

The rewards may look insignificant when compared to the excessively high price of the tickets in terms of daily life difficulties: appalling housing conditions, no sanitation, long hours of commuting for the lucky minority with stable employment, or high rates of underemployment or unemployment for the remainder. What are the rewards for people transformed into jacks of all trade who compete for an odd bit of scrap or occasional job? What is the reward for the 300 thousand peddlers crowding the streets of Lima?

The important thing is that there are rewards for some. The large cities are places of hope, while the drudgery of rural life looks hopeless. The first to emphasize this life-lottery aspect of the city was perhaps the 19th-century French historian Jules Michelet.[3] This interpretation of the urbanization drive in Third World countries which has already accomplished in a matter of a few decades a dramatic social transformation has a number of implications.

First, urbanization is likely to continue unless radically different strategies of rural development are urgently put into practice, resulting in rapid improvement of social, cultural, and economic conditions for people living in Third World villages and small towns.[4] We need to revise in several aspects the simplified concept of the division of labor between cities and the countryside, the former producing culture and industrial goods, the latter feeding the cities and supplying raw materials. At least two points should be raised in this connection.

Jane Jacobs (1984) may be wrong in her unilateral celebration of cities as prime movers of economic development. Braudel's interpretation (1979) is much more subtle. Whatever the wonders achieved by cities with the economic surplus they have been able to concentrate, one should not forget that primitive accumulation was largely through extracting this surplus from the peasantry of today's industrialized countries and their colonies.

Urbanization happened exactly the same way in the Soviet Union. The debate between Bukharin and Preobrazhenski (1922) centered on alternative methods, but not on the principle. Both agree there can be no other way of industrializing, urbanizing, and thus modernizing a backward, predominantly rural country than by making peasants foot the bill. Furthermore, there is today ample evidence of the "urban bias" in policies of many Third World countries, such as India (Lipton, 1977). While the surplus is extracted from the rural populations, the investment thus financed and most economic, financial, and foreign exchange policies are geared to interests of the urban minority, more exactly to those of the affluent urban upper classes. The recent urbanization processes which in a matter of 40 years inverted the ratio of rural to urban population from 2:1 to 1:2 in Brazil did not arrest urban bias. The reversed ratios only make it less efficient in spite of the rapid rate of technical progress in mobilizing the necessary resources to provide a minimum of services to urban dwellers, particularly the urban destitute.

A corollary to questioning the existing dichotomy between cities and countryside is the postulate of thinking in terms of rural development in which industries and even tertiary activities may play a major role. As Ismael Sabri Abdalla (1979) put it, densely populated and still predominantly peasant countries must seek a strategy of industrialization without depeasantization. China has made a great effort in this direction, but recent drastic changes in its development policies cast a doubt on the much-advertised success of the previous orientation (see Koshizawa, 1978; Murphey, 1980). More generally, what is needed is a carefully designed strategy of *aménagement du territoire* (land-use planning), aiming at an industrialization pattern that instead of disrupting takes advantage of the existing web of villages and small towns: the *industrializzazione senza fratture* (industrialization without fracture) of the kind that occurred in northeastern Italy is a good example (Fuà and Zacchia, 1983). Recent progress in telecommunication and computers makes obsolete all our concepts of positive externalities arising out of extreme spatial concentration of secondary and tertiary activities. We are only beginning to realize the implications of these breakthroughs.

If the life-lottery analogy is correct, we should expect that metropolitization will advance at an even greater speed than urbanization. People will tend to go to the capitals and large cities because of the illusion that more winning tickets are there. When news of an industrial boom spreads through a country, for every new job offered several newly-arrived candidates apply in addition to those already living in the town. When recession sweeps a country, moving to the largest city appears as a solution of last resort. So, whatever the ups and downs of the business cycle, migration continues. This has been clearly observed in São Paulo during the present crisis.

Large cities absorb above all the young and enterprising labor force. The distribution of costs and gains between the hinterland and the large city is thus once more biased in favor of the city: the countryside bears the social cost of bringing up this labor force. The benefit of their work accrues to the cities, except for the remittances of part of their meager pay to families left behind.

History and the Future

Demographers are not particularly good at forecasting, in spite of an entrenched prejudice in their favor because they were probably the first social scientists to discover the power of the compound rate of interest and the possibility it offers to produce striking projections which should not be, however, mistaken for previsions. Most demographic projections should rather be read as scenarios of the impossible. With this caveat, let us hope that present trends will slow down before the turn of the century, even though chances for a less dramatic turn of events are very thin.

Despite some superficial similitudes, the present situation in Third World large cities is quite different from the one experienced in the course of fast urbanization in Europe and the United States. Common features are the appalling living conditions of the majority of urban dwellers: in ancient Rome as documented by Carcopino (1939), in late 18th-century Paris as described by Mercier (1979), in 19th-century English cities evoked by Dickens and analyzed by Engels (1973), in Moscow which so much shocked Tolstóy (1886), or in early 20th-century Chicago, as it emerges from the novels of Dreiser (1900) and Farrell (1935). But the analogy ends there. History should never be used as an excuse to impose social inequity, nor as a source of models that ought to be repeated as if there existed only one universal development (or maldevelopment?) path (Sachs, 1984). On the contrary, the only progressive interpretation of historical experience leads us to consider past experiences as antimodels that can be surpassed.

At stake for Third World countries is the opportunity to transform their condition of latecomers into an advantage. Modern science and technology associated with a critical analysis of the impasses of industrialized societies should allow Third World countries to find alternative patterns of urbanization and to implement them at a far lesser social, ecological, and economic cost (see Blair, 1984). This is not to say that substantial progress could not be achieved in Third World countries through more traditional and Western-like methods. But the magnitude of financial effort required makes this proposition utterly unrealistic in the present political context, even if the volume of necessary investment does not exceed the theoretical possibilities of the world economy. The range of estimates varies from author to author. The figures below should be read only as an order of magnitude.

By and large, the urban population in developing countries is expected to grow by over 1.15 billion people between 1980 and the end of the century. This calls for building the equivalent of one Mexico City every three to four months! Owen's figures (1979:17) are even higher:

> In the last two decades of this century, the world will need housing, utilities, schools, hospitals, and commercial structures equivalent to building 1,600 cities of a million people each. To these growth requirements will be added the backlog of needs for the existing population, which numbers over a billion people more.

Satellite towns in Hong Kong were expected to cost in the mid-1970s $3,000 per capita, equally divided between public and private sectors. Figures from Singapore were about one-third lower. A very conservative estrimate of $1,000 of public investment per urban dweller, to be matched at least by an equal sum from private sources, would put the total necessary investment for 1.15 billion new urbanites at $115 billion per year, including provision for the backlog of necessary urban investment. This looks at first sight to be a staggering sum of money when compared to the present shortage of capital in Third World countries, and indeed it is. The shocking fact is that this figure is roughly only a third of the world's annual armament budget.

The draconian rules of the debt service imposed by the IMF and Western banks transformed Latin America into a net exporter of capital to the tune of 30 billion dollars per year, nearly enough to take care of the population of one new Mexico City. The planning secretariat of São Paulo has recently estimated that the equivalent of 12 annual city budgets—some seven billion dollars (less than $1,000 per capita)—would be sufficient to cope with the deficit in urban infrastructure (Wilheim, 1984). There is not the slightest chance of getting these sums, not because of their absolute magnitude but because of misplaced priorities in the world economy. In my opinion, Arthur Lewis (1977) is right in insisting on the heavy burden cities impose on Third World economies while arguments to the contrary advanced by a World Bank expert, Johannes F. Linn (1982), are less than convincing.

Another factor in urban expenditure, to my best knowledge not estimated for Third World large cities, is maintenance of the existing infrastructure. Judging by American standards where the backlog of infrastructural maintenance and repair investment—urban transport and rural—has been estimated at three trillion dollars (Beck et al., 1982), the figure must be quite high.

However, the main difference between Third World and European cities is the speed and scale at which they are developing. It took London 130 years to grow from one million people to eight million. Mexico City did it in 30 years from 1940 to 1970 and doubled again in 15 years (Pearce, 1984).

Robert McNamara (1984: 11) emphasizes the same point:

> Huge urban agglomerations are, of course, known in the West: the New York-northeastern New Jersey metropolitan area, or Tokyo-Yokohama, both with populations close to 20 million, or Los Angeles and London. These are now, however, growing slowly if at all. They have built up housing stocks, physical infrastructure, and public amenities over many decades of heavy investment—yet their maintenance problems are acute. The giant Third World cities—Mexico City . . . , São Paulo . . . , Shanghai . . . , Bombay and Jakarta . . . , and so on—will have doubled or more in the last quarter of this century. These sizes are such that any economies of location are dwarfed by costs of congestion. The rapid population growth that has produced them will have far outpaced the growth of human and physical infrastructure needed for even moderately efficient economic life and orderly political and social relationships, let alone amenity for their residents.

Our modest European experience is not of great value to Third World urban planners and managers who must work in much more demanding and difficult conditions.

The Failure of Urbanists

The postwar record of urbanism, with few honorable exceptions, is dismal. While dreaming of fantastic cities of the future, urbanists failed to foresee the patterns of postwar demographic trends and the preeminence, soon to be taken outside the United States, of the individual motor car. For example, much money was spent in cities like Osaka and Warsaw, completely destroyed by the war, to make up for the lack of vision of the first generation of urbanists who undertook their reconstruction. Urbanists, like economists and generals, were ready for the last battle they won—the few progressive realizations of Bauhaus in the late 1920s. The social rhetoric of the Charter of Athens served more as a screen to hide their fascination with new building materials, industrialized construction methods, and spatial and architectural aestheticism rather than as a pointer to look at the real person in the streets (see Woods, 1975). Urbanism continued to be taught and practiced more as a province of architecture than as a nodal point of all social sciences. In their conceptions of society and human needs, most postwar urbanists demonstrated the same mix of naivete, dogmaticism, and lack of interest in empirical evidence about people's life-styles as the protagonists of the discussions held in the Soviet Union in the early 1920s (Kopp, 1975).

As a result of all this, new towns planned for Third World countries became gigantic social failures, even though they may hold outstanding works of contemporary architecture. Thus, Chandigarh has been conceived as a very pleasant ghetto for civil servants, carefully segregated according to their rank. Ordinary people were simply forgotten and relegated to neighboring shantytowns. The same happened in Brasilia in a much larger way, with the difference being that Brasilia was conceived first for the bureaucrats' and their families' cars, becoming the model of an aesthetically attractive anticity, a place where people normally do not communicate except within their narrow neighborhoods and where there are fewer people in the streets than cars on the highways. At any rate, the bulk of the working people do not live in Brasilia but in the so-called satellite towns, which were not originally planned and whose populations now exceed one million, compared with the 300 thousand odd inhabitants of Brasilia proper.

The satellite towns are booming with life (and real estate speculation); their streets and shops are always full of people; they exhibit the vitality so characteristic of many Brazilian cities which just happened and were at best submitted to a measure of ex-post urbanistic intervention. The reverse of the coin is the high cost in time and money of commuting daily to Brasilia for work. Shop employees and junior office clerks spend a substantial part of their meager salaries on transportation. The bureaucrat's paradise is for them a five-day weekly purgatory where they expiate the planners' sins.

This being the situation in carefully designed and, in any sense, privileged cities because of their symbolic and political importance, it is easy to imagine in what conditions the bulk of Third World urban dwellers live and work. In practically all large cities the population increase outplaces the rhythm of expansion of infrastructures and services, of which a substantial part accrues to the central city—the rich residential quarters and the monumental government and large business office center. Most housing for the poor is built on the outskirts illegally, violating in one way or another the unrealistic legal requirements and building codes. Other people have no refuge except the shantytowns.

As for minimum public service provision, piped water and sewage systems lack the dimensions to meet rapidly growing needs. In Cairo, for example, 53 percent of the dwellings lacked sewage facilities, 46 percent had no water, and 25 percent no electricity in 1976. Cairo's average population density is 25 thousand people per square kilometer, but in some sections it exceeds 100 thousand. In Mexico City, one of the most affected by environmental disruption, drinking water supply poses major problems. Part of the water comes from sources a thousand meters below the altitude of the Valley of Mexico. In another ten years water will be brought from a distance of three hundred kilometers and lifted up two thousand meters at a staggering energy cost. Meanwhile some 40 percent of piped water is lost because of leakages.

Mexico City's transportation systems are faulty[5] and ill maintained—a serious mistake because better maintenance could save much money while creating jobs for maintenance workers.[6] Frequent traffic accidents and fires take a daily toll of human lives.

Epidemics from waterborne diseases are still very much a danger despite reduced urban mortality in recent decades. We should not forget that until the beginning of the 20th century, all cities in the world had a negative balance between births and deaths (see McNeill, 1976). When will the next wave of the Black Death come? Epidemiologists have no explanation for its sudden disappearance in southern China some 60 years ago (Boyden et al., 1981). Will it ever resurge at some point on the globe? What will be the capacity of modern medicine to cope with such an epidemic? Some scientists expect now a deadly blow of acquired immune deficiency syndrome (AIDS) on crowded Third World cities (Seale, 1985). Dwellers of giant Third World cities live dangerously indeed. The real price of the life-lottery tickets is much higher than it looks on face value.

Sometimes there are terrible accidents like the 1984 explosion of a gas storage facility in Mexico City or the Bhopal tragedy in India where 25 hundred people died and many more were disabled for life because of careless management of a pesticide factory by a multinational company. This dreadful industrial accident—the worst recorded up to now—raises many questions. Where should potentially dangerous manufacturing operations be sited? Under Indian conditions, it was evident that any factory in an urban or suburban area would be immediately surrounded by a populous shantytown. What kind of security measures should have been taken, first to prevent the accident and then to prepare the population for emergency?

Accidents like the ones in Mexico City and Bhopal can happen wherever potentially dangerous industries are sited, even in the most industrialized countries (cf. Lagadec, 1981). We should harbor no illusion about that. Our scientists and engineers tell us the probability of such accidents is infinitesimally small—this is their way of saying that with bad luck they *can* happen at any moment anywhere, even in the absence of such events as a major earthquake or sabotage.[7]

High Vulnerability and High Resilience

Because of their complexity, the life-supporting systems of large cities are highly vulnerable. Roberto Vacca, a systems engineer and fiction writer, imagined a scenario of doom for New York to emphasize the vulnerability of high-tech systems: two jumbo jets crash in a snowstorm and fall on the energy grid, plunging New York into cold and darkness for two weeks. Careless heating with old furniture transformed into fuel wood causes fires. But the streets are jammed with abandoned cars and fire fighters cannot intervene. Several million people die from fire, cold, epidemics, and violence. New York's Puerto Ricans manage to survive better because they are used to harsh life conditions (Vacca, 1973). Other sorts of catastrophies, including social upheavals, have been discussed in the scientific and science-fiction literature. For Soedjatmoko (1984), rector of the United Nations University, "events in the Western ghettos are but a rehearsal of more serious troubles that will affect all the agglomerations in the Third World. Violence, criminality, armed rebellion and revolution are considered as the only means of survival, the only response to intolerable injustice, oblivion, oppression, and a feeling of impotence" (my translation).

Reasoning along these lines, by any technological standards most Third World cities should have collapsed long ago. To what can we attribute that this is not so? Good luck (a true miracle) or some other reasons?

The UNESCO Man and Biosphere program made a meritorious effort to define the concepts of *vulnerability* and *resilience* in the urban context. The five areas of vulnerability, as seen by its authors, are:

(1) *Biological vulnerability:* susceptibility of human populations to disease, injury, and death, expressed in terms of the proportion of the city's population affected.
(2) *Structural vulnerability:* loss or damage to the physical existence of the city or its stock of real property, measured in proportional terms.
(3) *Life-support vulnerability:* the possibility of the absence, scarcity, or disruption in supply of essential life supports such as water, food, and (depending on climate) clothing or shelter from the elements.
(4) *Economic vulnerability:* the possibility of financial crisis through changed circumstances.
(5) *Functional vulnerability:* the danger that organizational responses required to manage urban problems may exceed functional capacity, causing a breakdown simply because of failure to conceive and implement appropriate strategies (White and Burton, 1983).

White and Burton go on to list the following five resilience factors:

(1) *Accumulated wealth:* the stock of financial or other resources on which the community can draw to meet emergency needs. To the extent a city or nation is less able to help itself, it is said to be less resilient.
(2) *Health:* the level of health as measured by incidence of disease and level of nutrition. Where a population is already at a low health level, it has less resilience on which to draw to cope with the impact and to recover.
(3) *Availability of alternative supplies:* the extent to which supplemental or alternative materials can be supplied or created.

(4) *Skills:* the level of training and education in a population that permits improvisation of responses. Skills are not necessarily the "modern" or "advanced" ones associated with high technology; "traditional" skills associated with less sophisticated forms of technology may also be helpful in certain circumstances.
(5) *Morale:* the spirit of confidence and capability that a community can produce in response to adversity. There are many examples of failure to respond to some cases and the enormous surge of energy and determination to overcome adversity that can occur in others.

I must confess, however, that I have the same kind of difficulty with the concept of vulnerability as with indicators of mortality and morbidity. Too much stress is put on extreme cases of death and statistically recorded illness (which means a physical or health center was consulted) and not enough on middle-of-the-road situations of people who are sick but do not resort to the health service (or else consult practitioners of traditional medicine whose existence is ignored, if not repressed, by official medicine). It is nonsense to measure the mental health of a society (or rather the lack of it) by the number of patients in psychiatric wards, without trying to estimate at least what is the proportion of neurotics. In the same way, one should expand the notion of vulnerability to encompass all the spectrum of locally-bound, small catastrophies and daily crises coped with by individuals, families, and informal solidarity networks (extended family, neighborhood, groups of friends, church congregations, etc.). To survive in a large city means expending day after day a tremendous energy, treasures of ingeniosity, and much time to secure the essentials of food, shelter, and income.

Viewed from this angle, *Third World cities are at the same time highly vulnerable and highly resilient.* But both the vulnerability and the resilience are *individualized.* Two comments are in order here. First, because of this nonadditivity of the most common vulnerability phenomena and people's response to them, there is little to expect except confusing matters from transposing to the city the concept of systems as evolved in science[8] unless explicit mention is made of the scope for human creativity, the dialectics between destabilization and reorganization, and adaptive learning processes.[9] Accordingly, the learning process should provide the main paradigm for creative urban development planning (see Friedmann, 1981).

Second, a parallel can be drawn between the situation analyzed above, in terms of simultaneous presence of high vulnerability and high resilience and the demographic patterns characterized in the past by high fertility and high mortality. By analogy with the concept of demographic transition, we ought to think in terms of *an urban transition resulting in lower vulnerability and stronger resilience.*

The lasting solution can only come from new approaches to land-use planning, resulting in overcoming the present differences between city and countryside and, ideally, a more balanced distribution of world population with respect to world resources. Land reforms, a carefully designed social, ecological, and technological strategy with respect to still existing economic frontiers,[10] and mass migrations on an unprecendented scale from South to

North ought to be the three pillars of such a long-range policy. But let us stop dreaming and try to go back to the hard realities of our world.

The medium-range approach passes through conceiving, designing, and building low-investment cities, which should not be confused with an unrestricted acceptance of so-called intermediate technologies. Richard Meier's work (1974) on what he calls resource-conserving cities is a good example of an approach blending social realism with modern technologies. Charles Correa's pioneering proposal (1973) for New Bombay is another very different instance of inspiring thinking, even though his project has not yet been properly implemented. At a September 1984 conference on Latin American metropolises facing the crisis, held under the joint auspices of the São Paulo municipality and the United Nations University (UNU), it was decided to prepare a state-of-the-art study of Latin American experiences on how to urbanize at low cost.

In the shorter run it should be possible to improve access to food, energy, and shelter for the urban poor by mobilizing the many underused, misused, or wasted physical and human resources of urban ecosystems. Starting from food and energy, the UNU initiated several research activities on alternative urban development strategies, centered on urban agriculture, waste recycling, energy recovery and conservation, self-managed housing construction, and social services production.

The following five themes emerge as central concerns of this program area:

(1) Analyzing the impact of urbanization trends on accumulation processes in the Third World in conjunction with the situation created by foreign indebtedness and the double-bind of urban bias and preeminence of the upper-class consumer city over lower-class residential quarters and the producer city.[11]

(2) Improving knowledge of the working of the urban economies going beyond the traditional dichotomy of formal/informal and trying to understand the complex interplay of all markets (ranging from official to criminal) with household economy and public sector intervention.

(3) Identifying the resource potential of urban ecosystems of utilization, keeping in mind the criteria of social usefulness, ecological sustainability, and economic survivability and looking to the formulation of specific policies aimed at improving the condition of the urban poor.

(4) Looking into the linkages between civil society (as represented by citizen groups, neighborhood associations, church groups, clubs, etc.), the state, and organized business with a double view at making operational, through this kind of institutional analysis, the concept of participatory local development planning, and at clarifying the kind of macropolicies necessary to open up and strengthen the spaces for local autonomy.

(5) Disseminating through a network of collaborating institutions information on concrete experiences in alternative urban development.

These themes and other alternative approaches discussed above represent a beginning. Much remains to be done (in a limited time span) to prevent the vulnerability in today's giant cities in the Third World from being transformed into catastrophic realities.

Notes

1. "For the most vulnerable urban groups, often made up of recent immigrants from the countryside, there exist of course no unemployment benefits. Prolonged unemployment periods, resulting from lack of work opportunities, bad health, or just bad luck, might lead to starvation and death, unless it is possible to have some access to welfare centers or hospitals (most often overcrowded) or to rely on begging, or on some other expressions of solidarity. The latter are not easily found in large cities. Credit is not easily obtained in the absence of security. As a result, deaths occur in large cities of poor continents, with each death surrounded by its own peculiar circumstances. The atomistic nature of these individual tragedies is such that their sum does not constitute a collective phenomenon equivalent to a famine in rural areas" (Spitz, 1984, p. 183).

2. See Theunynck, 1982. For other no less eloquent examples, see the chapters on Third World giant cities included in this volume. See also Hardoy and Satterthwaite, 1985.

3. "Le village est inhabitable. La ville, un abîme inconnu, est (vue de loin) une loterie; là peut-être on aura des chances, tout au moins la misère plus libre; l'atome inaperçu se perdra dans la mer humaine" (Michelet, quoted in Garden, 1975, p. 86).

4. The role of large cities is, as a rule, statistically overestimated. Planning should take into consideration the continuum encompassing large, medium, and small cities, as well as villages. See Hardoy and Satterthwaite, 1984.

5. To quote once more the example of Mexico City transportation, it is pandemonium with the notable exception of the subway. Twenty million mechanized journeys take place in the city every day—four million by private car, ten million on a fleet of seven thousand battered buses, the rest by collective taxis (the cheap and efficient *pesero*) and metro. The subway network carries more than three million people a day, twice as much as London's underground on a system one-seventh its size. The fare continues to be one peso, about one-hundredth of London's cheapest fare. But up to 2,800 people crowd each nine-car train, traveling at 115-second intervals; a train's maximum theoretical capacity is 1,530, including 1,170 standees. Average vehicle speed on the clogged roads of Mexico City is 12 kilometers per hour, or a crawl of 4 per hour at peak rush time. Workers spend an average of 2.5 hours every day traveling to and from work (Pearce, 1984).

6. In Brazil, a 5 percent reduction of fuel consumption by trucks and buses achieved through better engine maintenance would bring about savings sufficient to employ some 100 thousand additional mechanics—a considerable economy of foreign exchange, too.

7. As this manuscript was going to press, details of the Chernobyl nuclear accident began to surface—a disaster of great magnitude whose ultimate effects may not be known for many years. One can ask questions about nuclear facilities similar to those asked about the location of potentially dangerous industries.

8. For a pertinent critique of systems modeling as applied to cities, see Harvey, 1973, p. 303.

9. Although still quite abstract, Dupuy's article (1977) unfolds a very interesting perspective on our subject.

10. According to reliable estimates, the Brazilian *cerrado* (savanna) region alone has 150 million hectares of cultivable land still to open. If the population density there were to reach one person per hectare, this would mean doubling the present population of Brazil. If, on the contrary, extensive methods of land use prevail with one person per hundred hectares, at best the flow of migration to São Paulo would be slowed for a couple of years. Of course neither of these extreme solutions is good, but the range of rational choices remains wide.

11. In spite of its excessive schematization, Max Weber's distinction between the "consumer city" and the "producer city" is quite useful. Berthold F. Hoselitz (1960) rephrased it, proposing the terms *generative city* and *parasitic city*.

Bibliography

Abdalla, Ismael Sabri. 1979. "Dépaysanisation ou développement rural? Un choix lourd de conséquences." *IFDA Dossier* (International Foundation for Development Alternatives) 9(July).

Beck, Melinda, Phyllis Malamud, Jacob Young, Dan Shapiro, Joe Contreras, Susan Agrest, Katherine Koberg, Renee Michael, and Madlyn Resener with Jerry Buckley. 1982. "The Decaying of America." *Newsweek* 100 (2 August):12-16+.

Blair, Thomas L., ed. 1984. *Urban Innovation Abroad. Problem Cities in Search of Solutions.* New York: Plenum Press.

Boyden, Stephen, Sheelagh Millar, Ken Newcombe, and Beverley O'Neill. 1981. *The Ecology of a City and Its People: The Case of Hong Kong.* Canberra: Australian National University Press.

Braudel, Fernand. 1979. *Civilisation matérielle, économie et capitalisme, XVe - XVIIIe siècle.* 3 vols. Paris: Armand Colin.

Bukharin, Nikolaĭ Ivanovich, and Evgenil Alekseevich Preobrazhenskii. 1922. *The A B C of Communism: A Popular Explanation of the Program of the Communist Party of Russia.* Translated by Eden Paul and Cedar Paul. London: Communist Party of Great Britain.

Calvino, Italo. 1974. *Invisible Cities.* Translated by William Weaver. A Helen and Kurt Wolff Book. New York: Harcourt Brace Jovanovich.

Carcopino, Jérôme. 1939. *La view quotidienne à Rome à l'apogée de l'empire.* Paris: Hachette.

Correa, Charles. 1973. "Self-Help City: The Internal Organization of Metropolitan Areas." In *Symposium on Population, Resources and Environment, Stockholm, 1973.* E/CONF.60/SYM.III/9. New York: Economic and Social Council, United Nations.

Dreiser, Theodore. 1900. *Sister Carrie.* New York: Doubleday, Page.

Dupey, Jean-Pierre. 1977. "Autonomie de l'homme et stabilité de la société." *Économie appliquée* 30:85-111.

Engels, Friedrich. 1961. *La situation de la classes laboreuses en Angleterre.* Translated from German. Paris: Les Editions Sociales.

Farrell, James Thomas. 1935. *Studs Lonigan: A Trilogy.* New York: Vanguard Press.

Friedmann, John. 1981. *Retracking America: A Reissue of the Classic Book on Transactive Planning, with a New Preface.* Emmaus, PA: Rodale Press.

Fuà, Giorgio, and C. Zacchia. 1983. Industrializzazione senze fratture. Bologna, Italy: Il Mulino.

Garden, Maurice. 1970. *Lyon et les Lyonnaise au XVIIIe siècle.* Bibliotheque de la Faculté des Lettres du Lyon XVIII. Centre Lyonnaise d'Histoire Économique et Sociale. Paris: Societé d'Edition "Les Belles-Lettres."

Hardoy, Jorge Enrique, and David Satterthwaite. 1984. *Planning and Management of Small and Intermediate Urban Centres in National Development Strategies.* London: International Institute for Environment and Development.

———. 1985. "Third World Cities—The Environment of Poverty." *Journal '85* (World Resources Institute, Washington, DC).

Harvey, David. 1973. *Social Justice and the City.* Baltimore: Johns Hopkins University Press.

Hoselitz, Berthold F. 1960. *Sociological Aspects of Economic Growth.* New York: Free Press.

Jacobs, Jane. 1984. *Cities and the Wealth of Nations. Principles of Economic Life.* New York: Random House.

Kopp, Anatole. 1975. *Changer la view, changer la ville; de la vie nouvelle aux problèmes urbains, U.R.S.S., 1917-1932.* Paris: Union Générale d'Editions.

Koshizawa, Akira. 1978. "China's Urban Planning: Toward Development without Urbanization." *Developing Economies* 16:3-33.

Lagadec, Patrick. 1981. *La civilisation du risque. Catastropes technologiques et responsabilités sociales.* Paris: Le Seuil.

Lewis, William Arthur, Sir. 1978. *The Evolution of the International Economic Order.* Princeton, NJ: Princeton University Press.

Linn, Johannes F. 1982. "The Costs of Urbanization in Developing Countries." *Economic Development and Cultural Change* 30:625-48.

Lipton, Michael. 1977. *Why Poor People Stay Poor: Urban Bias in World Development.* Cambridge, MA: Harvard University Press.

McNamara, Robert S. 1984. *The Population Problem: Time Bomb or Myth.* Washington, DC: World Bank.

McNeill, William Hardy. 1976. *Plagues and Peoples.* Garden City, NY: Anchor Press.

Meier, Richard L. 1974. *Planning for an Urban World: The Design of Resource-Conserving Cities.* Cambridge, MA: Massachusetts Institute of Technology.

Mercier, Louis Sebastien. 1979. *Le Tableau de Paris.* Edited by Jeffry Kaplow. Découverte-poche no. 10. Paris: La Découverte.

Murphey, Rhoads. 1980. *The Fading of the Maoist Vision: City and County in China's Development.* New York: Methuen.

Owen, Wilfred. 1979. "Transition to an Urban Planet." *Bulletin of the Atomic Scientists* 35(Nov.):12-18.

Pearce, Fred. 1984. "Mexico, the City Unlimited." *New Scientist* 104(1426):42-44+.

Sachs, Ignacy, 1984. *Développer les champs de planification.* Paris: Université Coopérative Internationale.

Seale, John. 1985. "AIDS Virus Infection: Prognosis and Transmission." *Journal of the Royal Society of Medicine* 78:613-15.

Soedjatmoko. 1984. "Raz de marée sur les villes du tiers-monde." *Le Monde* (15 June):31.

Spitz, P. 1984. In *The Right to Food,* edited by Philip Alston and Katarina Tomaševski. International Studies in Human Rights. Boston: Martinus Nijhoff.

Theunynck, S. 1982. *Une capitale stérilisante: Nouakchott.* Mimeo.

Tolstóy, Leo. 1925. *What Then Must We Do?* Translated by Aylmer Maude. London: Oxford University Press.

Vacca, Roberto. 1973. *Demain, le moyen age. La dégradation des grands systèmes.* Translated by Louis Mezeray. Paris: Albin Michel.

Veyne, Paul. 1976. *Le pain et le cirque: Sociologie historique d'un pluralisme politique.* Paris: Le Seuil.

White, Rodney, and Ian Burton, eds. 1983. *Approaches to the Study of the Environmental Implications of Contemporary Urbanization.* Prepared in co-operation with IFIAS-Project Ecoville. MAB Technical Notes 14. Paris: UNESCO.

Wilheim, J. 1984. "Metropole e crise: O caso de São Paulo." Paper presented at the seminar, As metropoles latino-americanos frente à crise: Experiencias e politicas, São Paulo, 10-13 September.

Woods, Shadrach. 1975. *The Man in the Street: A Polemic on Urbanism.* Harmondsworth, England: Penguin.

13

Growth and Pathologies of Giant Cities

Henry Teune

In this chapter I discuss a variety of conceptual and methodological issues relevant to the study of urban forms and giant city growth. My approach is broadly comparative, looking at city growth (or decline) in the context of national development processes and the changing international division of labor. I shall be both retrospective and speculative, at times commenting on the forces that have shaped urban pasts as well as those likely to determine urban futures. Throughout my discussion, I touch on a number of thorny questions scholars must address if analytical and theoretical progress on problems of giant city growth (or decline) is to be made.

Large Cities and World Cities

The giant cities emerging in the latter part of the 20th century developed in a radically different environment from those in the latter part of the 19th. Large industrial cities were the cutting edge of national industrialization. They created surplus wealth to make entire populations relatively rich and secure in national welfare guarantees. Labor was harnessed to coal-steam power and moved to and from work by electrical power, which facilitated the process of suburbanization, aided by the mass-produced automobile. With the growth of large industrial cities the countryside was depopulated and a historical "system break" of about the same magnitude as the change from food gathering to agriculture occurred. Agriculture ceased to dominate economic and social life.

When manufactured goods began exceeding commodities in world trade in the latter part of the 19th century, the world economy was shaped into centers and peripheries with a few centers dominating the goods necessary for development and the peripheries dependent on them—a process some have argued that began with the rise of capitalism in the 15th century (e.g., Wallerstein, 1974). There were world cities and the "backward" towns. What some had seen happening within countries, described by the concept of central place theory, others extended to the world.

But what happened around the late 1960s was another kind of urban growth, a decentralizing global economy with exchange of manufactured goods being shared in all parts of the world—with almost all countries participating in the exchange of manufactured goods and increasingly in services. There were technological, political, and social reasons for the emergence of new components of the world system and changing relationships among the countries of the world, most of which were new politically but old socially.

Technological reasons deal with communication and control. Communication costs decreased and their speed dramatically increased. Some of these technologies stimulated the growth of standards for production on a global scale, allowing for interchangeability on an increasing scale. The main consequence was accessibility at decreasing costs to almost every area of the world; only a few dark corners remain.

Political factors are several. By intention after 1947 the United States designed a decentralized world order of independent states, abandoning efforts of President Franklin Delano Roosevelt to create a great power coalition of informal understandings with formal recognition in the United Nations and its agencies. One consequence was competition among many states to gain position in regions and the world. Although some tried the policy of closure to the world for national development, most accepted openness and its resulting penetration of international economic organizations. The Soviet Union created an alternative international economic order made up of strong states and cooperation based on shared political goals. The result was initial economic growth, peaking in the 1970s, with some countries requiring subsidies and later a few defecting to participation in the capitalist system of openness, the most notable being China.

The social factors in part derive from communications and the encapsulation of almost the entire world. A common world culture emerged with largely secular values and concern for material consumption, as well as for variety of products to consume, both in ideas and material goods. The consequence was a narrowing of values and a broadening of preferences and tastes. What was available in the world would not only be distributed but national capacities for its use expanded.

Other forces for change also had an impact on the newly emerging global political economy, such as immigration of Third World people to the world centers of North America and Western Europe. There were reactions to consequences of these changes, including some movements to return to sacred political systems—sometimes successful as in the case of Iran, and protectionism of the late 1970s, and generally unsuccessful efforts at international cartelization of world commodities, most dramatically petroleum.

If conditions for a new system of a decentralized world economy continue, then patterns of recent change should intensify. First, the relative share of production by older centers in Western Europe, North America, and Japan should continue to decline as a percentage of world production. This decline was gradual after 1945, and has been much quicker since the 1960s. Second, a greater spatial distribution of contribution to world production should continue, as it has in Asia and Latin America, but not in Africa for a variety of political reasons. Third, cities should be increasing their political autonomy from their countries and their independence as players in a global political economy. This situation has been the case in many parts of the world with the exception, of course, of the Soviet Union and Eastern Europe, although it could be argued that Yugoslavia, Baltic cities in the USSR, and a few other areas are being penetrated by the global economy and have achieved some measure of autonomy from the state.

Evidence for the softening of the state in favor of the autonomy of regions—some of which spill over national boundaries—and large cities, as well as the emergence of new global political and economic institutions, is relatively weak. Such a trend would include developmental policies of cities directed to international rather than national markets (e.g., foreign trade offices of cities); free trade zones within cities (e.g., as in Hamburg, Germany, and certain cities in the United States), regionally coordinated economic policies (e.g., in Northern Europe and North America), continued reduction in national tariffs (e.g., those proposed for 1980s negotiations), indeed, citizens' changed perceptions of cities' roles in a global political economy. There is at least some argumentative evidence for the institutionalization of global economic questions, such as regular meetings among leaders of major industrialized countries, increased importance of international professional societies in setting world standards, and expansion of multinational corporations.

If cities are taking on a new role in a global political economy, then there should be different kinds of roles for different kinds of cities organized in a structure of global economic and political developments. Those enriched with the technologies of communications and control should be dispersing their material goods production processes but centralizing their communications and access to all parts of the world. They would be the high-tech cities with smaller proportions of their populations enjoying the benefits of growth and a decreasing proportion engaging in older forms of production. In addition to deconcentration of economic activities, such cities should also be losing population as less and less depends on physical proximity for supervision and more and more on electronic messages. These world cities of the first order include many of the old industrial capitals, such as New York and London.

Cities with less capacity for communication and control have been taking on manufacturing roles, an increasing part of which is for export to particular regions of the world. Among the reasons for assuming this role are the incentives for economic organizations to escape national control by gaining threat power from international mobility. These cities are the second-tier cities. As their role in a global political economy improves, they not only should be increasing concentration of productive activities but also their population, especially those in poorer, agricultural countries where even marginal opportunities in cities are significant in comparison to the countryside and the declining ratios of land to population.

Comparing Cities and Countries

If cities are increasing their independence from national control and losing autonomy to a global political economy, then traditional forms of comparisons have become more complicated. Since the 19th century and the rise of modern social science, increasingly the point of departure for comparative research and analysis has been the nation-state. It was and is the primary unit of analysis. To what extent are observed structures of behavior—political organization or suicide—a consequence of general developmental processes—economic prosperity or class—and to what extent are older forms of human affiliations—religion, culture, or nationality? In methodological terms this is the level of analysis problem, which in effect is a general theoretical problem of assigning weights to different levels and forms of human aggregations and organizations. At what levels of human specificity or inclusivity can most similarities and differences be explained and why?

This general problem of social science was formulated in the discipline of anthropology as Galton's problem but in a different theoretical context (for an elaboration, see Naroll, 1965:32). To what extent are human organization and behavior attributable to diffusion of knowledge of values and skills, and to what extent to the uniqueness and specificity of human experience such as the spread of religions or ideas, such as those of the French Revolution, or technology, such as the factory? Are we seeing Russian culture or Russia because of its relatively peripheral position vis-à-vis Western Europe?

In more recent debates, this same question has been formulated as "convergence" theory (see Laquer, 1967). Will the Soviet Union and the United States become more alike because of common forces of industrialization, secularization, and modernization? Or will they diverge because of a different heritage, political system, and society? What this question expresses, as do the previous ones, is again the theoretical problem of assigning weights to different kinds and levels of human organization, in this case the more universalistic ones of modernization and the more specific national experiences.

Certainly, comparative research since 1945 has been self-consciously cross-national. In comparing local governments, explanation of differences has focused on national policy, the political system, especially decentralization, and history. Comparisons of public policies, of course, are central to the political system (Hiedenherimer and Heclo, 1978). Trade unions in France, England, and the United States must be understood in terms of the countries where they operate. Indeed, there are arguments over which is the best entry point to choose in order to understand a particular political system—elites, the general population, or specific political institutions, such as parliament?

In industrialized countries, the main cutting points of development were, and to a considerable extent still are, the nation and the city; they were engines of change. In the West cities represented the new; the locality and region, the old. Cities could be compared either within a country or to those of other countries. If cross-national comparisons were made, the national system was taken as controlling the contextual variables.

Cross-national research on cities has been especially sensitive to national political systems. Is housing a national matter (as in the Netherlands), a local political issue (Italy), or a local market matter (United States)? Is transportation a national or local political question? In general, to what extent is there urban local autonomy, or specifically what are national-local linkages?

The foregoing is a relatively static set of issues. They are the relatively standard set of theoretical questions in comparative research addressed to the levels of analysis problem. The more important theoretical question is not what are the coefficients of importance in explaining structure and behavior at a given time, but how and why do those coefficients change? If different levels of human aggregation and organization for cities are changing, what are they and why and what will they be? Should unemployment rates in Amsterdam be interpreted in terms of Holland, Northwestern Europe, the Common Market, the industrial world, or the global political economy, or indeed North America or Southeastern Asia? And how and why will these rates change?

If only two factors are considered, the emergence of a global system with some impact on national components, including large cities, and technologies that affect dispersion and aggregation, then the levels of analysis problem for comparing cities empirically or understanding change theoretically must proceed with at least four different points of human activity: the city itself and what is unique to it; a region whether nationally circumscribed or internationally defined (Southern U.S. or the Pacific Basin); a nation; and the global political economy. If the region and global economy are growing, then, their coefficients should be projected to increase in explanatory power while those of city and nation decrease. If world cities will be increasingly differentiated in terms of their roles as organizers and controllers of access and production centers, then traditional measures of urban growth must be reinterpreted. Cities losing industry and population should be seen as reflecting a qualitatively different kind of growth—those gaining in population as reflecting a changing role in a global economy. The different directions of urban development in the advanced and developing countries are both part of a process of global development. What we are doing in comparing cities is comparing components in a newly emerging system influenced by developmental processes and the diffusions of technologies of communication and control. The national "communities of will" remain. The nations and their choices remain important but their relative decisiveness for urban phenomena will diminish.

Comparing anything states something theoretically. Since it is true practically and perhaps theoretically that any object, social or physical, has an infinite number of characteristics, if only those of distance and time related to all other known objects, then comparisons of any object with any other on any particular characteristic are highly selective. More general comparisons are based on understandings or judgments concerning the "significance" of clusters of generally related similarities and differences. This theoretical statement is almost always difficult to interpret.

Regional clusters of cities such as those found in this and other comparisons of cities, express a variety of discontinuity hypotheses. The region has cultural similarities; it has common historical experiences—for example, having similar colonial heritage; or relations to others in global economic developments are alike—for example, relationships to the European centers as peripheries. These may be in combination, such as is the case in Eastern European countries, with something shared in culture and language, some commonality of political institutions, and similar historical experiences with Western Europe. Outliers, such as Yugoslavia, are recognized for specific kinds of comparisons.

Types of national political systems, level of economic development, relationships with other cities (the historical case of the Hansiatic League), size, role in a regional or global economy (Northern European cities or those engaged in capitalistic trade), structure of city or national government (autonomy), are among the characteristics having theoretical importance for comparative analysis of cities.

City size, independent of national setting, culture, history, or place in regional or international configurations or relationships, is pervasive in comparisons of cities. Not only is there interest in growth, in the breaking of limits, whether those of Malthus and his constraints of population size or of buildings surpassing the limits of heights, but also of the ecological limits of density. This density—numbers and space ratios—makes the large city an attractive target for comparisons.

Size, however, carries other interesting theoretical relationships. One of these, long understood, is demand for government control not only to offset dangers of fires and crime but also to provide for amenities. Another is the relationship between density and access to variety. More people not only generate more activities but also more diverse ones. Finally, size is associated with certain pathologies (or at least that has been part of Western social analysis since the emergence of the large industrial cities) associated with crime, drug abuse, and a host of problems emanating from human congestion.

Theories of Cities and Change

Urbanization and urban forms of human settlement and organization have many theoretical explanations. One such is social ecological—putting together population, growth, death (largely from epidemics), and decay. Another is that technologies of production led to the demise of European feudalism and the rise of capitalism. Such theories are often suspect because of their limited scope regionally and historically. Were the large cities of the various Chinese dynasties, for example, a consequence of ecological forces or political decisions enforced by the benefits or losses from centralized control of irrigation and an imperial political system?

Cities, however, have generally been associated with human development, including urban forms of settlement and their political organization. This general but often controversial theoretical perspective on the emergence of large cities in the 19th and 20th centuries is based on the industrial revolution. If human development is defined in terms of advances in innovation, then the city in the modern era has certainly been decisive in integrating diversity of peoples, things, and ideas. If innovation is a key to developmental change, then what conditions or variables are likely to increase the probability of putting existing variety or ideas together to create new variety? First, the amount of variety available should increase the probability of any two or more different things, such as the cart and the combustion engine, together. Second, the lower the costs of moving things or confronting ideas, the greater the probability that combinations will occur. Third, the certainty of searching for something to solve some unanticipated "problem" should reduce the risks of experimentation and its resultant innovation.

Whatever else may be defining characteristics of urbanization, a core one is sheer size and density—a social-spatial concept. Large aggregations of populations in a small space increase the potential for conflict, reduce costs of access in terms of the constraints of distance, and strengthen the probability of labor force specialization. Within an urban area, central location functions in much the same way as the city for the countryside. Urban centers with easier access to the variety of other cities should innovate more than those isolated. Trade marked the development of cities in Europe, most of them small, and their economic development took place in regional clusters (De Vries, 1984).

Innovation is one cutting edge of human development; efficiency is another. Efficiency emanates from new technologies—know-how—and control or its application, at least most of the time. The main technological advances in production were in the generation of inanimate power to replace animal and human power and the supervision of human effort to achieve increasingly complex goals. The former is tied to the use of coal (later oil), transformed first as steam and then electricity. Coal-steam has a poor ratio of access costs to the distance of its dispersion. Steam quickly cools but electricity only slowly depreciates in power as it is transmitted. Although oil pollutes when produced, it does so very little when consumed and its pollution levels are acceptable when used to transport people through heavily populated areas.

It has been estimated that production costs in 19th-century American cities were about five times lower than in dispersed production using various forms of power, including water (Higgs, 1977). Concentration in the production of power produced efficiencies in production at a margin substantially greater than the overhead costs of density.

Access costs were also radically reduced within 19th-century industrial cities. First was access to information. A number of technologies provided for decreasing costs of information. The mass-produced newspaper, the telephone, and city directories dropped access costs in cities, and spread them beyond their political boundaries. Second, hardened steel and elevators allowed more people to be packed into smaller geographical areas in taller buildings. Third, electrically powered transportation systems gave easier access to a greater variety of skills among a larger labor force.

Several other technologies yielded greater access to more variety at lower costs, including lighting streets, thereby expanding the time people could gain access to things. In general, control and access required concentration, and the size of cities in the industrializing world grew until the first quarter or half of the 20th century.

Costs of density, however, are considerable and pollution from concentration exceeds physical capacities to absorb it. Density also increases the probability of contagious diseases, a significant historical constraint on urban population growth, resulting in costs for socializing rural immigrants and upward pressures on the price of labor. Density also generates the need to substitute government for the social controls of family and kin. A number of improvisations and technologies were required in sewerage, sanitation, and disease control, as well as efforts to control fire and behavior.

The foregoing is largely an accounting of cost-benefit-technological relationships that contributed to the creation of large cities. It does not include factors that push people from the countryside: land-holding policies, challenges from agricultural imports, or changes in weather patterns. Also not addressed is the more general theoretical question about massive increases in the sheer scale of human organization in regions and nations and why cities became integrated into national and international relationships of unprecedented size and scale to a point where today most large cities have become part of a global political economy. In less than one generation, even one of the most economically self-sufficient nations in the 20th century, the United States, increased exports and imports from about 4 percent of GNP to 6 percent in 1959 to 17 percent in 1984 and is projected to surpass 25 percent by 2000 (Schulz, 1985). But what is perhaps more significant about the international engagement of the world's largest nationally circumscribed economy is the spread of those imports and exports around the world. Trade has increased more rapidly with Asian, Latin American, and other less-developed economies contributing to a decentralization of world economic activities.

Cities have always provided others access to their variety and their people sought that of other cities. Exchange or trade was not only necessary for the survival of their populations but also for their status as attractive centers. Politically, cities compete not only to replicate the variety of others but also to offer something distinctive from other cities. The design of cities—in the modern era, urban planning—is one way of achieving uniqueness. Historically, artistic artifacts and expressions have marked cities as special. Material things may be redundant; their arrangement provides almost infinite possibilities for uniqueness.

For cities to compete they must access variety, limited by capacity and cost. First is the cost of information about existing variety as well as expenses and risks of exploration. Second are costs of transporting and integrating variety. Both have to be measured against anticipated value, either yielding new values for consumption or possibilities for efficiencies and control.

Limits on the kind and amount of variety also come from the character of a city and its organization and level of variety. Importing variety must be controlled for it to be integrated and made compatible with particular forms of political control. Cities act as a political unit to facilitate and fit in what is new and different, hence the conflicting impulses of those within a city to import variety and locate it optimally in ways that minimize disturbing the existing relationships.

Development emanating from growth and new variety is an uneven process; certain domains and areas change faster than others. Development is also more than the logic of technology and production. In the 1960s a few large cities in Western Europe, North America, and Northeast Asia were transformed from national manufacturing centers into world cities. In part this means that their transactions with the rest of the world are at least as important, or in some cases more important, to their economic life than their national hinterlands. These cities became involved with global finance, administration, risk sharing (insurance), information, and today, world culture (art, literature, and music, as well as popular culture); this is the first developmental tier of cities.[1]

As this shift took place, industrial labor became less necessary in these cities and all but a few, and those mostly because of national political decisions, lost population in the 1970s and experienced urban decay, despite efforts to offset it by subsidies or urban renewal. The two general consequences were deconcentration of industrial production and some decentralization of political decisions. Manufacturing was standardized, power for production dispersed, and control exercised through advanced communication technologies. Production could be scattered physically and integrated organizationally. Political control of conflicts among a work force based on class and ethnicity became a matter of indifference. Not only was the local work force not needed but "urban" economic organizations acquired mobility. Economic organizations had the option of moving away from unsatisfactory environments at relatively low costs in contrast to "in place" industries, such as steel, in which investments in particular places had national economic advantages.

The incentives to urban elites for maintaining these first-tier cities as political communities for internal control diminished.

A second tier of cities has grown up in a global political economy characterized by population growth and concentration. These cities are centers to which industrial production is being transferred. Seoul, Singapore, and São Paulo have developed along with the post-1945 global political economy and are subject to its fluctuations in prosperity. Such cities are easily penetrable and hence part of a "world capitalistic system," influenced not only by the first-tier cities but also by international cartels, such as the International Monetary Fund. Some of these cities act as intermediaries between first-tier cities and certain geographical regions or other cities within their countries.

A third tier of cities is those tied to national political and economic development. These cities are not necessarily national capitals, but their economy is tied to national economic goals in much the same way that cities in the West were in the 18th and 19th centuries. Some of these cities are strongly protected by national boundaries, such as those in Eastern Europe.

A fourth tier of cities is those with the role of traditional regional markets. These, of course, are smaller in size than other types of cities but can play a vital role in national development by linking segmented societies into a national economy. Examples of these are middle-sized cities in Thailand or Indonesia. At present, they have a minor role in the global political economy.

The four tiers of cities generate conflict. One central conflict is between the imperatives of national political development and competition in a global, economic system. Another occurs when the institutions of world cities bypass second-tier cities, undercutting their potential for economic survival. But, as stated, developmental processes are not only uneven but also contradictory.

Examples of the relationships among tiers of cities are several. One is the process of transferring technology downward. The transistor is invented in one or perhaps two places and its properties standardized, including its projected value in terms of finance. Its early use in manufacturing (its combinatorial possibilities) occurs near the place of its invention. As its properties become standardized, its use in manufacturing can be decentralized and yet controlled. Manufacturing moves to a regional capital or another country. The computer is invented and a similar process follows. But the regional capital now shifts its manufacture of the transistor to capitals in its region and begins to participate in computer manufacturing. Yet, certain forms of hierarchical control remain. Although there are different national and regional overtones to the products in these transfers, the units of manufacture remain standardized. The transistor is replaced by the microchip, at a higher level of technology. This process continues at an ever-increasing rate and what results is a hierarchy of places, but, as always in human affairs, one beset with imperfections and contradictions.

The automobile is an example of decentralized production and marketing on a world basis. Automobile manufacturing is now approaching a global network of production and control. This phenomenon is not merely a matter of substituting cheaper labor for capital, if only because production units must be standardized and that generally requires similar types of technology based on a common or easily translated language.

In the West, national urban policies and city decisions encouraged growth of manufacturing and services in the 19th century, providing surplus wealth for national taxation and expansion of the welfare state. Cities, as has been indicated, adopted policies to advance their competitive position among others. Managing growth is one thing; limiting growth is quite another. Despite national and local efforts, often increasing disparities among jobs, services, and population (with perhaps the exception of that unique case of Singapore) practically no combination of national and city policy is able to limit growth today. This situation is true in almost every Third World country. As has been documented, the gulf between population and housing in several Eastern European countries has remained since the 40s, whether or not its city housing stock was destroyed in World War II (see Friedrichs, chapter 5, in this volume).

Managing decline in cities of the West is a relatively new problem. Changing the use and functions of physical space often imposes heavy costs. This problem was especially severe in the United States with vast areas of urban blight in most of its older industrial cities. Indeed, the language of U.S. national policy concerning cities shifted in the late 1970s from urban renewal to conservation of cities. Many compete now as museums of history rather than as environments for innovation and productivity.

Pathologies of Large Cities

Growth, especially rapid growth, generates visible pathologies. At the very least what was once new becomes the old and ratios among aggregate components change, for example, housing stock and people. These phenomena derive from sheer quantitative growth. Growth and its consequences have always been normatively controversial, whether planned or not.

To assess growth the concept of niches is necessary. Growth of anything requires a particular niche. Pathologies occur if growth is faster than the growth of the boundaries of the niche—in the case of cities, physically and politically defined ones. The niche is a container and a container has some upper limit. A bathtub is a niche and increase or decrease in the amount of water in it is a simple function of water going in and out over time. If what goes in is more than what goes out, the niche breaks, the water spills, the room is ruined. Thus, the concept of growth is necessarily tied to some notion of equilibrium with maximum and minimum flows and by implication some idea of optimum based on such principles as the largest size with the least damage to the holding power of the niche. Since the maximum niche for an urban area or city has been radically increased by technological advances during the past two or three centuries, research on optimal size of cities has not made much progress in recent decades.

The health or durability of a niche, however, depends on other nearby niches (we consider niches conceptually as places). These other niches constitute the ecology, social organization, or environment of the niche. At some point growth of the elements within one niche will either expand its size and shrink the size of niches around it, or perhaps in response to declines of other niches will bloat its size. Niches that are growing in the number of their elements are destabilizing. The increasing members can spill over and invade other niches or the expanding niche can crowd others. Growth thus may not only be destabilizing but also may produce negative outcomes for other niche compartments.

Growth within a particular niche can further lead to changes in the structures of relationships among niche occupants. The extent to which the diversity among members of a niche is expanding increases the probability of structural change within that niche. Growth is always quantitative; its consequences typically are structural or qualitative. Qualitative changes include a variety of microfactors—probabilistic events (innovation) of learning and adaptation at the microlevels of individuals, groups, and organizations—or macropolitical, economic, and social "revolutions."

Pathology, of course, is a pejorative, complex, and relative concept with political connotations. Its manifestations are attached to particular values, niches, and systems of niches. Cities can be bad for human health but good for national economic development, destructive of family life but an effective means to assure human services, harmful to plant life but good for individual opportunities.

Western intelligentsia have long held ambivalent views about urban growth and large cities. Those holding positive views see cities as aggregating energy for economic activity, freeing people from the dismal chain of population growth and scarcity, mixing diverse populations for assimilation, stimulating thought in art and science for new achievements, liberating individuals from control of local social and economic elites, providing a base for new collective identity, and unleashing human creativity from routines of agricultural production. Those convinced of the negative consequences of cities advocate controls on urban growth, improved design of small towns, and the virtues of rural living.

Whatever conclusions were reached about the virtues of cities by dominant political elites, efforts to control urban growth in the West were not historically effective as is the case in the Third World today. Some policies have assumed the pathology for urban growth. Growth of U.S. cities was seen as not compatible with democratic governance and its reliance on personal relations, trust, and independence. But perceptions of pathologies are relative and often focused on the most salient consequences, which in the case of mammoth cities have been only recently seen as a new international phenomenon, shattering historical expectations.

First, most analyses of pathologies of large cities are directed to individuals. The inventory of pathologies is long, varying across population categories. It includes noise, air pollution, infection, unhealthful water, uncertainty, psychological overstimulation, lost time in travel, ugliness—all contributing to physical and psychological maladies. In saying large cities are more threatening to individuals than rural villages or small towns, comparisons must be disaggregated, not based on national data relating urbanization to such indicators as infant mortality and longevity. Assessments of urban pathologies for individuals are often comparisons with some idealized environment; yet these, too, may be flawed. Long-term and short-term effects must be distinguished. Often, long-term effects are poorly documented, partially as a result of recent awareness of the urban environment and its relationship to physical and mental health and partially because of the lack of systematic population surveys over periods of time. Also there are confounding urban-related developments of improved diets, medical care, and sanitation. Modern medicine, for all practical purposes, has confined death by contagion to small percentages.

Second, cities are linked to various social pathologies: aggression, crime, social disintegration, and personal callousness. To what extent cities of large size generate a step function in these social maladies remains a matter of debate. It has frequently been proposed that density makes it easier to rape, steal, and otherwise behave aggressively toward others because of weakened societal controls and greater opportunity. But these factors must be discounted by greater detectability in cities. Reporting crime, suicide, and incest is more likely in cities with their diminishing social controls over such revelations and improved capacities for governmental surveillance.

Third, there are costs for cities—their overhead in maintenance. The problem with this concept, however, is one of determining costs and benefits to whom. To conclude that cities are too big, it is necessary to have some baseline for optimal city size, a focus of some social science research. But the concept of optimum size is subjective unless defined in terms of the whole human race or some normatively stated idea of human development. What we know is that large Third World cities have been growing and most probably will continue to do so into the 21st century. Locational decisions away from and into cities, it is believed, have some "rational" basis, but the outcome is often either decay or problems which in some cases, such as Chicago or Calcutta, cannot be addressed by an indifferent national government or a shadow city government.

Fourth, aside from costs and benefits—the pathologies of individuals and groups living in large cities—there may be benefits to regions, the country, a transnational region, or the world economy. The benefits of Los Angeles may accrue to the residents of Southern California, not the citizens of the city; the costs to the citizens of Paris may redound to France; Amsterdam may drain the Dutch economy but serve as an essential component of Northwestern Europe; and Tokyo, whatever the losses to its residents for its large size, may be vital to the world economy. These different levels of costs and benefits may induce the larger to subsidize the city in a variety of ways, including those by multinational corporations. Pathologies of large cities may be worse were it not for the interest of outsiders who use them for their purposes.

Fifth, some pathologies of large cities must be sustained for long-term development. Viewed as separate entities in the 1980s, large industrial cities may be judged as economic losers, but they were critical in transforming the U.S. economy from manufacturing to services (around the early 1950s) and from that of traditionally defined services to those of information and control (around the middle 1970s). Large cities in their period of growth were the tax base for improving the countryside, for example, via roads and rural electrification; today the cities require subsidies. The "developmental" residuals in agriculture, the small farmer, required subsidies to maintain order in the face of social and economic change. Today large cities are homes to those needing social welfare and may become relatively efficient systems for the delivery of services, providing for domestic tranquility, and a national sense of fairness. Growth and development always generate problems for some and a developmental perspective is necessary to evaluate them, which as has been argued previously, means the contradictory trends of decline and dispersion of some cities and growth and concentration for others.

Setting aside the foregoing qualifications about the concept of pathologies, we find that cities carry obvious impediments to human development and dangers to human health. Cities have fewer social controls than towns or rural settlements. Air pollution, contaminated water, and threats to life from traffic and criminals are intensified as cities increase in size. The hard evidence for these differences between the large city and real or idealized alternative forms of settlement may be difficult to find. But few would argue that large cities in most Third World countries are good for all but a small number of their residents. And few would dispute that the single largest threat to urban life and growth—epidemics—disappeared in Western European and a few other countries in the early 20th century and is now largely contained in Third World countries.

Perception of a relationship between human pathologies and giant cities results in part from the rapidity of their growth. Chicago grew as an industrial city from a few hundred thousand in the latter part of the 19th century to over three million in the 1920s. The urban population of Mexico was about 20 percent in 1930, increasing to close to 40 percent in 1960, and 60 percent in 1980 (Gil Díaz, 1984). Various estimates of Mexico City in the middle 1980s range from about 15 million to nearly 20, and the number of cities of over one million is increasing exponentially.

Aside from forced deurbanization and dislocations because of war, urbanization and growth of large cities are primarily the result of individual choice. Certainly in the last two centuries urbanization has been associated with increased prosperity, and economic and population growth. Alternative forms of human settlement have been abandoned as a result of millions of individual decisions. The question as always with human choices is to what extent they improve the individual's condition in the short run but destroy the collective potential in the long run.

Cities as Systems of Access

Urban in Western terminology comes from the Latin *urb* (city); *city* derives from the Latin *civis*, connoting citizenship. Social and economic organization that is *urban* should be distinguished from *city*, which is political. The relationship between the city and urban forms concerns the problem of controlling access, especially in free market or capitalistic countries. Where access is priced according to markets with free migration of labor and capital, the city has limited capacity to control. Lack of such capacity, some would argue, because of political alliances between economic interests and the state, leads not only to the phenomenon of urban sprawl but also to the diminishing ability to control access by political means, demeaning the importance of public goods, such as clean air, or creating collective pathologies, such as impossible traffic.

Controlling access is achieved through "markets" and political authority—the historic pillars of social control. One must do something to get something; or one must do or not do something because one agrees it is good ("positive legitimacy") or one will lose something because of the application of sanctions. The question is whether urban places add another factor to social control, the organization of physical space—Who occupies what place relative to others? The access of particular pieces of physical space depends on technology and what role a particular niche-settlement has in a broader environment including relative rates of growth of the various niches in the environment—this is the ecological perspective on urban places and urbanization. The perspective is tied to market forces, technology, and physical ecology or the natural advantages of location. But cities are also built and designed according to political goals and values.

It is difficult to assess a single function or dominant function for cities. Apart from centers of political control of larger areas, whether through religion as in cities of ancient empires, such as Egypt, or through administration, as in the case of Rome, or by designated trade centers, as were many in medieval Europe, most cities were either in conflict with other larger political groups, or made political alliances with others or with emerging centers of political control. Applications of the concept of function ignore contending political purposes and aspirations concerning cities as political entities.

Today almost no countries fail to allow for their urban places to adapt ecologically through market responses and adjust politically through some locally defined forms of urban governance. Within similar structural forms, such as mayors and elections, there is a wide variety of city governments resulting from the importance of the city, its particular physical or economic ecology (such as ports), its history, and its cultural traditions. A renewed old tradition of cities is to seek exemptions from national regulations. Functions and political organization, the markets and politics of cities, are the traditional means to differentiate cities. Another is developmental, involving technological change, extending capacity to access (locality, region, nation, and world), and the kind and amount of variety accessible.

As the world has been increasingly encapsulated into a single system, cities that were once centers that housed factories for production are becoming control centers and others are becoming production centers, the factories of the world. In a sense, the new large cities are becoming part of a "put-out" economy, contractually based units in a world production system that is becoming more inclusionary, locally specialized, and decentralized.

If so, have these developments led to a political consciousness and political decisions of city governments to deal with and take advantage of these changes much in the way that cities in the early part of the 19th century in the U.S. responded to urban growth by creating more effective governments? These changes are just beginning and if they continue will challenge the authority of the state.

Cities as Political Communities

To what extent do and can cities function as political communities, providing for individuals and groups a sense of psychological identity and belonging that overcomes economic and ethnic divisions and responds to political leadership? The same question, of course, is focused on the nation. Both city and nation were the main thrusts in national development in the industrialized West. They made political alliances against other social and economic elites, often based on land and locality. Urban governments had to address the problems of concentration to use the technologies of industrial production; the nation, the problems of unity to govern.

In general creating political identity under conditions of growth is easier than under those of decline, despite consequent pathologies and losses to some for whom future opportunities may be psychologically satisfying. This relationship depends on other factors, including political capacity and leadership—strongly argued in the first wave of writing on political integration (see Deutsch et al., 1957).

Ethnic, social, and economic conflicts that become political, that are consciously pursued for authoritative resolution, can be controlled or dampened by threat of coercion, segregation, benefits, participation, or removal to other political arenas. Of course, there are other means, especially those directed to psychological satisfactions, for example, pride in local or national citizenship.

Growth of large industrial cities in the West in the 19th century required importing labor not only for simple economic expansion but also to replace those lost through disease.[2] The faster the growth, the farther the reach for labor; the farther the reach, the greater the differences among the city's population.

The industrializing cities strengthened surveillance capacities, expanded their police forces, established jails, and created various petty courts, often directed to crimes associated with new residents and the lower classes. In rapidly growing cities, ethnic, cultural, and religious differences became increasingly identified with those of income or class. One of the most pervasive means of social control, especially in the United States following massive immigration from workers from Europe, was spatial segregation.

With increasing wealth to tax in the cities, they were also able to provide social welfare to reduce conflicts among economic groups. Although national welfare was provided in most countries of Europe beginning in the 19th century, it was a municipal and state function in the United States by and large until the 1930s. Indeed the political movement for social welfare in the U.S. had exhausted itself by about 1915 because of the necessity of targeting each state and city separately. The success of those advocating welfare depended on the industrial wealth of the states and put them into competition to lower taxes to attract industries.

Also beginning in the 19th century political participation in city government gradually expanded, related to the degree of political decentralization of the nation. In the United States, elections of mayors were followed by the growth of city political machines—the urban political parties—involving the lower classes and new immigrants in what the middle classes considered to be an unacceptable system of corruption and which they attempted to change through a variety of reforms commencing as a national movement after 1900.

The political system of the West had in general evolved by the 20th century into at least three tiers of territorial government—national, provincial, and local, with varying levels of autonomy, sectorized into urban and rural. Cities in Europe in many cases had established earlier special status as political entities. In the United States such status was sought as "home rule."

To the extent to which a political system is decentralized, there are incentives to avoid conflicts by shifting them to other levels of government; for cities this meant the provinces or the nation. Whether class conflicts in Europe or labor-management disputes in the United States, once they were politicized, cities could shift the problem of control upward to avoid the costs.

Between 1870, beginning in Germany, and the 1920s, welfare in the West became a national responsibility. Most Western European countries had national welfare policies and regulation of labor conditions. The welfare state was slower in the United States because of its decentralized political system, size, and diversity. But in the 1930s, the federal government assumed responsibility for welfare and regulation of the labor force by nationalizing it, setting conditions of entry and exit (retirement), pay rates, and rights of unions. Following this, the national government extended its control to ethnic and social conflicts, especially important in the U.S. with its diverse population. But similar national efforts to ensure civil rights occurred in Europe, some of which today partially involve the European community.

These trends often resulted from pressures for cities to get rid of the conflicts their growth generated. It was also in the interest of national governments to acquire more control, if only to achieve greater international status. But the overall consequence was to diminish the importance of city politics. Economic and social elites were no longer interested in municipal government. What affected them most occurred in the nation's capital.

With deconcentration of production in the industrialized countries and their relative economic decline in the world, fewer economic interests are concerned with cities and their problems. Many national governments did try to offset decline with urban renewal programs in the 1950s; some, of course, had to rebuild cities after their destruction in the war. But many now have used up their late 19th-century urban infrastructures of transportation, housing, and water supply systems. Cities evolving into first-tier world cities are building on top of the old, bifurcating the city physically and socially into parts devoted to the global political economy and the residual of the old industrial order.

In addition, when national governments began to shift from transfer payments to localities to regulating them, pressures mounted from localities, increasingly more varied and dispersed nationally in ecomonic activities, to decentralize. National regulations tend toward uniformity; they accumulate into priorities progressively at variance with local ones. Public transport is not a priority for Dallas; water is. Northeastern U.S. cities have centers that cannot survive if the automobile is the only means of access; water is abundant. These local differences along with the problem of national government overload led most Western countries to begin a process of political decentralization in the 1970s, reversing the centralizing trend of over a century.

What are the prospects of political community and effective municipal governance in the industrialized countries of the West? Managing decline as noted is not a positive condition for creating political community. Recently nations again are increasing the number of new immigrants, guest workers, and "undocumented" aliens, who have replaced a larger number who have moved out. The economic elites often do business in the city; they do not reside there. Economic elites do not need much labor at all; in contrast to "in place" industries, they have mobility. It may cost less to move out of a giant city than to stay and there are more and more alternatives. An accountant can leave New York and serve clients from Denver. Interests of various groups in large Western cities are no longer so interdependent that they must accommodate their differences politically.

In socialist states, the role of the city as a political community with varying capacities to resolve conflicts between groups is difficult to assess. First, these political systems assert that economic differences are not significant sources of conflict. Second, social and ethnic conflicts are treated as residual, not central. Of course, there remains pride of identity with the city and its history and accomplishments.

Because resource allocations come largely through the political system, many problems of cities in market systems are not as aggravated as those in Western countries. Since most resources are allocated by the central government, however, there are enormous pressures to locate in the national capital for access to political decision making. These pressures have led to locational decisions in political capitals that the national center did not and cannot control.

Local politics in controlled economies plays a role, and cities compete for resource allocations and adjustments in plans made by those at higher levels of authority. But group conflict has only minimal political space for free play, and it is generally addressed at levels higher than the traditional points of local political identity.

Differences among Third World political systems are greater than those of either the First or Second; and so are the problems of their large cities as political communities. Some, such as Mexico, are highly centralized; others, such as India, have considerable provincial and hence municipal autonomy. Comparisons with wealthier countries are difficult because of generally weak local and national governments. The GNP allocation by governments in almost all Third World countries is considerably less than those of others and on a per capita basis even lower.

Conflicts between poor and rich and various ethnic groups is perhaps less than could be expected from the range and depths of differences. But many new populations take the city as a place of temporary opportunity to improve their and their relatives' lot in their place of origin. This situation is true in many African as well as Asian cities that have rapidly increased their populations. But identification with place of origin rather than city may be only a short-term condition, and conflicts may intensify as groups settle in.

Urban elites in large Third World cities with some exceptions are allied with national political forces, as indeed are the preponderant parts of the national government. Low food prices in some countries are forms of urban subsidies, which may have led to deteriorating food production, especially in parts of Africa. A few governments, such as India and the Republic of Korea, appear to have shifted their main political alliance to the countryside, as indicated by votes for the ruling parties and relative improvement in rural income.

The basic dilemma, however, is that either political position would stimulate rural migration to the cities: low prices for agriculture makes cities more attractive; better income for farmers requires larger farms and fewer farmers.

Most large Third World cities, especially those with massive populations, grew in the context of an international political economy; most industrialized giant cities grew during the processes that led to the formation of strong nation-states. Industrial cities of the 19th and 20th century made alliances with their national governments to resolve conflicts emanating from their growth, to escape the confines of their regions and controls of provincial governments, and to gain increased opportunities from the national government's efforts to create national markets through regulation of labor, transportation and communication infrastructures, and reliable currency. After that phase, some of those cities entered a global political economy but with a relatively firm national base. Many large cities of the Third World grew up with national bureaucracies imposed on them, either as a legacy of colonialism or from new regimes seeking to revolutionize society. Opening these cities to the global economy, as recently happened dramatically in the

Peoples' Republic of China, provides an alternative to the nation. That alternative makes credible claims to special status to compete in the world economy, by-passing the nation and making difficult the creation of strong, effective states that can claim control of their economies equivalent to others.

Prospects for political community and effective government in most large cities of the world are generally not positive. There are exceptions. The decline of large cities demographically and economically also means decline politically even though there may be some regional areas of economic growth.

Strong national governments and a growing national economy help urban governance; so does ethnic homogeneity as is the case for several large cities in Northern and Southeastern Asia. Also, as that most exceptional case of Singapore might illustrate, absence of conflicts from a national government and its interests in territorial consolidation and balancing claims of a variety of cities, should provide conditions for strong municipal political leadership to control and reduce pathologies.

But most large cities in the developing world have poverty, insufficient growth in their national economies, ethnic diversity, and weak national governments. The ecological consequences of urban growth add to governmental ineffectiveness, which allows for more pathologies and yet more, at least perceived, ineffectiveness of government. The national government is usually of limited support. Growth and its pathologies expand, perhaps to the old notion of a "natural" ecological limit.

The Ecology of World Cities

Satellite communications, jet passenger-cargo planes, containerized freight, and breakthroughs in geology (opening new areas to extraction of resources, especially oil) have altered the spatial advantages of older cities and given rise to new ones. The impact of these technologies, and the control capacity provided by new generations of computers linked to satellite communication, became obvious in the 1970s. The shrinkage of time in communications and control changed distance from earlier conceptions as friction and constraint to inconvenience. Interest on money is sometimes calculated in days rather than weeks, months, or years. Standardization of information and international languages facilitated and contributed to, both as cause and effect, these technologies, further reducing costs of search for resources and variety. New intermediaries emerged. Not only were there new international governmental institutions but private organizations, such as international banks, professional societies, and regional associations.

One key element to making cities large, productive, and accessible is energy. Various political crises (1973 and 1979) triggered recognition that known configurations of geological distributions of energy resources were inadequate. A global political awareness was created by shortages of oil whose discovery and use grew up with the application of modern science to production (about the 1860s and 70s) in contrast to coal, used for centuries for energy and heating. Oil had become embedded in many transformations into useful materials, including clothing, now as important for mankind as the use of cotton was to clothe millions. Coal, although in principle not different chemically from oil, remained tied to burning-heating. Oil became in the 1970s a political weapon. Its use is not only for the sustenance of large cities and their economic production but also for transfer of resources among nations. The big issue of the 1970s was whether politics or markets would in the short and long run dominate oil distribution.

One concern in extracting resources is cost per unit—in particular sites increases in proportion to the amount extracted over time. The implication here is that new sites are sought to lower costs to a point where eventually in the very long run all are used up. Wherever new sites are opened, there will be concentrations of population, some basis for urbanization, and redistribution of wealth geographically. Some national governments will lose; others will gain. Those gaining can more readily support urbanization costs through subsidies. This situation clearly has happened in parts of Africa and the Middle East; it may be happening to China with vast areas where resources have only been scratched rather than dug out. The winners are potentially in the Third World. Counterpoints for advanced technological societies are the oceans whose potential up to now has been largely animal (fishing), minimally plant, and almost not at all mineral.

The combination of vastly reduced access costs and depletion of old sites of resources has been a considerable force in the growth of large cities in the Third World—a spread of the percentage of world wealth away from older industrialized countries, and the incorporation of a larger percentage of the world's surface into the global political economy. One might ask whether the noted higher rate of economic growth in the Third World in the 1970s and first part of the 80s in comparison to the First World is because of increased costs of extraction from older areas rather than increased Third World productivity, "decline of the West," or international politics of income distribution. The traditional ecological factors remain part of the dynamics of urban growth: rivers and oceans, terrains, atmosphere, land type, as well as the subtle ones, absorptive capacity of land and availability of natural underground "storage areas." But many of these features have been increasingly discounted by technology, although the concern now is whether it too has some limits. Los Angeles has overcome its limited access to water in its immediate vicinity through engineering and politics. But can it go much farther in increasing its import of water? Can air pollution be absorbed without nearly unacceptable costs of installing public transportation? What is the ultimate holding power of the area on which Mexico City sits?

It is now understood, without benefit of computer, that annual growth rates that most governments of the world aspire to as reasonable double things in less than two decades. Most of that growth is urban centered (Ophuls, 1977). The problem is that no one knows the capacity of the earth's surface and its varied population niches, such as countries. Surely concepts of these limits have been radically expanded more than once in the 20th century.

World cities also live in social-cultural and political environments which may be more explanatory of the nature of large cities and their growth than the physical-technical environments. Are Eastern European cities similar because of the political structure of their region or because of their shared pattern of economic development? Are cities of Southeastern Asia similar because of their common culture and its resultant attitudes toward what people can or should do? Although these are basic and difficult questions, they often provide the set of assumptions for grouping large cities of the world into regions.

Patterns of Pathologies

Implicit in discussions of problems or pathologies of cities are their patterns and types. Are there some general and significant ones, and with what are they associated?

First, urban poverty can be found in some Asian, most Latin American, and almost every African country. Poverty here is defined not in terms of some national average or goal but rather by objective standards of physical and social manifestations. The World Bank estimates about 800 million to one billion people are in a state of "objective" poverty. The number in poverty may be declining relatively as a percentage of the total, although it is increasing absolutely. To what extent is urban poverty caused by the city and how does it compare to rural poverty? The former is visible internationally; the latter is difficult even to estimate. Whether urban poverty is related to rapid urbanization or simply attracts those already impoverished in the countryside, its presence in large cities, particularly capitals, detracts from the legitimacy of governments that claim international parity.

Second, corruption is pervasive in most Third World cities, although present in all. *Structural corruption,* defined as significant discrepancy between what a government asserts as justice, for instance, in law, and what it actually does, not only leads to loss of its effectiveness and legitimacy but also to destabilization of the entire political system. There is *day-to-day corruption*—bribery. Although this type is not a distinctive feature of the city, often it is a way of life, especially in large cities for primarily two reasons. There are more transaction points to corrupt than in the countryside—permits, taxes, entitlements, and so on—and new groups and minorities, both significantly present in large cities, seek access to government but often do not know how to get it or are excluded. Buying favors is a form of "participation," leading to stabilization; an alternative is threat and violence. But petty corruption may destroy even the limited effectiveness of urban government (see Scott, 1972).

Third, socially deviant behavior, perhaps more closely associated with large cities than the previous two, is at least perceived to be a problem for all large cities. It is also the most problematic in terms of concept and measurement. The theoretical question is to determine the extent to which pathological behavior, again a pejorative concept and presumably defined politically as crime, can be explained by factors in the general social, economic, and political environment, in this case those that constitute the nature of large cities or other factors, generally hard to isolate, such as family. Nonetheless, deviant behavior, some threatening to the well-being of others, is apparent in all large cities. Although general comparative analysis, such as Japan versus Western Europe versus the United States attributes differences in crime to strong local and central control, it is likely to be only one of the variables leading to observed differences.

Other pathologies of cities already have been mentioned. All are problematical theoretically as the parameters for their assessment are unspecified. All had historical counterparts before the advent of the large cities in the 19th and 20th centuries. It may be that most pathologies of large cities are short-term effects of growth rather than inherent in them as a particular form of human organization.

Perspectives on the Future

Whether or not world population growth declines or increases and despite potential population decline in parts of Africa and other areas of the world, the number of cities over one million will continue to expand, and in the Third World their populations will grow immensely. Most policies to keep people on the land, especially land reform, have failed.

The European and North American "system break" of retiring agriculture as a way of life—perhaps as important in the long run as agriculture's replacing hunting-gathering—has spread throughout the world, replacing agriculture with urban civilization as the dominant form of life, rather than having urban pockets in vast fields and forests. The problem faced in this transformation was how to get people off the farm and into the factories, the most productive of which were in cities.

In the Third World most giant cities grew up in the past three decades and are made up of immigrants or their children. They are no better prepared for urban life than their Western counterparts were to adapt to the industrial cities. The costs to some, whether or not pathological, are enormous. There are exceptions in Asia and Southeast Asia where the population grew mostly from natural increases across generations, such as Hong Kong. To get new immigrants into the factories means developing life-cycle or cross-generation transmission belts of low-skill services to better skills in other occupations. To provide for manufacturing and other economic opportunities, however, it is necessary to become part of the global political economy and the vulnerabilities that entails. To choose national closure is to enter intellectual combat with the contemporary history of China and Albania. The fear of openness is collapse; of closure, stagnation.

The problem the Western industrialized world faces is quite different: how to get people out of the factories and cities and in front of a telephone or computer terminal. Although not challenged with physical harm or assaults on health, the psychological consequences of this change can be as great as the trauma of moving from the Midlands or Iowa to Liverpool or Chicago or of a peasant seeking a new life in a big city.

A few major cities might have made the basic adjustment to a new global political economy, such as New York, despite the problems remaining. Several such as Shanghai have taken major steps to what they anticipate will be their future. Some remain victimized by the changes that have taken place, such as Lagos, and must make substantial adjustments, being caught in a global economy that fluctuates. Although the Eastern European cities are increasingly being affected by the global political economy, they remain in a more or less separate enclave of the world.

Whatever else the impact of integrating diversity on a global scale and the new technologies of access have, one consequence is the near dissolution of distance as a significant factor in the ecology of concentration and dispersion. Except where concentration still is part of a productive future, such as in the making of certain goods, the large city does not provide the incentives necessary for creating and maintaining a political community. What are the new urban forms of organization? Here only speculation is left. Some alternatives are suggested.

First, in every large city and to a greater extent in the older ones, definable enclaves of people engaged in creative activities and the information industries will grow, but never be a large number. Unstandardized information comes at high costs; information must be in dense patterns; it is difficult to channel. One way of intensifying information is through face-to-face contact, a traditional means of introducing the new and reducing uncertainty. The possible gains of accessing unstandardized information are high but the probabilities of such gains low. The innovative industries of fashion, art, and taste need large cities for easy access and those of science need the economies of relatively dense science centers and research parks. This new urban form is already with us in a few parts of the world.

A second urban form is based less on space but on stable social structures of activities governed by clear values and strong elites. They have no particular place, but the people in them grew up together in those sectors rather than in the towns and neighborhoods of the past. The medical, scientific, bee-keeping, car-manufacturing groups have known each other for years, see each other for some time each year, attend to each other's families and personal fortunes, and provide for the psychological needs of affiliation and familiarity.

Third, are the new centers of production which comprise many of the large world cities. Rather than being places with political community and relative self-sufficiency when combined with a hinterland, they are made up of a variety of components with various levels of linkages to their hinterlands, including transnational regions, their own and other nations, and the world at large. These economic and social segments occupy a single place, although with diminishing potential for social political integration and with their neighbors.

Then, of course, there are those that remain in the old nation-city political economy of national development, targeted to industrial output. In addition to these, are a variety of other urban forms of recent history, including neighborhoods, towns, small cities and parts of larger cities, some of which are maintained to remind us of our history. These are the centers of leisure and entertainment, a different kind of place of production.

Those occupying the first of these must reside in a first-tier city and be able to move around to others as well as those of the second tier. Those of the second of these urban forms, dealing mostly in standardized information and services, can reside anywhere and be linked to about everyone in their sector, working together to achieve status. Many of them have and are moving from the major cities and in a few countries from the metropolitan areas. Spatial community, as the dominant form of political community of the past, does not have much meaning except perhaps as nostalgia. The third occupy locations within a big place, perhaps responding to attempts to make the area work as a city, a political community, but not able to achieve security from it.

Beyond this it is difficult to go. There will be other kinds and ways of defining new urban forms. But urbanization as process, and city as a political organization tied to place and based on an economy and social order, has always been undergoing change. The historical manifestations of concentration and dispersion, new organizations replacing the old, reflects the process of putting things together and taking them apart and is essential to development. The historical urban form of industrial cities in the West and the giant cities of the Third and Second World countries is only one component of the long process of human and societal development.

Notes

1. This particular point, the organization of space, is an idea of Oliver P. Williams who has also helped in formulating other arguments made in this chapter. (See also Teune, 1987.)

2. This section is biased toward the U.S. experience and my chapter on political development of the welfare state in the United States (Teune, 1986).

3. Although these problems were discussed earlier, corruption and government effectiveness were not. It is not an urban phenomenon but an age-old practice to deal with authority.

Bibliography

Deutsch, Karl W., Sidney A. Burrell, Robert A. Kann, Maurice Lee, Jr., Francis L. Loewenheim, and Richard W. Van Wagenen. 1957. *Political Community and the North Atlantic Area. International Organization in the Light of Historical Experience.* Publications of the Center for Research on World Political Institutions no. 6. Princeton, NJ: Princeton University Press.

De Vries, Jan. 1984. *European Urbanization, 1500-1800.* Harvard Studies in Urban History. Cambridge, MA: Harvard University Press.

Gil Díaz, Francisco. 1984. "Mexico's Path from Stability to Inflation." In *World Economic Growth,* edited by Arnold C. Harberger. San Francisco: ICS Press, Institute for Contemporary Studies.

Heidenheimer, Arnold J., Hugh Heclo, and Carolyn Teich Adams. 1983. *Comparative Public Policy: The Politics of Social Choice in Europe.* 2d ed. New York: St. Martins Press.

Higgs, Robert. 1971. *The Transformation of the American Economy, 1865-1914; An Essay in Interpretation.* The Wiley Series in American Economic History. New York: Wiley.

Laqueur, Walter Z. 1967. *The Fate of the Revolution, Interpretations of Soviet History.* New York: Macmillan.

Naroll, Raoul. 1965. "Galton's Problem: The Logic of Cross-Cultural Analysis." *Social Research* 32:428-51.

Ophuls, William. 1977. *Ecology and the Politics of Scarcity: Prologue to a Political Theory of Steady State.* San Francisco: Freeman.

Schulz, George. 1984. "Economic Cooperation in the Pacific Basin." Current Policy Series no. 658 (21 Feb.). Washington, DC: Editorial Division, Office of Public Communications, U.S. Department of State.

Scott, James C. 1972. *Comparative Political Corruption.* Prentice-Hall Contemporary Comparative Politics Series. Englewood Cliffs, NJ: Prentice-Hall.

Teune, Henry. 1986. "The Political Development of the Welfare State in the United States." In *Nationalizing Social Security*, edited by Douglas Ashford and E. W. Kelly. Greenwood, CT: JAI Press.

———. 1987. "Cities in a Global Political Economy." In *Theoretical Frameworks of the Contemporary World in Transition.* Vol. 2, *Transnational Issues and Global Politics*, edited by Kinhide Mushakoji and Hisakazu Usui. Tokyo.

Wallerstein, Immanuel M. 1974. *The Modern World System: Capitalist Agriculture and the Origins of the European World-Economy in the Sixteenth Century.* Studies in Social Discontinuity. New York: Academic Press.

Index

About the Editors

MATTEI DOGAN is scientific director at the National Center for Scientific Research, Paris, and professor of political science at the University of California at Los Angeles. He is author, coauthor, editor, or coeditor of 15 books and a contributor to 38 others. He has written more than 50 articles in French, English, Italian, and German academic journals.

JOHN D. KASARDA is Kenan professor of sociology and chairman of the Department of Sociology at the University of North Carolina at Chapel Hill where he is also a Fellow of the Carolina Population Center. His research focus is urban economic restructuring and public policy. He has published widely in the United States and abroad in the area of urbanization processes.

About the Contributors

MANUEL CASTELLS is professor of city and regional planning at the University of California, Berkeley. He holds a Ph.D. in sociology from the University of Paris where he has taught. Formerly on the faculty of the universities of Montreal, Chile, Wisconsin, Copenhagen, Boston, Mexico, Hong Kong, Southern California, and Madrid, his 12 books include *The City and the Grassroots,* winner of the 1983 C. Wright Mills Award.

XIANGMING CHEN is a James B. Duke fellow at the Center for International Studies, Duke University. He is working on a dissertation which examines China's recent development strategy of operating Special Economic Zones and coastal city trade in a comparative context of East Asia. He has published in social science journals.

JÜRGEN FRIEDRICHS is professor of sociology and director of the Center for Comparative Urban Research at the University of Hamburg. His research deals with urban demographic and economic processes. He has been widely published in Europe and the United States. His most recent book is *Introduction into the World of Work.*

GUILLERMO GEISSE is professor of urban planning at the Institute of Urban Studies, Catholic University of Chile; director of the Center for Environmental Research and Planning in Santiago; and president of the Inter American Planning Society. He was a guest scholar at the Wilson Center in Washington, DC, in 1985.

SIDNEY GOLDSTEIN is professor of sociology and director of the Population Studies and Training Center at Brown University and a past president of the Population Association of America. Among his recent publications are *Migration and Fertility in Peninsular Malaysia; Migration,* Subject Report no. 2, *1980 Population and Housing Census of Thailand;* and *Population Mobility in the People's Republic of China.*

PETER HALL is professor of geography and chairman of the School of Planning Studies, University of Reading (England) and professor of city and regional planning, University of California, Berkeley. He is author or coauthor of many books on urban geography and planning, most recently, *The World Cities* and *Silicon Landscapes.*

HANS NAGPAUL is associate professor of sociology at Cleveland State University. He formerly taught at the University of Delhi. He has worked with the Indian National Planning Commission and has published four books and numerous articles.

DENNIS A. RONDINELLI is senior policy analyst in the Office for International Programs, Research Triangle Institute, in North Carolina. He has taught at the universities of Wisconsin, Vanderbilt, and Syracuse. He has written 8 books and more than 80 articles on urban and regional development, most recently, *Applied Methods of Regional Analysis.*

FRANCISCO SABATINI is assistant professor of urban planning at the Institute of Urban Studies, Catholic University of Chile. He is currently a doctoral student in urban planning at the University of California at Los Angeles, where he is writing his dissertation on urban popular movements in Santiago, Chile.

IGNACY SACHS is on the faculty of the Ecole des Hautes Etudes en Sciences Sociales, Paris; director of the International Research Center on Environment and Development, Paris; and director of the Program on Alimentation and Energy at the United Nations University, Tokyo.

HENRY TEUNE, a political scientist, has taught at the University of Pennsylvania since 1961. His comparative research on local political systems has involved collaboration in Poland, Yugoslavia, India, and the Philippines. He is coauthor of *The Integration of Political Communities, The Logic of Comparative Social Inquiry, The Developmental Logic of Social Systems,* and *Values in the Active Community.*

YUE-MAN YEUNG is professor of geography and Director of the Centre for Contemporary Asian Studies at the Chinese University of Hong Kong. His publications include *National Development Policy and Urban Transformation in Singapore, Changing Southeast Asian Cities, Hawkers in Southeast Asian Cities,* and *A Place to Live.*